T0181044

Design Theory

Pascal Le Masson · Benoit Weil
Armand Hatchuel

Design Theory

Methods and Organization for Innovation

 Springer

Pascal Le Masson
MINES ParisTech—PSL Research
  University
Paris
France

Armand Hatchuel
MINES ParisTech—PSL Research
  University
Paris
France

Benoit Weil
MINES ParisTech—PSL Research
  University
Paris
France

ISBN 978-3-319-84364-3   ISBN 978-3-319-50277-9   (eBook)
DOI 10.1007/978-3-319-50277-9

Printed on acid-free paper

This Springer imprint is published by Springer Nature
The registered company is Springer International Publishing AG
The registered company address is: Gewerbestrasse 11, 6330 Cham, Switzerland

Acknowledgements

A course manual summarizes established knowledge. This manual is no exception. However, in the field of design, this summary also required significant efforts to arrange, complete, and adapt a complex, multidisciplinary corpus to today's challenges. This manual is, therefore, the result of a research and instruction program that was launched at the School of Mines (École des Mines, today MINES ParisTech—PSL Research University) more than 20 years ago. In 1994 the Paris School of Mines created a field of study dedicated to "design theory and engineering". Under this unusual name at the time, a team of professors and researchers was formed under the responsibility of Armand Hatchuel. The team consisted of Benoit Weil, Jean-Claude Sardas, and Christophe Midler, and a few years later of Pascal Le Masson as well.

That program revived an old tradition of training engineers on technology and product design, all the while addressing contemporary innovation challenges. It aimed to unite the subject fields of innovation, project management, and design and, eventually, bring design sciences to a level of maturity equal to that of decision or programming sciences. Within a few years the program gained surprising momentum. The set of subject fields would progressively be organized around a single theoretical framework, the C-K theory (Hatchuel 1996; Hatchuel and Weil 2001, 2009, etc.), taught in many schools today. This program also responded to a significant need for change, namely the need to strengthen companies' innovation capacity. These advances thus made it possible to establish valuable partnerships with multiple companies.

This led to the formation of a research and instruction platform that has not ceased to grow, especially with the establishment of the Chair of Design Theory and Methods for Innovation in 2009, sponsored by five and subsequently seven companies, and renewed in 2014 with currently more than ten corporate sponsors.

The theories and methods developed 20 years ago in this framework gradually enriched an already comprehensive corpus, including innovation management, innovation economy, engineering design, industrial design, as well as history, mathematics and logic, cognitive sciences, etc. They made it possible to create a synthetic and reasoned instruction of design theories, methods, and organization including contributions from these multiple disciplines. Many courses were thus created. Since 1999 at the School of Mines, Vincent Chapel and Benoit Weil created the Product and Innovation Design course, which was further developed later on with Pascal Le Masson. In 2007 a version of that course was developed for ParisTech doctoral students. Those corpora were also included in several other institutes, not only in engineering schools (e.g. Ponts et Chaussées, École Polytechnique, AgroParisTech, INSA, ENSAM, CNAM, etc.) but also in business schools (HEC, ESCP, Grenoble École de Management, etc.), universities (Université Paris Dauphine, Université Paris-Est Marne-la-Vallée, etc.), and schools of design (Strate College, ENSCI, etc.). The courses were also adopted abroad, at Chalmers School of Entrepreneurship, Stanford, Carnegie Mellon, as well as in Delft, Tel Aviv, and Tokyo, among others.

We would like to thank first and foremost the School of Mines for the support it has consistently provided to these programs from the start and to this day. We are thankful to the professors of Design curricula, and particularly to Christophe Midler (CRG, École Polytechnique) who participated from the very first day and who contributed significantly to this work. We would like to express our gratitude to Maurille Larivière for his remarkable contribution to our work with his industrial design experience. Thanks to Blanche Segrestin, Franck Aggeri, Philippe Lefebvre, and Michel Nakhla, whose contributions often served as a reference in our research. Marine Agogué, Akin Kazakçi, Sophie Hooge, Cédric Dalmasso, Mathieu Cassotti, Anaëlle Camarda, Annie Gentes, Anne-Françoise Schmid, Kevin Levillain, Olga Kokshagina, and Georges Amar, who joined the research program of the Chair a few years ago and whose work is a fundamental contribution to Design theory and psychology. We are thankful to all of our students who chose to study engineering design and whose master's theses were sources of advancement and motivation for professors.

We also wish to thank Yacine Felk, Ingi Brown, Olga Kokshagina, Pierre-Antoine Arrighi, Elsa Berthet, Frédéric Arnoux, Aytunc Ün, Hassen Ahmed, Kenza El Qaoumi, Louis-Etienne Dubois, Morgane Benade, Mario Le Glatin, Daniel Carvajal, Maxime Thomas, Agathe Gilain, Honorine Harlé, Juliette Brun, Hicham Ezzat, Kevin Levillain, Camilla Freitas-Salgueido, Laura Le Du, Milena Klasing-Chen, Fabien Jean whose theses and master's dissertations constitute remarkable advances in Design Science and whose work contributed significantly to this book.

This book, however, would not have been possible without the support of numerous partners in the companies that support the Innovative Design Theory and Methodology department. We are especially grateful to Dominique Levent, Philippe Doublet, Lomig Unger, Pascale Nays, Rémi Bastien (Renault), Dominique Laousse, Eric Conti, Jean-Jacques Thomas, Cédric Brogard (SNCF), Guillaume Lapeyronnie, Denis Bonnet, Christian Delaveau, Joseph Huysseun, Philippe Benquet (Thales), Pascal Daloz, Patrick Johnson, Anne Asensio, Manuel Gruson (Dassault Systèmes), Patrick Cogez, Jean-Luc Jaffard, Loic Liétar, Jean-Marc Chéry (ST Microelectronics), Laurent Delattre, Céline Sches, Nicolas Bréziat, Alain

Dieulin (Vallourec), Stéphane Cobo, Pierre Becquart, Paul Gardey de Soos (RATP), Denic Clodic (Ereie), Vincent Chapel and Jean-Pierre Tetaz (Helvetia), Guirec Le Lous and Hervé Le Lous (Vivasanté), Anne Bion, Isabelle, Michel and Adeline Lescanne (Nutriset-Onyx). Guillaume Bulin, Michel Comes, Gilles Dussault, Bernard Marmann, Thierry Pardessus, Axel Flaig, Alain Fontaine (Airbus).

The courses benefited greatly from the experience of consultants who deployed these methods: Yvon Bellec (TAKT) who was the first to embark on the KCP adventure, Patrick Morin, and Patrick Corsi, Dominique Lafon, Fred Arnoux, Benjamin Duban and all the teams of TAKT and STIM.

For agreeing to give lectures and share their experience with our students, we wish to thank Jean-Hervé Poisson (Renault), Nils Saclier (Renault), Bernard Vaudeville (T/E/S/S), Simon Aubry (T/E/S/S), Yves Parlier (Beyond the Sea), Marc Maurer (Saint-Gobain), Jean Schmitt (Jolt Capital), Hervé This (INRA), Benjamin Duban, Charlotte Leleu (Regimbeau), Philippe Picaud (Carrefour), François Pouilloux and Stéphanie Jacquemont (Ixxo), Guillaume Bulin and Thoerry Pardessus (Airbus), Patrick Cogez (ST Microelectronics).

For their work and their stimulating discussions with us we owe much to many researchers in various fields: Olivier Houdé (Sorbonne), Albert David (Université Paris dauphine), Gilles Garel (CNAM), Sylvain Lenfle (CRG, Ecole Polytechnique), Olivier Hirt (ENSCI), Patrick Llerena (Beta Université de Strasbourg), Patrick Cohendet and Laurent Simon (HEC Montréal), Susanne Ollila, Maria Elmquist, Tobias Fredberg, Anna Yström and Mats Lundqvist (Chalmers University), Carliss Baldwin (Harvard Business School), Victor Seidel (University of Oxford), Alan Mac Cormack (MIT), Rafael Ramirez (University of Oxford), Annabelle Gawer (Imperial College), Francesco Zirpoli (University Ca'Foscari de Venise), Markus Becker (University of Southern Denmark), Franck Piller (Aachen RWTH), Yoram Reich (Tel Aviv University), Ade Mabogunje (Stanford University), Eswaran Subrahmanian (Carnegie Mellon), Chris McMahon (Bristol University), Jean-François Boujut, Michel Tollenaere and Eric Blanco (Polytechnique Grenoble), Toshiharu Taura (Kobe University), Yukari Nagai (Tokyo University), Ken Starkey (Nottingham University), Ehud Kroll (Technion), Kathrin Möslein (Erlangen-Nürnberg University), John Gero (Krasnow Institute for Advanced Sciences), Udo Lindemann (Technische Universität Munich), Patrick Fridenson (EHESS), Anne Françoise Garçon (Université Sorbonne) and Hélène Vérin, Jean-François Basserau (ENSAD), Thomas Gillier (Grenoble Ecole de Management), Amaresh Chakrabarti (Bangalore Univ), Roland De Guio (INSA Strasbourg), Tom Howard (DTU), Suan Imholz, Olivier Irrman (Lille), Jean-Baptiste Mourer (INRIA), Vincent Bontems and Vincent Minier (CEA, ENS), Dorian Marjanovic (Zagreb Univ), Panos Papalambros (Michigan Univ), Olivier Potier (Nancy Univ), Vivek Velamuri (Erlangen Univ), Jan Dul (Rotterdam Univ), Klaasjan Visscher (Univ Twente), Gwenaelle Vourch (INRA), Amos Winter (MIT), Peter Vermaas (Delft Univ), Offer Shai (Tel Aviv Univ).

Multiple projects were developed thanks to seminars and conferences organized by leading scientific communities in the field: the Design Society, which we thank for the creation of the SIG (Special Group of Interest) on design theory; EURAM (European Academy of management) which regularly offers tracks on Design; IPDM (International Product Development Management), a pioneer group in innovation management, especially its president Christer Karlsson and its Board, to whom we are grateful for kindly welcoming original projects on design and innovation.

Finally, we wish to express our most sincere gratitude to all researchers of the Center of Management Science of MINES-ParisTech—PSL Research University for their contributions. We thank Céline Bourdon, Stéphanie Brunet, and Martine Jouanon for their help.

We are also very grateful to Atenao Translation Agency, particularly to Emma, and to Atenao's translators, Chris and Maria, for their rigorous and respectful translation. We warmly thank Chris McMahon (University of Bristol) for his intellectual support in preparing the translation project. We are most grateful to Anthony Doyle, who accepted to include this book in his collection at Springer.

Paris, France   Armand Hatchuel
September 2016   Pascal Le Masson
                                                                              Benoit Weil

References

Hatchuel A (1996) Coordination and control. In: International Encyclopedia of Business and Management. Thomson Business Press, pp 762–770
Hatchuel A, Weil B (2009) C-K design theory: an advanced formulation. Research in Engineering Design 19 (4):181–192.

Contents

Chapter 1
Introductory Chapter: Contemporary Challenges of Innovation—Why New Design Theories

1.1 Introduction: Training for Design Today

In 1993 a new Engineering Design course was created at the Ecole des Mines, France. This addressed a double need: preparing future engineers for the changes under way in the organization of innovation in large companies, and developing a training course incorporating the most recent advances in the theory and methods of design. Developed in partnership with major innovative firms and closely linked with a research program focused on managing design activity, this training program gradually built up a set of original courses covering the theory, methods and organization of design. This is the purpose of this publication.

What does training in design mean? There are those who have claimed that the creative design process is typically something that can be learnt only through practical experience, learning via a project. In reality, this received wisdom is not based on fact. Historical experience has shown that training in design was indeed possible, including designs with a powerfully creative content - with some remarkable effects. We recall two prominent examples where training for design was one of the major factors in an economic and social policy of innovation: in the 19th century, the German policy for catching-up in the industrial world relied in particular on creating design training programs in the brand-new technical colleges, thereby contributing to the emergence of a scientific and technical managerial class for the design offices of future large companies such as AEG, Siemens, Bosch, etc. (see the historical case study in this chapter of this article). At the end of the 19th and start of the 20th centuries in Germany, intellectual movements and industrial policies sought the means to design products emblematic of a new style for a changing society, and which would show that the "made in Germany" label could be a yardstick for quality. These ideas led, in particular, to the creation of new schools such as the Bauhaus, whose courses are held up to this day as the gold standard for industrial design training (see Chap. 5). Hence these two examples show that training for design is that of fitting within the prestigious traditions that this article will attempt to extend.

These examples also show us that the question is not so much that of knowing whether we can train for design and innovation but rather *how* we train for design. We shall emphasize several lessons to be drawn from these training programs. First lesson, the importance of conceptual reasoning: the greatest artists taught at the Bauhaus (Klee, Kandinsky, etc.) but they did not demonstrate their practical way of working—they taught the theories that they had themselves developed over time to train creative minds; similarly, teachers at the technical schools might have been great practitioners very much au fait with industrial issues, but that drove them precisely to formulate theories of conceptual reasoning to get their students to overcome the stumbling blocks and fixed ideas they encountered in the industry of that time. Second lesson: that of teaching that reasoning is not sufficient; it is also necessary to teach the methods enabling the designer to make use of an often very demanding conceptual reasoning. Third lesson: design, even that from creative minds, is not individual, and design organization as well has to be taught, thus allowing division of work, coordination and cooperation. Fourth lesson: the design process involves taking account of the logic of performance and the management of specific resources after the designers have been trained. Fifth lesson: any training for design must also make the designers aware of trends and dynamic properties of objects, uses, collective bodies and contemporary society, i.e. of the fields of innovation they might have to explore as designers. Five lessons, and the same number of facets of training for design: reasoning, methods, organization, performance, fields of innovation. These are the five facets this course proposes to tackle, applied to contemporary design situations.

Covering these five facets are currently many highly extensive bodies of knowledge, originating in particular from the management of innovation and technology, the economics and sociology of innovation, and engineering or industrial design. Over the last few years these bodies of knowledge have been evolving radically. Hence the management of technology and innovation gradually incorporates methods which extend beyond incremental innovation management in allowing the management of disruptive innovation; it also describes forms of leadership in which "design reasoning" complements traditional decision-making capabilities; it discusses the use of external resources within the logic of "open innovation". With regard to the economics of innovation, the question is no longer simply that of the optimum allocation of resources; the logical processes of resource creation through growth and learning, and new forms of relationships between participants allowing novel ecosystem dynamics must also be included. Within the field of engineering design, new theories and new methods of design incorporate contemporary creative logical reasoning and set out powerful methodologies for the generation of ideas.

The aim of this course is to present these different bodies of knowledge and their evolution within a summarized and reasoned framework which, on the one hand, will provide a fruitful dialogue between the various approaches, and on the other, will be sufficiently open to allow the students to include, or even develop themselves, future design methods.

To this end, training relies on a pedagogic (and theoretical) course of action, with teaching based on formal design theories and models. Complete mastery of these will then provide a better comprehension of the other facets (methods, organization, economics and fields of innovation) (see Fig. 1.1). Thus we can represent the logic of the teaching process by means of the diagram below.

These theoretical foundations also enable the management of technology and innovation (which tend to favor organizational and strategic aspects) to be combined, along with engineering approaches (which tend rather to emphasize questions of theory and method) and "lab" type industrial design approaches (e.g. the MIT media lab, Stanford Design Lab, etc.) which favor an approach using objects and fields of innovation.

Finally, these theoretical foundations provide a common basis for design professionals who sometimes feel, incorrectly, very distant: R&D, design offices or consultancies base their expertise (and their legitimacy) on their mastery of method and organization; experts in program management or strategy consultancies on their mastery of strategy and the economics of innovation; and product or design managers on their knowledge of the fields of innovation. Starting on theoretical foundations provides the elements of these three major skills and encourages collaboration between them.

The general organization of this work is based on the distinction between two historical design regimes: the rule-based design of the large company with its traditional R&D, and the innovative design emerging these days in all manner of shapes and forms in today's industry. Chapters 2 and 3 deal with rule-based design, while Chaps. 4 and 5 cover innovative design. Each of these regimes corresponds initially to very contrasting modes of reasoning, making use of specific methods, organizations, performance logic and types of innovation.

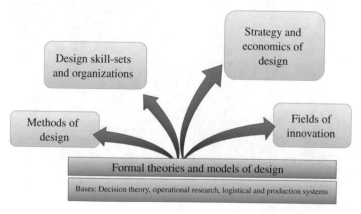

Fig. 1.1 Pedagogic approach: base the teaching on formal design theories and models

Dealing with each regime, we start off with some elementary activity, that of a design project, a rule-based design project (in Chap. 2) or an innovative design project (Chap. 4). For this project, we start by studying the type of reasoning we need to follow, then the nature of the performance to be attained and the organizations involved in the project to attain these objectives. Secondly, we investigate the general conditions under which a set of such projects can develop. We then tackle questions related to the company and its associated ecosystem under a rule-based design regime (Chap. 3) and under an innovative design regime (Chap. 5). We start with the economic logic pertaining to each of these regimes, then move on to study the cognitive conditions required to attain this performance (types of knowledge) and finally the associated organizations.

The course synopsis is outlined in the diagram below (Fig. 1.2).

Each chapter contains different material: a main course, contemporary case studies, historical cases and workshops. There is a review of the main ideas at the end of each chapter and suggestions for further reading.

Before covering these four chapters, we shall indicate in this current section the general logic of an approach by design regimes. We first show that the preconceived ideas and simplistic approaches to the management and economics of innovation cannot explain the puzzles and paradoxes at the heart of contemporary design. Hence we pose several questions to be resolved over the course of the book. We then establish the first theoretical elements for developing our response: we show the reasons for an approach by design and by design regimes, and we set out a canonical model for a design regime. We shall see that a design regime can be characterized by a conceptual reasoning model, a performance model and an organizational model.

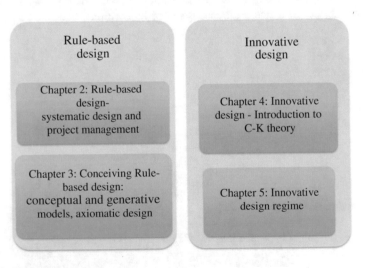

Fig. 1.2 Synopsis of this book

1.2   A Few Puzzles and Paradoxes

We begin with the received wisdom, pervasive as it is when we speak of innovation today. Why take on a design course while we have ready answers for obtaining the innovations apparently essential to the economies of today? After all, is it not enough (1) to make a substantial effort in R&D? (2) to organize innovative and daring projects based on original ideas and driven by competent expert-managers? (3) to demonstrate creativity and decision-making rigor?

These facts permeate the spirit, indeed they lead to remarkable paradoxes. Reviewing this received wisdom will enable us to set out the reasons justifying a course on "design" today.

1.2.1   The Paradox of R&D: Investment Does not Mean Innovation—Design is Increasingly Difficult to Organize

The first accepted idea: No innovation without concomitant investment in R&D. "Invest 3% of GDP in R&D",[1] "increase efforts in R&D" are the standard catch-phrases trotted out by nations trying to encourage innovation or by financial analysts seeking to assess a company's potential for innovation. They draw on a two-fold commentary:

- on the one hand there is no significant correlation between the commitment to the R&D efforts made by the biggest global businesses and their growth (Jaruselski et al. 2005, 2012, Jones 1995); see our own study on DTI figures, as illustrated below (Fig. 1.3). These suggest we should seriously revise this catchphrase, given that some businesses, such as Apple, built as they are on innovation, have shown impressive growth while nonetheless requiring an R&D effort rather less than the average for their particular sector;
- on the other hand the effort in R&D, or more generally in design, is steadily increasing, making France, for example, a country with nearly 50% of her engineers involved in design.

As of now it is important to give a few orders of magnitude. When we speak of design, or a design team, we no longer mean an engineer surrounded by several technicians or a little squad of inventors. There are nearly 20,000 engineers and technicians at the Renault Technocentre at Guyancourt in the Paris suburbs, and more than 800 PhDs at the Astra-Zeneca research centre in Göteborg, Sweden. When we speak of design, what we mean today is new factories and white-collar workers.

---

[1]This was the objective Europe set for itself for 2010 at the Lisbon summit, and assigned in 2010 to the new policy of innovation for 2020.

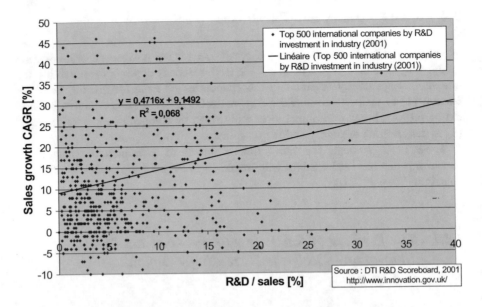

Fig. 1.3 The paradox of R&D. *Top*: R&D intensity (R&D/Turnover) versus smoothed annual growth over 5 years for the 500 largest global industrial investors in R&D (DTI figures, 2001)—there is no significant correlation. *Bottom*: R&D investment by the 10 companies judged to be the most innovative (figures taken from Booz Hamilton 2011)—these companies are not the biggest R&D investors in their sector

We note that these activities are often poorly identified in statistical studies, tending as they do to focus on economic activities such as production - for which design would be seen as a service. These "services" have assumed massive proportions in today's economies, which have thus become economies of "design" and no longer just "production". We shall study these design economies in the remainder of this publication: their scope, dynamic, effectiveness, etc.

1.2.2 Paradoxes in the Organization of Innovative Design

The second piece of accepted wisdom: innovation equates with having original ideas, knowing how to find the resources to develop them and knowing how to show tenacity in a necessarily long, costly and risky process of trial and error - this would be the "discipline" of innovative organizations (see Drucker 1985; Tidd et al. 1997).

Some businesses (see Le Masson et al. 2010) show more complex organizational traits: among some of them, crazy ideas are not left to a few creative enterprising individuals, but are subject to *rigorous and precise management* by the company's senior directorate. With other businesses, it is not so much a question of assembling ad hoc the skills within the project teams as ensuring that the *skills of the business are renewed* in accordance with a complex metabolism (Chapel 1997). With yet others, trial and error is not random but in fact follows *the logical processes of rigorous exploration*.

This is very much an organizational model which poses the question: that of a funnel for the selection and development of ideas. Might the process of innovation consist of going from idea to product? To claim such a thing today is to forget 150 years of industrial history, to ignore two industrial revolutions, the R&D organization of large companies, and the essential elements of the skills and bodies of knowledge within the great design professions (engineers, scientists, business, the design process, etc.). To claim that is to be mistaken on at least four points:

1. *Should there be more ideas?* In fact, it seems that it is not the quantity of ideas which gives rise to problems—it is not generally the ideas which are lacking, it is the "good" ideas; and *the great numbers of them do not compensate for fixations*: unconstrained and with sometimes immense resources, collectives will make orphans of any exploration of broad fields of innovation (see later on in this work notions of fixation and orphan innovation). Innovative concepts do not come about by chance, but are the result of a rigorous and demanding process.
2. *Should there be more selection?* Is this actually a question of quantity? Recall that the most demanding processes of selection with selection ratios of one in a hundred (consumer marketing), or even one in several million (pharmaceutical high throughput screening), allow products through for launch on the market, of which a very large proportion will fail. It is the process, rather than the selection ratio, which raises the question: not to select, but to guide and learn.
3. *Would it be better to make use of the skills available*, available internally and, especially in this era of open innovation and big data, externally? But is this really a question of access and usage? On the websites of the biggest players in open innovation, challenges with titles such as "smart cities", "cancer therapy", "new mobility", "sustainable energy", "energy storage", "autonomy for vulnerable individuals", etc. cannot be posed simply! And what if contemporary design also needed new forms for *creating* knowledge?

4. *Should there be more trial and error (and more means for doing it)?* However, are we not guilty of the "Shadoks" syndrome (creatures created by French cartoonist Jacques Rouxel in the 1960s & 70s)? Knowing that their rocket has only one chance in a million of working, the Shadoks scurry off to "make sure the first 999999 tests fail" "The more often it fails, the better its chances of working" they declare proudly. Joking aside, the organizers of the design process are faced with a fundamental question of statistics: either the organizer is content to ask for the means to play sufficiently long and often (but will our companies and economies allow it?); or the task of the organizers is exactly that of inventing a design process for which, even when there is one chance in a million of winning the first time, the process can be guided such that the probability gradually changes, increases, and tends to 1. In other words, *the work of the organizer of a design process is not to operate in a given probability space, but to change the probability space itself.* As we shall see, it is not a matter of deciding within the uncertain but rather of giving structure to the unknown.

It is often thought that innovation is matter of laissez-faire or looking for that one great blockbuster, but for the last two centuries the major design organizations have not adopted such an approach, preferring greater rigor in the exploration and logic of metabolic growth involving the entire organization in a constant process of renewal of skills and products. We shall be studying these perhaps poorly understood logical processes in this book.

1.2.3 Paradoxes of Reasoning in Innovative Design

Third accepted idea: one has to be creative or clever to innovate, to have original ideas and/or draw on leading-edge scientific discoveries. But what exactly is creative thinking? And what is the role of knowledge in design reasoning?

Creative thinking seems to be the key to innovation; paradoxically, this is one of the representational blind spots. "Innovation is knowing how to develop good ideas". However, what constitutes a good idea? What is an original idea? Strictly speaking (Boden 1990), the notion of originality suggests breaking certain rules of design, i.e. breaking certain attributes defining the object. An original idea forces us to revise the axioms of the object. The demanding nature of this process should not be underestimated!

"Innovation is applying the results of fundamental research". However, what does "applying" mean? What are the results to be applied? And conversely, can we not suppose that it is those "good ideas" which stimulate us to revisit the questions of fundamental research? Do nanomaterials or nanofluids not pose questions of new crystalline structures or of new states of matter?

Beyond the obvious, creative thinking raises fundamental questions on revising the axioms of objects and the role of knowledge in this process.

Is this about a paradigm shift? managing—was that not making decisions? Are managers not decision-makers? In fact, the reasoning useful for managing innovation could be of an entirely different nature: no longer making decisions but generating (imagining) alternatives; no longer do theories of decision provide the basis for innovation management models, but theories of design. A design course these days is therefore an introduction to the post-decisional paradigm which characterizes contemporary society.

These few paradoxes therefore pose just as many enigmas: how to study and improve the capacity for innovation? How to organize innovation? How to model reasoning in innovative design? These are the questions we shall seek to answer in the remainder of this book. First of all, we shall define certain key terms (innovation, design regime), and then we shall give a general framework (canonical model) to gradually enable us to set out the various models we intend to study.

1.3 The Issue of Definitions: Capacity for Innovation and Design Regimes

We start by establishing a few definitions.

1.3.1 Traps in the Term "Innovation"

We are not looking to write the history of the word "innovation". The literature in general and the economic sciences in particular are in agreement over the definition derived from econometrics, according to which "an innovation is an invention transformed into a product sold in the marketplace" (see e.g. Schumpeter's disciple McLaurin, who was among the first to develop an econometry of innovation (Maclaurin 1953)).

From a management perspective, this definition presents a difficulty: regarding innovation as a judgement on an *existing* product or service (since it can be found on the market), this leads inevitably to *ex post* assessments where the management of innovation demands an ex ante capacity for action.

A second difficulty relates to the contrast between this definition and the history of innovation: by favoring a singular vision of innovation, the definition has a tendency to mask the phenomena of repetition, diffusion and expansion characteristic of the innovation-system as, for example, the history of the railways demonstrate. Far from the figures of some singular achievement such as created by the Renaissance engineers (e.g. the Duomo of Santa Maria del Fiore in Florence built by Brunelleschi without scaffolding), innovation in the railways presupposes a world of coordinated innovation: new forms of energy, new materials, new infrastructures, new sciences, new uses, new urban planning, new architecture, a new aesthetic, new business models and new forms of collective action (e.g. the limited company or société anonyme).

Finally, this definition leaves intact the enigma of innovation *today*: much time is spent talking about innovation, but are we being unduly influenced by fashion? Is this craze for innovation merely cyclic or superficial, and are we discussing innovation just because we have forgotten the old "recipes"? But more significantly, we may be discussing a new industrial revolution.

Initially at least, this course will avoid debating notions of innovation, but will rather describe the logic of action that may *lead to* innovation. Thus the notion of capacity for innovation, described as a capacity for design and steered more towards collective action, seems more appropriate.

1.3.2   The Notion of Design Regime

Although less ambiguous and more promising in describing the process of innovation, the notion of design is loaded with restrictive interpretations. The former AFNOR (*Association Française de Normalisation* - French standards organization) definition is an example of such restrictions. According to standard X50-127 of January 1988, design consists of a *creative activity which, on the basis of expressed needs and existing knowledge, results in the definition of a product satisfying these needs and which is industrially manufacturable*. This definition, since revised and extended, is a good illustration of the classical restrictions: existing needs might exist, and the necessary knowledge is available. Conversely, we will gradually work out a definition of design (see C-K theory, course 3) combining the creation of knowledge, the generation of alternatives and new worlds.

Replacing the term innovation by that of design, and expanding the definition of design, remains insufficient to describe the *logic of collective action* in design leading to innovation. From a *management* perspective, several facets may be added. Hence, in the remainder of the course, we will be introducing the original notion of **design regimes**, characterized by three dimensions: a particular design reasoning, a form of collective organization, and a performance logic (specific to the company or ecosystem). The notion of regime, common in economics, has the advantage that it emphasizes the long-lasting nature of collective action (a design regime is a form of industrial design characterized by a certain stability over the course of time), varied forms (a priori, there are several regimes) and transitions (regimes succeed one another and even overlap).

However, the specific features of this notion of regime as applied to design cannot be overemphasized. To clarify this point: remember that in a production regime products and skills appear as stable structural elements—product architecture and distribution of knowledge can explain certain productive regimes. By contrast, design involves a *renewal* of the world of objects and the universe of skills—what, therefore, are those stable elements which constitute a design regime? Could there even be stable elements for a regime of action characterized by deviance and subversion? The precise advantage of the notion of design regime will be that of setting out those structural elements forming the basis of the specific dynamics of the trends in skills and products.

1.4 Canonical Model for a Design Regime

In the remainder of the course we shall deal with several design regimes, each characterized by forms of reasoning, forms of organization and particular logics of performance (Le Masson and Weil 2008). These regimes are contained within a very general non-specific canonical model, a sort of minimal framework with which to obtain a sufficiently specific description for a regime. To have a complete and consistent representation of a regime, the canonical model must set out the dimensions to be described.

Describing a regime will consist of describing the reasoning processes, organization and performance of the design regime in a canonical framework. We shall do this next.

1.4.1 Canonical Model of Reasoning X, K(X), D(X), P(X)

A design regime is first characterized by a particular model of reasoning. This idea of a model of reasoning may appear rather abstract. However, an understanding of the specific nature of design reasoning is critical, in particular with respect to the thought processes around scientific modeling or decisional reasoning (see enigma 3 above).

1.4.1.1 Intuitive Approach

We can give a few examples of questions requiring a form of design reasoning see table 1.1:

- find an even number between 1 and 3,
- find a statistical estimate of an unknown parameter for a known family of distributions,
- find the solutions to an equation for which an analytical solution is not known,
- find the shortest route in an unknown country
- design a brake to bring a vehicle at 100 km/hr to rest in 20 m.
- design magic lighting for the interior of a car.

Disregarding the entertaining nature of these examples, note that they also correspond to some specific aspects of industrial design. The designer is "visible" in cases 5 and 6; let us show a few analogies for cases 1–4: from a catalogue of components suggested by a supplier, a designer will have been able to choose the component corresponding to his criteria (case 1), he will have to choose a certain procedure taking account of the performance spread in the samples (case 2), he will have to find the optimal dimensioning for systems of complex equations (case 3), and he will have to discover better alternatives at certain stages of design (identifying a less expensive supplier for a key component—case 4).

Via a quadruple continuum, these examples illustrate the dimensions of what can be designed. Hence we move on:

1. from the logic of selecting an object within a list of known objects X_i to the design logic of a new object X_x;
2. from a logic in which all knowledge is available to a logic in which new knowledge must be produced (the unknown country, etc.);
3. from a logic in which the parameters required to design the object are known (a number or path such as a set of ordered geographical points, etc.) to a logic for which the design parameters are unknown (the braking technologies required are not necessarily those of today—no known technology can provide the braking distance demanded);
4. and finally from a logic in which the expected properties of the object and its performance parameters are known (parity, quality of the estimator, the "shortest" braking distance) to a logic wherein the expected properties of the object are to be expanded upon or specified (the "magic" of the lighting).

The ambitions of design questions depend on the expected explorations over four dimensions: the definition of the object X, the knowledge K(X) associated with the object (its environment, use, governing "laws", etc.), the possible decisions to be taken to make the object D(X) exist (technologies, design choices, etc.), and the expected properties P(X) of the object to be designed (Table 1.1).

1.4.1.2 Decision Theory Approach

The canonical model X, K(X), D(X), P(X) (Hatchuel 2003) is a generalization of decision theories (see Savage 1972; Wald 1950; Raïffa 1968). The most general decision theories model decision reasoning in the following manner: given a set of objects X and a knowledge of these objects K(X) (a random variable modeling certain undecidable parameters for this object - probability density of the size of a product's market, for example), possible decisions about this object D(X) (choice of a particular technology for the product) and a level of performance P(X) associated with the decision (minimizing the expectation of a cost function), decision theories can define the best decision function to minimize the cost expectation.

This result is very powerful: it incorporates, for example, the notions of uncertainty (random variable over states), notions of belief (subjective probability over states), forms of learning by sampling the state space (Bayesian decision theory in an uncertain environment), etc. However, a few restrictive conditions remain as to the structure of X, K(X), P(X), D(X)—these conditions are critical for design issues: the possible decisions D(X) about object X are known from the beginning $(D(X) \subset K(X))$ and a solution exists only if P(X) is possible in K(X) (X_0 exists in K(X) such that $P(X_0)$ is true). In other words, decision theory allows us to choose the best alternative among the known alternatives, but does not allow the creation of new alternatives.

Table 1.1 Intuitive approach of the various forms of design reasoning

Design/find		X	K	D	P
An even number between1 and 3	The set of objects is known (an even number takes the form 2n, where n is an integer); known choice function: browse through all integers between 1 and 3 and test	Integer	Peano (successors of 1); parity test, "<" and ">" tests	Algorithm: 1, 2,… and test integer division by 2	Parity
Statistical estimate (of m for an LG(m, σsigma) (e.g. find the average price per m² given n transactions)	For simple distributions, P and D are in K	A statistic, i.e. a function of the X_i;	Estimation theory (Minimum variance unbiased estimator, Maximum likelihood estimator); Sampling;	Follow one of the known procedures (unbiased estimator then exhaustive stats; or maximum likelihood estimator, etc.); calculate the value for sampling	Minimum variance, unbiased, calculable, etc.
Solutions to an equation for which an analytical solution is not known	D and P in K (if we have proved that there exists a solution in a known set of e.g. real numbers; otherwise it is more complicated!)	A real number	Check that the equation is soluble with a real number; algorithms to identify real numbers, etc.	Successive approximation algorithms, etc.	Check the equation
The shortest route in an unknown country,	The set of routes is unknown. Hence D is not in K; P is in K	A path made up of sections, some of which are known, others not yet	Certain sections; calculation of length	Follow the routes; but the know routes are not sufficient (new routes can be discovered, but without knowing the results beforehand)	Path as short as possible (we know how to evaluate P)
Braking system	The set of systems with this braking distance is unknown. D is not in K; P is in K	A braking system (the nature of some components is unknown)	Known (but inadequate) braking systems; ability to calculate the braking distance at 100 km/hr	Use components from past braking systems (or more generally, the language). However, that is insufficient for the braking distance required; D for P is therefore partially unknown	Braking distance of 20 m at 100 km/hr: known, can be calculated (we know how to calculate P)
Magic lighting for a car	Neither D nor P are in K	A lighting system (the nature of the components and the nature of the performance are partially unknown)	Past lighting systems (and their associated languages—but the "magic" bit is irrelevant…)	Use the components (or languages) of past lighting systems; however, this is not sufficient to make a magic lighting system. D for P is partially unknown	"magic": some meanings are known, others are not (P partially unknown)

With respect to the standard decisional framework, design starts off with given values for X, K(X), D(X) and P(X) but is concerned with cases for which the propositions P(X) are not true in K(X), i.e. where K(X) does not imply P(X); this means that there are no decisions D(X) included in K(X) allowing P(X) to be constructed. We are interested in a new, unknown X_x for which certain properties can only be stated gradually; in particular, $P(X_x)$ must be true.

To clarify (see the "modeling-optimization-design" workshop): not only is X_x verifying $P(X_x)$ *unknown* but, unlike certain forms of modeling, it is also considered to be *unobservable*: scientific modeling may be confronted by cases where the object is unknown (X-rays, Pluto, planet X, etc.) but it postulates these objects as observable, i.e. it assumes that these objects can be characterized by known variables (mass, position, energy, wavelength, etc.); in design, the object X is not only unknown but unobservable, meaning that the dimensions characterizing it may be the result of exploration, and that they may take unexpected forms.

The canonical model enables us to suggest a more precise definition of design:

Definition of design. The design of X_x is the set of decisions D to be taken and knowledge to be created (δK) so that X_x is be known in the new extended K(X), K'(X). Initially we have $D(X_x) \not\subset K(X)$ (we do not know what decisions to take); after the design, it comes down to the situation $D(X_x) \subset K'(X)$.

1.4.1.3   Canonical Model and Regimes

Next we shall study models of design reasoning characterized by the particular nature of the conjoint expansions of D and K. A priori, in the above model, design reasoning may require an infinite number of learning processes and decisions, and there is nothing to guarantee convergence of the process to a new X_x. We shall see that design regimes are based on theories guaranteeing a certain robustness of reasoning. In particular, we shall show that **theories of rule-based design** preserve as far as possible the initial knowledge base K(X) in K'(X) and minimize the expansions, at the price of restrictions on acceptable $P(X_x)$; conversely, **theories of innovative design** consist of redefining certain properties of the object in K(X), i.e. sometimes profoundly redefining K(X), then enabling the new $P(X_x)$ to be tackled.

1.4.2   Canonical Model of Performance

The above model invites us to specify a general performance model of a design regime. Drawing our inspiration from classical efficiency factor models, we propose to construct a canonical performance model by considering the inputs and outputs of the design process.

The inputs will be the design resources associated with a particular regime. For example, staff or R&D efforts. More generally, we are interested in types of knowledge and the capacity for producing knowledge.

Fig. 1.4 Diagram of the
canonical performance model
of a design regime

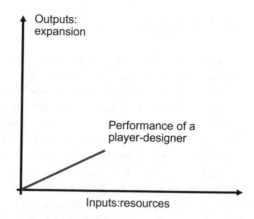

The outputs will be the expansions that have been made (the new X_x and the new associated $P(X_x)$. We will ponder the nature and variety of what has been designed (incremental or radical innovation, continuous or disruptive, etc.).

One of the particular features of design regimes, as compared with production regimes for example, involves the nature of inputs and outputs. While a production regime tends to count the quantities of a known item and known production factors (at output and input respectively), a design regime will be characterized by the particular nature of the resources and expansions it takes into account. For a particular regime this is precisely one of the issues, i.e. defining what will be the nature of these inputs and outputs. The expansive power of design regimes will depend on these inputs and outputs, as we shall see.

The performance of a regime could be studied as the possible expansions from a set of given resources. We will then find various forms of effectiveness (Fig. 1.4).

1.4.3 Canonical Model of Organization

The canonical model of the organization of design comprises two aspects: forms of coordination between the design capabilities, but also forms of cohesion (Segrestin 2006). Describing the forms of coordination consists of focusing on the division of work, responsibilities, forms of prescription, tasks and their interdependencies, resource management, etc.

Describing cohesion consists of examining the reasons behind making a collective a collective: we examine the potential "common purpose" (Barnard 1938), shared interests and forms of solidarity. We shall see that, in rule-based design regimes, the question of cohesion is supposed not to arise (we design the X_x which correspond with the $P(X_x)$ associated with a common purpose and with the company name). With an innovative design, exploration may lead to exploring those $P(X_x)$ likely to readdress the common purpose, either because they contradict certain strategic directions of the company, or because they are an invitation to define new directions.

1.5 Conclusion

In conclusion we can say that a design regime will be described when we have defined each of the three dimensions, namely design reasoning, performance and organization. For each of these we now have a canonical model, a formal framework allowing us to give this description an operational basis.

On the basis of these models we can distinguish two families of regime: rule-based design regimes in which a stable corpus of rules builds reasoning, organization and performance; and innovative design regimes, characterized by an ability to create regular disruptions in the systems of design rules. The first of these lies at the heart of the great contemporary R&D based enterprises, and more generally, at the heart of the organization of industrial development. The second is very much to the fore nowadays, and provides a completion of rule-based design, forming the basis of the transformation of today's firms and their ecosystems.

These are the regimes to which we now turn our attention.

1.5.1 Main Ideas of the Chapter

- Paradoxes: the paradox of R&D, the "good ideas" paradox, the paradox of rigorous creative thinking.
- The notion of design regime.
- Canonical model of reasoning: X, K(X), P(X), D(X); unknown character and supposed unobservable of the X_x to be designed.
- Canonical model of performance: design resource and design output.
- Canonical model of the organization of design: coordination and cohesion.

1.5.2 Additional Reading

For the most part, the ideas in this chapter are taken from the work of several authors. For greater depth of understanding, the following reading is recommended:

- On decision and design: decision theory (Savage 1972; Wald 1950; Raïffa 1968); problem design and solution (Simon 1969, 1979; Hatchuel 2002; Dorst 2006); decision—modeling—design (Hatchuel et al. 2013)
- On the notion of design regime: (Hatchuel and Weil 2008; Le Masson and Weil 2008); see also the notion of innovation regime (Godoe 2000); ideas of Product Life Cycle, Dominant Design, Path Dependency, Path Creation, etc. will be discussed in the next chapters.
- On the paradoxes of innovation: see from this chapter to Chap. 3 in (Le Masson et al. 2006, 2010).
- On cohesion and coordination see (Segrestin 2005, 2006; Barnard 1938; Hatchuel 1996).

References

Barnard CI (1938) *The Functions of the Executive*. Harvard University Press.

Boden MA (1990) *The creative mind. Myths and Mechanisms*. George Weidenfeld and Nicolson Ltd.

Chapel V (1997) La croissance par l'innovation intensive: de la dynamique d'apprentissage à la révélation d'un modèle industriel, le cas Téfal, Thèse de doctorat en Ingénierie et Gestion (thèse), Ecole des Mines de Paris, Ingénierie et Gestion, Paris. 274 p.

Dorst K (2006) Design Problems and Design Paradoxes. *Design Issues* 22 (3):4–17.

Drucker PF (1985) The Discipline of Innovation. *Harvard Business Review*.

Godoe H (2000) Innovation regimes, R&D and radical innovations in telecommunications. *Research Policy* 29:1033–1046.

Hatchuel A (1996) Coordination and control. In: *International Encyclopedia of Business and Management*. Thomson Business Press, pp 762–770.

Hatchuel A (2002) Towards Design Theory and expandable rationality: the unfinished program of Herbert Simon. *Journal of Management and Governance* 5 (3-4):260–273.

Hatchuel A (2003) Théorie de la Conception. Cours Ecole des Mines, Option Ingénierie de la Conception. MINES ParisTech.

Hatchuel A, Reich Y, Le Masson P, Weil B, Kazakçi AO (2013) Beyond Models and Decisions: Situating Design through generative functions. Paper presented at the *International Conference on Engineering Design, ICED'13*, Séoul, Korea.

Hatchuel A, Weil B (eds) (2008) *Les nouveaux régimes de la conception*. Vuibert-FNEGE, Paris.

Jaruselski B, Dehoff K, Bordia R (2005) The Booz Allen Hamilton Global Innovation 1000: Money Isn't Everything. *Strategy and Business* Winter 2005 (41):15.

Jaruzelski B, Loehr J, Holman R (2012) Making Ideas Work. *Strategy + Business* 69 (Winter).

Jones CI (1995) Time Series Tests of Endogenous Growth Models. *Quarterly Journal of Economics* May 1995:495–525.

Le Masson P, Weil B (2008) La domestication de l'innovation par les entreprises industrielles : l'invention des bureaux d'études. In: Hatchuel A, Weil B (eds) *Les nouveaux régimes de la conception*. Vuibert-FNEGE, Paris, pp 51–67 (chapitre 53).

Le Masson P, Weil B, Hatchuel A (2006) *Les processus d'innovation. Conception innovante et croissance des entreprises*. Stratégie et management. Hermès, Paris.

Le Masson P, Weil B, Hatchuel A (2010) *Strategic Management of Innovation and Design*. Cambridge University Press, Cambridge.

Maclaurin WR (1953) The Sequence from Invention to Innovation and its relation to economics growth. *Quarterly Journal of Economics* 67 (1):97–111.

Raïffa H (1968) *Decision Analysis*. Addison-Wesley, Reading, MA.

Savage LJ (1972) *The foundations of statistics*. 2nd edition (1st edition: 1954) edn. Dover, New York.

Segrestin B (2005) Partnering to explore: the Renault-Nissan Alliance as a forerunner of new cooperative patterns. *Research Policy* 34:657–672.

Segrestin B (2006) *Innovation et coopération inter-entreprises. Comment gérer les partenariats d'exploration ?* Editions du CNRS, Paris.

Simon HA (1969) *The Sciences of the Artificial*. M.I.T. Press, Cambridge, MA, USA.

Simon HA (ed) (1979) *Models of Thought*, vol 1. Yale University Press, New Haven.

Tidd J, Bessant J, Pavitt K (1997) *Managing Innovation : Integrating Technological, Market and Organizational Change*. John Wiley & sons, Chichester, England.

Wald A (1950) *Statistical Decision Functions*. John Wiley & Sons, New York.

Chapter 2
Designing in a Rule-Based Regime—Systematic Design Theory and Project Management

The first class of design regime we shall analyze in this book is that of rule-based design. This relies on a set of rules for the efficient design of new products (or services), whence the name (Le Masson and Weil 2008). Historically, several rule-based design regimes were gradually established, culminating in 1970's in systematic design, surely the most common and perhaps the most effective: it is this organizational model (often implicit and seemingly natural) which is adopted by the major R&D companies, and it is this model that provides the structure for the logic underlying project management in product development (NPD).

In this chapter we examine the design logic of "rule-based design", i.e. when well formed design resources (i.e. rule systems) are available. Hence in this chapter we shall answer the question: how do designers conceive a new object within the framework of a rule-based design regime? In the next chapter we shall address the design of the rule-based system itself. We shall see how to design a rule-based regime and at the same time discover some of their fundamental properties. In addition, examination of an historical case will provide a description of the gradual build-up of rule-based design in companies; the historical perspective will also show how theories of rule-based design have progressively developed.

Following the analysis framework of the design regimes outlined in the introductory chapter, we shall examine in turn the reasoning processes, performance and organizations, highlighting the tools associated with them.

2.1 Reasoning in Systematic Design

Systematic design forms part of a German tradition of design theory and method that arose at the start of the 19th century (König 1999; Heymann 2005; Le Masson and Weil 2010, 2013). The international work of reference is the manual written by Gerard Pahl and Wolfgang Beitz (Pahl and Beitz 1977; Pahl et al. 2007; Wallace and Blessing 2000), "Engineering design, a systematic approach", published first in

© Springer International Publishing AG 2017
P. Le Masson et al., *Design Theory*,
DOI 10.1007/978-3-319-50277-9_2

German in 1977, translated by Ken Wallace into English in 1984 and republished several times since. This work and its successors are widespread to the point that the theory of systematic design (and its variants) is today commonly taught in engineering design courses the world over.

2.1.1   Expectations of Systematic Theory

Pahl and Betz stipulate that a good design method must satisfy several requirements:

- It must be applicable to the most varied of problems, regardless of the field of specialism—it must therefore be independent of the objects to be designed; however, it must also be compatible with the concepts and methods of the disciplines involved (engineering sciences in particular).
- It does not rely on discovering solutions by chance but contributes to the inventiveness and understanding of the whole, and facilitates the finding of "optimal" solutions; however, wherever possible it must also facilitate the application of known solutions.
- It must be easy to learn and to teach and, "taking account of advances in ergonomics and cognitive psychology", it must be capable of reducing the workload (including the mental load), saving time, avoiding human error and sustaining the interest of the designers. It must also be compatible with the instruments available (in particular the use of computers for data processing).

The systematic design they propose meets these criteria, as we shall see.

2.1.2   Fundamental Principles

According to Pahl and Betz, design follows a linear process which can be broken down into phases.

Reasoning in systematic design comprises four main phases (see Fig. 2.4), with each making use of a specific object language to the exclusion of any other language:

- **Functional design** involves clarifying the design task and of setting out the functional specifications for the future product. Only *the language of functions* is used to describe the object. During the course of this phase, several functional specifications may be developed, several possible requirements sheets may be discussed, and at the end of the phase, just one requirements sheet is retained.
- **Conceptual design** on the basis of the requirements sheet of the previous phase, involves formalizing the functional structure (interdependencies, functions and sub-functions, potential modularization) and mobilizing conceptual models, i.e.

the main techniques and technologies required to fulfill these functions. Only *the language of the main techniques* (the laws of engineering science) is used to design the object—this is no longer a matter of discussing the functions or indeed of discussing the components. Several conceptual alternatives must be developed, with one of these being selected by the end of the phase.

- **Embodiment, or morphological design** on the basis of the technological scheme obtained in the previous phase, consists of proceeding to the "organized assembly" of the various component parts. Only *the language of components and their inclusion in a coherent whole* is used; no alternative technologies are discussed, and neither are the exact dimensions of the components. This is an architectural design phase, where we speak in terms of components, modules, parts, assembly procedures, etc. Again, several embodiments are developed during this phase, with one of them being selected.
- **Detailed design** which, on the basis of the previous embodiment, involves dimensioning all the free parameters (sizing of parts, procedure configuration, material identification, suppliers, etc.). This is the language of dimensionalization and product reference.

It is possible to design in one of the phases using inputs from the previous phases, independently of subsequent phases (Fig. 2.1).

From a formal point of view, systematic theory takes account of the $P(X)$ (functional specifications), checks whether they are unattainable with known solutions ($K(X)$ does not imply P) during selection from the specifications sheet (if they are, then the design becomes a form of optimization); the process is then a series of "decisions" d_i taken from $D(X)$ gradually defining a family of objects verifying the initial (PX). The list of d_i is not completely known at the start but the process does facilitate their development, on the one hand by structuring a priori the types of d_i (functional, conceptual embodiment, detailed), and on the other hand by enabling each stage to re-use, post hoc, the known d_i at each of the levels of language.

The reasoning structure in different languages for the object can be interpreted as a division of $\{X, K(X), P(X), D(X)\}$ into four sub-spaces $\{X_{func}, K(X_{func}), D(X_{func}), P(X_{func})\}$, $\{X_{conc}, K(X_{conc}), D(X_{conc}), P(X_{conc})\}$, $\{X_{emb}, K(X_{emb}), D(X_{emb}), P(X_{emb})\}$ et $\{X_{det}, K(X_{det}), D(X_{det}), P(X_{det})\}$; in each of the sub-spaces the available knowledge may, or may not, allow the proposed concept to be achieved (again, if it can, we find ourselves back in a decision-optimization situation), and if it cannot, the decisions to be taken about the object will result from a learning process associated with the level of language used and restricted just to this level of language. In other words, the segmentation into language types guides the search for new d_i: even if at each level we have $D(Xx_{level\ j}) \not\subset K(X_{level\ j})$, the expansions necessary to have $K'(X_{level\ j})$ such that $D(Xx_{level\ j}) \subset K'(X_{level\ j})$ are confined to this level, thus limiting the "distance" between $K(X_{level\ j})$ and $K'(X_{level\ j})$.

Several essential elements are introduced by systematic design. Without going into too much detail here, we can say immediately that (1) systematic design combines the logics of convergence and divergence; (2) it tends, paradoxically, to "slow down" the process of design by avoiding complete, pre-existing solutions but

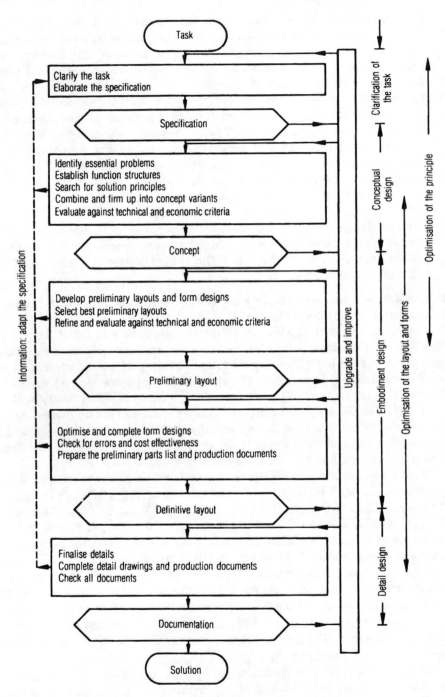

Fig. 2.1 The four main languages of systematic design in the sense of Pahl and Betz

in fact it is there to preserve forms of "technical creativity" and avoid "fixations"; (3) systematic design relies in particular on the key stage of the conceptual phase where abstract technical languages can be used to design the object—a "conceptual wedge" (like a log splitter!) for separating functions and components, thus ensuring a richer and more complex relationship between these two terms.

2.1.3   Illustrative Examples of Language

We give an example of the different languages for an object in the case of a refrigerator (the reader can practise by trying to do the exercise himself before reading on) (see Figs. 2.2 and 2.3).

- Functional language: this is the language of the object's conditions of value (e.g. sales criteria) and existence (e.g. commercial standards). Hence we find: "volume (in liters) at a temperature (a few degrees or tenths of a degree, more or less)", standards (for safety and power consumption, fluids and recyclability, etc.), reliability and robustness (in specific usage scenarios: opening & closing the door; air-tightness in a standardized humid atmosphere, etc.), shelves, ease of upkeep, noise, automatic defrosting, temperature indicator, etc (Fig. 2.2).
- language of embodiment: this is the language of components, often presupposing that the object itself be analyzed (disassembly). For the refrigerator we find the "cabinet" (enclosure, door, etc.) and then, on the other side, a grille (technically called a "condenser"), a black bowl-shaped object which on

Fig. 2.2 Some elements for the functional language of the refrigerator

Fig. 2.3 Some elements of the embodiment of the refrigerator

dissection turns out to be a compressor, pipes (including the expansion capillary), insulating foam inside the walls of the enclosure, and tubes inside the insulating foam (technically called an "evaporator"). A more detailed analysis shows that the pipes all connect the compressor, condenser, expansion capillary and evaporator (Fig. 2.3).

- The conceptual language is perhaps the least evident for the refrigerator non-specialist, but is essential for understanding the design of the object. In this case there is no direct correspondence between function and component (a component can be assigned a function or a function a component); it is the conceptual language, which explains the complex relation between functions and components. The main conceptual model of the refrigerator is a two-phase thermodynamic cycle which "makes cold" (the language of function) via the change of phase of a liquid to a gas, this phase change being organized in a complete cycle starting with evaporation (the "cold production" phase), going on to compression, condensation and expansion before returning to evaporation. Such a model may, for example, be represented by a (T, S) diagram (temperature and entropy) with greater or less precision in the model, each phase in the cycle then corresponding to a physical entity (the language of embodiment) (see Fig. 2.4).

Note for the moment that all we have done is describe the object (already known, already designed) in each of the languages. The systematic design reasoning applied to the design of a new refrigerator takes a different form. Let us assume now that we have to design a new range of refrigerators for the elderly. The process follows the logic below:

1. we start by skimming through all possible functions before selecting those deemed relevant for the new product (conservation of medicines, pre-prepared meals, etc.)
2. we then consider all possible conceptual designs for the product. At this stage, designers must in theory look at all possible technologies: cold via the two-phase cycle, but also cold via expansion of a gas or the thermoelectric (Peltier) effect, etc.
3. Once one of these principles has been adopted, we then look at the embodiment, and so on.

(a)

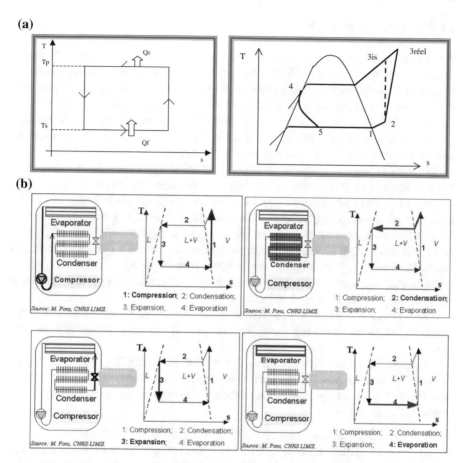

(b)

Fig. 2.4 Some elements of the conceptual language of the refrigerator. Note that the model can have several levels of precision. The first representation (Fig. 2.4a, **top left** on the diagram) provides a crude idea of the magnitudes of the energy exchanges (Qf = quantity of heat extracted; Qc = quantity of heat produced); the second (Fig. 2.4a **top right**) locates the cycle with respect to the phase diagram of the fluid used and in particular checking that the cycle is "good" (i.e. avoiding liquid "blips" in the compressor by ensuring that all the fluid is completely gaseous at the inlet to the compressor); on the bottom (Fig. 2.4b): the diagram makes each phase of the cycle correspond to a component of the embodiment: **top left**: compression requires a compressor, then on the right condensation uses a condenser, then on the **bottom left** the expansion device uses a throttling valve and on the **bottom right** evaporation takes place in an evaporator

2.1.4 Tools and Associated Techniques

The way in which reasoning is structured according to the systematic design model has given rise to the development of numerous tools associated with each of the phases. We mention three types of tool: functional analysis, knowledge management catalogues, and Computer Aided Design (CAD) .

2.1.4.1 Functional Analysis

Functional language is one of the critical languages of systematic design. Thanks to functional analysis an object can be discussed without evoking any constructional techniques, i.e. talking about the object even though it may not yet exist, and with the aim of preserving degrees of freedom for those engineers who will have to design it. The functional language also carries a concomitant logic of validation essential for design. It must be possible to validate a function, i.e. it must be associated with a protocol allowing such a validation.

We shall define a function as an object's condition of existence or value, from the perspective of a stakeholder, or the object's environment. This language must be as abstract as possible in describing the conditions of existence or value to avoid a priori technical solutions ("inform", "communicate", "beguile"), but it must also be as specific and as concrete as possible when it comes to describing the environment or the observer ("inform handicapped persons with motor disabilities who cannot reach the counter…"), to be pertinent in terms of validation.

Note that the same object may perform several functions and that the same function may mobilize several objects.

The fundamental assumption of functional analysis is that there exists a minimal group of functions qualifying the object and which are independent of how it is made.

Example: "Functional description of a system for controlling traffic at a road crossing". It must be possible to describe the functions without knowing beforehand whether to use traffic lights or a roundabout. The reader can do this exercise for himself. For example:

- F1: the system must inform drivers of the existence of the crossing.
- F2: the system must allow each driver to know what vehicles are at the crossing.
- F3: it must allow each driver to take the correct decisions when approaching the crossing.
- F4: it must clearly describe the rules of the road and any contraventions that may result in a (police) ticket.
- F5: as far as possible, the system should avoid any risk of collision.

In practice, we refer to the functional analysis workshop for more detailed elements.

This type of analysis can be refined using additional tools. Hence a functional analysis may give rise to a value analysis leading to a ranking of the functions according to their customer value (see the recent ISO standards on "value management"). Hauser and Clausing (1988) proposed building functional analysis into a "house of quality", providing a relationship between functions (functional requirements, FR, in columns) and "Customer attributes", weighted with respect to the competitor's bid (see Fig. 2.5).

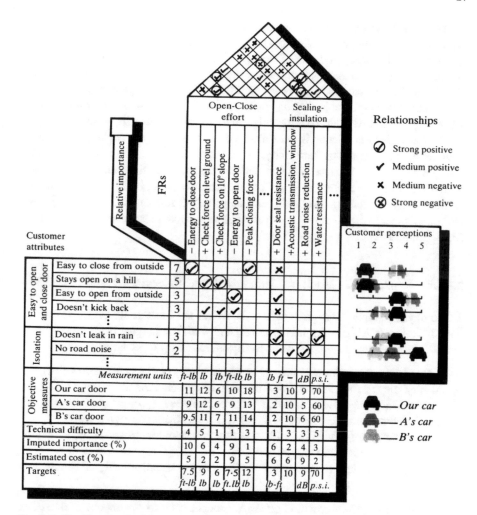

Fig. 2.5 The "house of quality" according to Hauser and Clausing

2.1.4.2   Catalogs and Knowledge Management in Systematic Design

Corresponding to systematic design reasoning is an equally systematic knowledge management: for it to be readily activated, knowledge is organized in accordance with the languages of design (knowledge of functions, conceptual design, embodiment, etc.) and conversely, gradually acquired knowledge is accumulated in each of these languages.

Thus work in systematic design can draw on a prepared knowledge base. Hence catalogs of conceptual models can be found for "energy storage" or "change of energy mode" (see Fig. 2.6). The language of embodiment will also be structured, based in particular on the list of "recommendations" set out (see the recommendations below on ease of assembly or disassembly for recycling purposes) (see Fig. 2.7).

The way in which reasoning is structured provides a knowledge management which overcomes the two classical obstacles to knowledge management as a matter of course: difficulty of use and cost of storage. This is because of the strict relationship between design stage and type of knowledge (functional, conceptual, embodiment and detailed level) which on the one hand allows the knowledge user to understand immediately what type of knowledge he should make use of, and conversely that the knowledge produced is stored appropriately to enable its effective re-use.

Note also that knowledge is not presented as is usual in the engineering science courses of an engineering university, where knowledge is grouped under major disciplines such as thermodynamics and electricity, etc. Systematic design will organize knowledge according to an entirely different logic, by listing, for example, all the energy conversion modes (in each engineering science discipline it will seek the phenomena within those sciences that are capable of energy conversion) (Figs. 2.6 and 2.7).

2.1.4.3 CAD, Digital Mock-up, Simulation and Validation Tools

These days systematic design is able to draw on the tools provided by Computer Aided Design (CAD), being highly consistent with the languages of embodiment and detailed design. Today's CAD and digital mock-up tools have gradually:

- enabled complex shapes to be mastered (freeform surfaces in aeronautics)
- enabled the resizing and local optimization of certain parameters to be facilitated
- enabled integrated technical data to be managed (product-process-resource management over the product's entire life cycle from concept to scrap)
- aided validation (simulation tools and tests on numerical mock-ups; validation process of "workstation ergonomics" type)
- enabled the product at the upstream phase to be visualized (digital demonstrators, virtual reality)

For more in-depth information on these aspects, refer to (Daloz et al. 2010). For an historical approach to CAD see also (Fridenson 2015).

2.1.5 Contemporary Trends in Rule-Based Design

We mention three of today's trends in rule-based design.

2.1.5.1 TRIZ, or the Temptation of Universal Conceptual Models

The TRIZ method provides a more systematic treatment of the conceptual design phase, enabling the transition from a specific to a general problem: by studying the bases of patents, work carried out by G. Altshuller and his teams showed that all the

Type of energy / Working principle	mechanical	hydraulic	electrical	thermal
1	Pot. energy (m, h)	Liquid reservoir (pot. energy)	Battery	Mass (m, s, Δv)
2	Moving mass (transl.) (v, m)	Flowing liquid	Capacitor (electr. field) C	Heated liquid
3	Flywheel (rot.) (J, ω)			Superheated steam
4	Wheel on inclined plane (rot. + transl. + pot.) (J, ω, v)			
5	Metal spring (d, F)	Other springs (compr. against fluid + gas) Δp; ΔV		
6		Hydraulic reservoir a. Bladder b. Piston c. Membrane (Pressure energy)		

Function	Input	Output	Physical effects						
$E_{mech} \to E_{mech}$	Force, pressure, torque	Length, angle	Hooke (Tension/compression/bending)	Shear, torsion	Upthrust Poisson's effect	Boyle-Mariotte	Coulomb I and II	...	...
		Speed	Energy Law	Conservation of momentum	Conservation of angular momentum	...		...	
		Acceleration	Newton's Law	...	...	...	...	...	...
	Length, angle	Force, pressure, torque	Hooke	Shear, torsion	Gravity	Upthrust	Boyle-Mariotte	Capillary	
			Coulomb I and II	...	...	...	...	...	
		Speed	Coriolis force	Conservation of momentum	Magnus-effect	Energy law	Centrifugal force	Eddy current	
		Acceleration	Newton's Law	...	...	...	...	...	...
$E_{mech} \to E_{hyd}$	Force, length, speed, pressure	Speed, pressure	Bernoulli	Viscosity (Newton)	Toricelli	Gravitational pressure	Boyle-Mariotte	Conservation of momentum	...
$E_{hyd} \to E_{mech}$	Speed	Force, length	Profile ↑↓	Turbulence	Magnus-effect	Flow resistance	Back pressure	Reaction principle	...
$E_{mech} \to E_{therm}$	Force, speed	Temperature, quantity of heat	Friction (Coulomb)	1st law	Thomson-Joule	Hysteresis (damping)	Plastic deformation	...	
$E_{therm} \to E_{mech}$	Temperature, heat	Force, pressure, length	Thermal expansion	Steam pressure	Gas Law	Osmotic pressure	...	...	
$E_{elect} \to E_{mech}$	Voltage, current, magn field	Force, speed, pressure	Biot-Savart-effect	Electro-kinetic effect	Coulomb I	Capacitance effect	Johnson-Rhabeck-effect	Piezoeffect	...
$E_{mech} \to E_{elec}$	Force, length, speed, pressure	Voltage, current	Induction	Electro-kinetics	Electro-dynamic effect	Piezoeffect	Frictional electricity	Capacitance effect	...
$E_{elec} \to E_{therm}$	Voltage, current	Temperature, heat	Joule heating	Peltier-effect	Electric arc	Eddy current	...	...	
$E_{therm} \to E_{elec}$	Temperature, heat	Voltage, current	Elect. conduction	Thermo-effect	Thermionic emission	Pyroelectricity	Noise-effect	Semiconductor Super-conductor	...
$E_{mech} \to E_{mech}$	Force, length, pressure, speed	Force, length, pressure, speed	Lever	Wedge	Poisson's effect	Friction	Crank	Hydraulic effect	...
$E_{hyd} \to E_{hyd}$	Pressure, speed	Pressure, speed	Continuity	Bernoulli	...	...	...	...	...
$E_{therm} \to E_{therm}$	Temperature, heat	Temperature, heat	Heat conduction	Convection	Radiation	Condensation	Evaporation	Freezing	...
$E_{elec} \to E_{elec}$	Voltage, current	Voltage, current	Transformer	Valve	Transistor	Transducer	Thermogalvanometer	Ohm's law	...
...	...	...	...	...	...	...	...	...	...

Fig. 2.6 Various principles for "energy storage" (*top*) and energy conversion (*bottom*) *Source* (Pahl et al. 2007)

Oper.	Guidelines	Type	Wrong	Right
Simplify interfaces				
Jo	Avoid hindering caused by air cushions.	MA AA		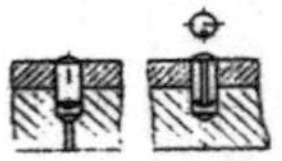
Jo	Provide tapering to ease joining.	MA AA	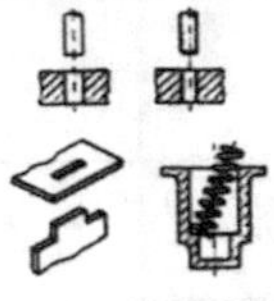 after [7.42, 7.258]	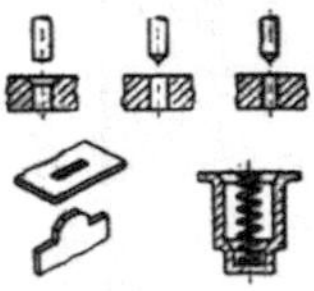
Jo	Divide large interfaces into several smaller ones.	MA AA	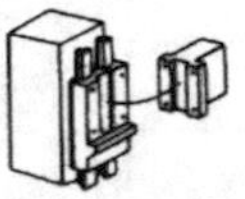	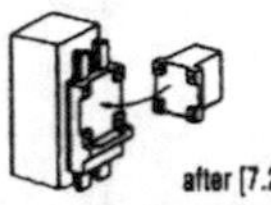 after [7.238, 7.258]
Jo Ad	Avoid simultaneous operations that influence each other.	MA AA	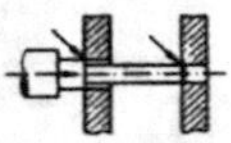	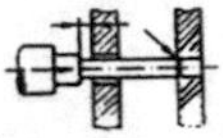 after [7.42]
Jo Ad	Provide access for assembly tools.	MA AA		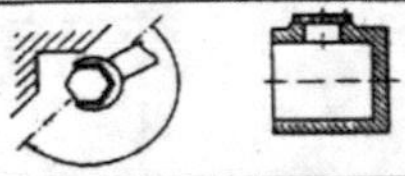
Jo Ad Se	Prefer connecting elements with elastic, elastic-plastic or material tolerance compensation.	MA AA	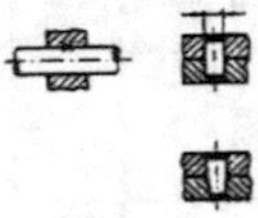	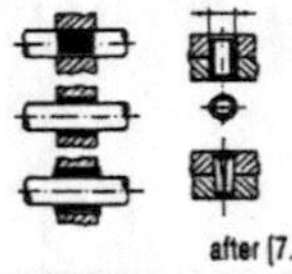 knurled / tolerance ring / cast after [7.42]
Jo Se	Allow for large tolerances through assembly parts that are flexible.	MA AA		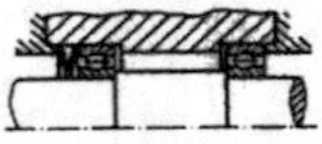
Ad	Adapt using standardised matching parts without disassembling.	MA AA		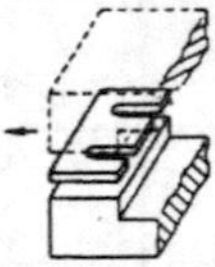
Se	Apply locking elements that are easy to assemble.	AA	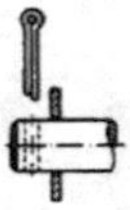	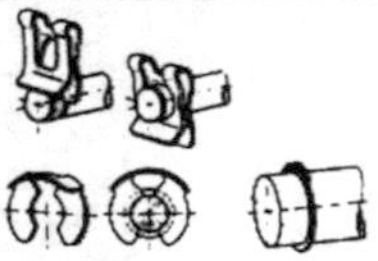

Fig. 2.7 Recommendations for ease of assembly *Source* (Pahl et al. 2007)

problems resulting in patents involved resolving a contradiction between two primary techniques among a finite list of 39 principles. Hence there are just 39×39 possible contradictions. They also showed that these contradictions were always resolved by using one of the 40 inventive principles they listed (Altshuller 1984).

Hence the TRIZ method offers a rich catalog of conceptual models and a method for activating them.

Note that the TRIZ method is often invoked in tackling questions of innovation in the broadest sense. We should emphasize here that the method has two special properties: it presupposes a good knowledge of the object ($K(X)$ already important) so that any critical contradiction(s) can be characterized; this leads to a decision about what technical principle to choose among a finite list – it does not model the learning process, apart from the 39×39 contradictions and the 40 inventive principles. These two special features make it particularly consistent with a systematic design reasoning which relies precisely on an extensive knowledge base and on a constant effort to limit the production of new knowledge. On the other hand, these two conditions make it less effective for innovative design situations such as those we shall examine in Chaps. 4 and 5 (for a more in-depth discussion of the method, see (Rasovska et al. 2009; Reich et al. 2010)).

2.1.5.2   Extension of Systematic Design to Other Designers and Other Objects

Today's numerical modeling and PLM (Product Life Cycle management) tools are able to extend the design to other participants previously marginalized by the traditional processes of engineering design. Hence systematic design processes can be deployed for product distribution, maintenance or after-sales: these participants, often treated as simple "producers" charged with carrying out routine tasks, may however contribute to the design not only by outlining their own functional requirements (as they may already have done in the systematic design model), but also by bringing in their own specific design variables (logistical scheme, original promotional campaign, after-sales contract), etc. The new tools leave these parameters "free" over the course of the process (or manage the gradual stress they are put under) and enable skilled designers in their turn to take part in the process.

Systematic design and its associated tools have gradually conquered numerous fields. These days it is not just "machines" (cars, aircraft, machine-tools, telephones, microprocessors, etc.) that are designed on these principles, but also software, medicines, buildings, urban areas, insurance contracts or banking services.

2.1.5.3   "Parameter Analysis" Approaches

Some authors have shown that, in certain cases, systematic design can slow down the design of an individual object (Ehrlenspiel 1995) by committing the designer too early and too comprehensively to dimensions which are, of course, necessary for the final object but not always necessary in the exploratory phases. Hence the author of a specifications sheet may list obvious standard functions which are easy to realize along with those that are hard to attain requiring more intensive explorations, without distinguishing between them; moreover, the exploratory leads suggested by systematic design might fall within complex superabundant combinations, sometimes a little sterile (see the combinatorics opened by the morphological matrices of Zwicky (1969)).

This is why, all the while preserving the logic of rule-based design (use of a knowledge base restricting exploration and minimizing any challenges to knowledge), more exploratory processes have been proposed. Such is the case for the "parameter analysis" method (Kroll et al. 2001, 2013), in which exploration focuses on several critical parameters before then reverting to a systematic design logic. This reasoning is based on the fact that parameters not instantiated at one level (conceptual alternatives can be explored without having all the functions, for example) are not necessary for exploration (even though they may ultimately be necessary for the final object) and that taking them into account would render the exploration less effective (contrary to the normal assumption of systematic design which avoids too broad an exploration through the constraints of the specifications sheet).

For a detailed study of the Parameter Analysis method see (Kroll 2013). It can be shown that the logic of Parameter Analysis is an extension of the Branch and Bound logic applicable to design situations (Kroll et al. 2013, 2014).

Note that the three developments mentioned satisfy the fundamental assumptions of systematic design reviewed below in conclusion:

- a linear process that can be broken down into phases
- each phase makes use of a language specific to the object and to the exclusion of others
- there are four object languages: functional, conceptual, embodiment, and detailed
- it is possible to design in one phase independently of subsequent phases

2.2   Performance in Systematic Design

2.2.1   Fundamental Principle: Maximizing the Re-use of Knowledge

The performance of a project under systematic design is based on a fundamental principle: maximize knowledge re-use to design X_x such that $P(X_x)$ is true.

This general principle appears as two guiding criteria over the course of the process:

- Limitation of explorations. Several aspects of Systematic Design contribute to this:

  - Recurrent test of unknown nature: as soon as it has been shown that the task corresponds to a known set of specifications, or that the specifications corresponds to known technical principles, systematic design tends to an optimization-decision regime.

- If the necessity for exploration has been demonstrated, restrict it to one of the levels of language without pushing the exploration beyond that level (making a complete prototype just to assess certain technical principles is pointless).
- Make an early selection of the alternatives: exploring all the branches of the tree diagram created by all the examined specifications sheets and then all the technical principles for all the specifications and so on is pointless.

- A knowledge aggregation logic:

  - The knowledge produced is incorporated within K(X) and hence can be re-used for subsequent projects.
  - Design reasoning implicitly avoids any challenge to the knowledge base. Explorations of the type "a refrigerator with the same functions but a technical principle different from those with which the company is familiar" run the risk of being quickly brought to a halt.

These principles are therefore able to reconcile exploration and the generation of alternatives with the re-use and maximization of knowledge; they can also ensure forms of divergence while maintaining overall convergence at each phase of the process, thanks, in particular, to a gradual process of validation.

2.2.2   Practical Assessment

This general principle lies at the heart of measuring the performance of a systematic design project. In practice, the performance of a systematic design project is defined by a target and a drift with respect to the target.

2.2.2.1   Project Target—the Idea of NPV

A cost-quality-time (CQT) objective is clearly identified at the start. C represents the cost of development, Q the product or service target (functions, production cost, etc.), and T the development timescales (generally the time initially set between the start date and the intended date for launching the product; in the automobile sector in particular, this date corresponds with the marketing agreement).

This initial CQT objective is validated at the start of the project, and corresponds to an economic equation which characterizes its value. Value is assessed in the same way as the profitability of an investment is assessed, i.e. on the updated earnings or "net present value", NPV. The profit curve is characterized by an initial "entry ticket" (essentially costs) and a production service-life phase (essentially revenue, with deduction made for direct manufacturing costs). The transition into the positive

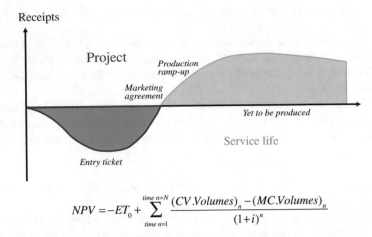

$$NPV = -ET_0 + \sum_{time\ n=1}^{time\ n=N} \frac{(CV.Volumes)_n - (MC.Volumes)_n}{(1+i)^n}$$

Fig. 2.8 The economic logic of a systematic design project and formula for calculating the NPV. ET = entry ticket, CV = customer value (anticipated sales price), MC = Manufacturing costs. In this formula, time n = 1 begins with the marketing agreement and the ET starts at time zero, n = 0

zone occurs with the marketing agreement defining the "timescale" T; the quality criterion Q corresponds to the expected turnover in the market and the anticipated production costs, while the cost criterion C is the "entry ticket" corresponding to the initial production investment (tools, etc.) and design costs (see Fig. 2.8).

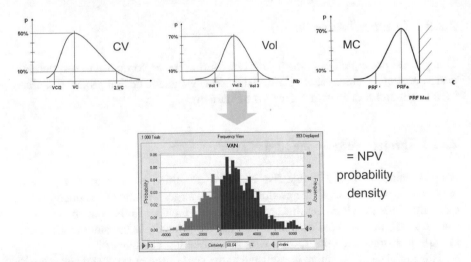

Fig. 2.9 Stochastic simulation for estimating the probability density of the NPV for an innovative project (*Source* (Hooge 2010)). Instead of taking a point value for the customer value (i.e. the forecast sales price), volumes and manufacturing costs, a random variable is taken for each of these terms. The density of the variable is given by experts (the *upper curves*) who can, for example, estimate the lowest price below which there is a 1 in 10 chance of a sale, the median price, and the highest price above which there is a 1 in 10 chance of a sale. Simulation gives the NPV's probability distribution

Note that this economic equation can be expressed in probabilistic terms (turnover, costs and time can be considered as random variables); the expectation of the NPV can then be calculated (see an estimate of the expectation of an NPV using stochastic simulation) (Hooge 2010) (Fig. 2.9).

If the NPV or the expectation of the NPV exceeds the company's conventional thresholds, the project is launched with the aim of ultimately reaching the target CQT.

2.2.2.2   Drift Assessment

Assessing a project consists of assessing a deviation from some initial target. In practice, this assumes that:

- The target is qualified in terms of CQT
- The resources allocated to the project have been identified (initially budgeted for, and reassessed over the course of the project)
- A follow-up is made and a final assessment in terms of CQT: what specifications have been attained? Within what timescales? With what costs?

In the canonical performance model the project "outputs" appear as a minimized drift with respect to a fixed target (realized—expected) (possibly weighted by the updated profit); the "inputs" are the design resources agreed upon. The "efficiency" of a project is therefore measured in terms of the ratio of drift to resources.

This measure invites a few remarks:

1. performance is indeed measured by a *deviation* from an objective: a systematic design project is a priori intended to "best" fit the target, and its objective is not to explore new targets. Note that doing something "faster" or of "higher quality" is not always expected of the company: rather, the product or service should be released "when expected" and at the predefined level of quality. What is being assessed is compliance with the target, not the target itself.
2. It is understood that the success (or failure) of a project depends on the one hand on how well (or badly) the project is guided through the course of the process but also on the definition of the initial target and its deviation from the skills and competence the allotted budget allows. In other words, the initial difference between $P(Xx)$ and $K(X)$ largely determines the future performance of the project. It is this definition of the "difference" between project target and available resources which lies at the heart of the project "contract" that the project leader concludes with the company. The critical points of this negotiation are the definition of reasonable but cost-effective objectives, and allocation of the necessary resources.
3. Measurement of performance must take account of unexpected events, technical and market-related uncertainties. The standard principles do this in two ways:

- vagaries of the market are generally included when calculating the expectation of the updated profit; the CQT target adopted can then be robust against external vagaries (even if the scenario of externally generated hazards is highly unfavorable, the CQT target allows the company to limit its losses). However, that generally leads to a "hardening" of the target (shorter timescales, more demanding specifications, lower costs to enable unfavorable scenarios to be dealt with). Once the CQT target is agreed, the project leader is assessed without always taking account of any externally generated pitfalls that might affect the project. In other words, the project leader is not expected to include these hazards during the course of the project (under systematic design it is assumed that the specifications will not change over the course of the project). Or further, a project leader who has reached his CQT target cannot be held responsible for commercial failure (which would be due to an error in the initial target).

• On the other hand, any technical hiccup is the responsibility of the project leader and his team. Performance is very tightly bound with the ability to gradually reduce the initial uncertainty in order to reach the set target using the allotted resources.

In terms of risk management it can be said that the project leader manages risk related to "market pull": the market uncertainty is under control (by initial market studies and the margins taken on the initially set CQT target), and all that remains is the technical uncertainty, being reduced gradually by the rule-based design project. However, certain projects (coming out of research) may be governed by "technology push", i.e. the technical uncertainties are reduced and there only remains the "market" uncertainty, which market research or commercial predictions gradually aim to reduce.

To conclude: we have a logic in which *uncertainties have been decoupled*, and any reduction in uncertainty is built on *known elements* (there is no technical exploration without market assumptions and no commercial exploration without the available technology).

2.2.2.3 Risk Management in Rule-Based Design—Decision Theory Under Uncertain Conditions and Real Options

Project Selection Tools in an Uncertain Situation

The decision model can be enhanced to take account of externally generated hazards. In decision theory under uncertainty we reason according to the models introduced by Savage, Wald and especially Raïffa (Savage 1972; Wald 1950; Raïffa 1968) (Models for calculating real options are derived from these models (Trigeorgis 1996, 2005)) (see Fig. 2.10).

The figure below outlines the tool's underlying principle, which can be effectively analyzed within the {X, D(X), K(X), P(X)} framework. K(X) is constituted thus: the decider knows a set of alternatives D_i, a set of states of random E_j independent of D_i and whose (subjective) probability $P(E_j)$ is known (the weather, for example: the probability that it will be rainy or fine, on which sales of a certain product may depend), given a certain value V_{ij} associated with the pair (D_i, E_j) (we assume that there exists a utility function U which allows an optimal decision to be constructed—although it may

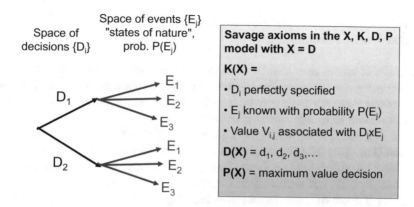

Space of events {E$_j$}

Space of
decisions {D$_i$}

"states of nature",
prob. P(E$_j$)

**Savage axioms in the X, K, D, P
model with X = D**

K(X) =

• D$_i$ perfectly specified

• E$_j$ known with probability P(E$_j$)

• Value V$_{i,j}$ associated with D$_i$xE$_j$

D(X) = d$_1$, d$_2$, d$_3$,...

P(X) = maximum value decision

Fig. 2.10 General framework of the theory of decision under uncertainty (Savage axiom)

be necessary to prove the existence of an optimal solution, we will not discuss this idea here; the interested reader may refer to (Wald 1950)); this decider then maximizes the value of the decision D (the performance function P(X)) taking that decision D$_i$ which maximizes the expectation of utility S_i^* (Fig. 2.10):

$$S_i^* \sum_j U(V_{i,j}) \cdot P(E_j)$$

This type of technique enables a project to be selected from among a set of alternatives taking account of the uncertainty of all possible states of the world. This is a standard tool for choosing product development projects.

The figure below gives an example (a very classical case of decision-making under uncertainty) (see Fig. 2.11). There are two decisions: either to go for a walk with a raincoat (D$_1$) or to go for a walk with a hat (D$_2$). The states of nature are: there will be rain during the walk (E$_1$) or there will be sun during the walk (E$_2$). The probabilities (which are actually the beliefs of the decision-maker) are P(E$_1$) = P(E$_2$) = 50%. The values are: V(D$_1$, E$_1$) = 100 (pleasure of the walk in the rain with a raincoat); V(D$_1$, E$_2$) = 10 (pleasure of the walk in the rain with a hat); and conversely V(D$_2$, E$_1$) = 10   and   V(D$_2$, E$_2$) = 100.   The   value   of   the   decision   D$_i$ is: V(D$_i$) = $\sum_j P(E_j) \cdot V(D_i, E_j)$. Here we get: V(D$_1$) = V(D$_2$) = 55 (See Fig. 2.11).

Fig. 2.11 Simple example of
a decision-hazard tree

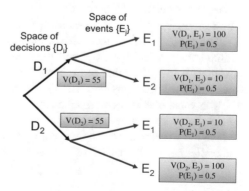

Project Assessment Tools for Reducing Uncertainty
(Market Research or Analysis)

The previous model may also be used to calculate the value of certain additional explorations prior to choosing whether or not to develop a product, if such explorations allow uncertainty to be reduced. This is so, for example, with research projects aiming to reduce commercial uncertainty.

It is assumed that the decider can proceed to a test, this being for him a new decision D_{n+1}. This test will enable him to learn, and this learning process will alter the subjective probabilities. Following the test, he will still be able to proceed to the prior decisions D_i, but evaluated this time using the new subjective probabilities, i.e. taking account of the reduction in uncertainty due to the learning process (and also taking account of the cost of the test, which will be deducted from the value V_{ij}). The learning process is modeled by assuming that the test gives us an "index" as to the state of the nature but does not completely reveal the "true" state. The experimenter knows that if the state is E_j then the test will give a result U_i with probability $P(U_j/E_i)$.

In the case of the weather seen above, it may rain with probability $P(E_1) = 50\%$ or be fine with probability $P(E_2) = 50\%$. The test involves consulting the weather forecast, which is known not to be completely reliable, and hence it is known that when it rains, the forecast predicted 80% i.e. $P(U_1/E_1) = 80\%$ and $P(U_2/E_1) = 20\%$; conversely, when it is fine the forecast predicted 4 times out of 5, i.e. $P(U_2/E_2) = 80\%$ and $P(U_1/E_2) = 20\%$.

Using Bayes' formula and the total probability formula it is then possible to calculate $P(U_1)$ and $P(U_2)$:

$$P(U_j) = \sum_i P(U_j/E_i) \cdot P(E_i)$$

Then we calculate $P(E_i/U_j)$ $(P(E_i/U_j) = P(U_j/E_i) \cdot P(E_i)/P(U_j))$ (see the practical calculation in the figure below) (see Fig. 2.12).

The products s of the initial Di are recalculated with the new probabilities for each result U_i of the test.

It is important to emphasize that the test does not alter the states of the nature: whether it rains or shines does not depend on the weather forecast. The test only changes the belief as to these states: the decider believed in 50% for state E_1 before the test; after the test, he believed 80% if it was U_1 and if U_2, he believed 20%.

This process enables the value of the test to be calculated, and is the difference in value between decision D_{n+1} and the best of the decisions without the test. In the example below, the value of D_3 is subtracted from the value of D_1 (or D_2, since the two decisions are equivalent in this example), i.e. 82–55 = 27 (Fig. 2.12).

The value obtained corresponds to the value of tests whose intention is to reduce uncertainty. This approach to uncertainty reduction is one of the first modes to which industrial research and marketing studies aim to add value; this was clarified, in particular, by Peirce in 1879 (Peirce 1879); complete models had to wait for

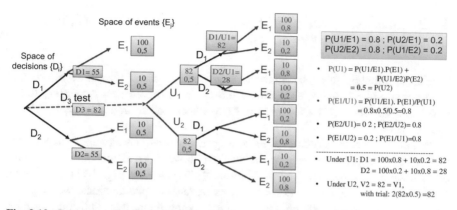

Fig. 2.12 Calculating the value of a test using the decision theory model with learning

developments within the theoretical framework of statistical decision theory (in the 1950s). Work on evaluating research projects via real options emerged directly from the above formal framework.

2.3   Organization of the Systematic Design Project

2.3.1   General Principles: Project and Project Leader

The model of reasoning in systematic design makes the latter compatible with a project management mode: there is a clear objective, resources are identified (internal or external), possible milestones, forms of division of work with respect to this objective, a partial delegation of responsibility and possible oversight by the corporate hierarchy or a project backer (CQT target).

We immediately note that in the canonical organizational model, the systematic design project is characterized by a form of cohesion with a clear "common purpose" (CQT objective endorsed by the company strategy); we shall see that systematic design encourages relatively simple forms of coordination based on prescribed and controlled tasks.

We emphasize the paradoxical side of this organization: as we have seen in the introductory chapter, design is quite generally distinct from traditional optimization and decision situations, and is liable to revise the associated knowledge and interests, rendering obsolete the skills of the company and calling its strategy into question—but systematic design can fit into a simplified, not to say bureaucratic, organizational form where the reasons to act are clearly established at the outset and stable over the course of the process, hence avoiding constant changes in company skills and strategy.

2.3.2   Division of Labor in Systematic Design: Stage-Gate Process and "V" Cycle

The linear structure of systematic design reasoning first of all allows a temporal sequencing with phases of exploration and decision milestones. According to the terminology introduced by Cooper (1990) we speak of an organization in "stage" and in "gate".

In a finer sense, the structure of reasoning in systematic design allows a complex and a priori exploratory task to be broken down into elementary tasks. This breakdown can be described by the famous "V" cycle (see Fig. 2.13): the overall task of the project is broken down in terms of ranked specifications; we begin by specifying the services offered, these services being themselves split into systems and sub-systems finally giving rise to the elements of machines. To these machine elements correspond the elementary tasks of detailed design. Once this has been done, we then move on to an integration process of validation and synthesis: whatever can be validated is validated at component level, then at sub-assembly level and then at the systems architecture level to finally validate the services provided. The interlocking validations correspond to an economic principle: don't include a defective component if the faults can be detected beforehand; simplify the quest for causes when a sub-assembly or system is faulty. The "V" cycle also aims to guarantee that the project converges, avoiding any late discoveries of defects in quality or the failure to meet technical objectives. The coherence between the breakdown of the specifications and the available validation protocols is a determining factor in the performance of a project under rule-based design (Fig. 2.13).

This task breakdown enables a division of design labor thus: prescribing some elementary tasks to competent experts; coordination between tasks (definition of interdependencies and precedence constraints); allocation of responsibilities and resources, etc.

Again we must make it clear that it is because systematic design involves reasoning, that stage-gate type processes are possible along with the division of work into prescribed elementary tasks.

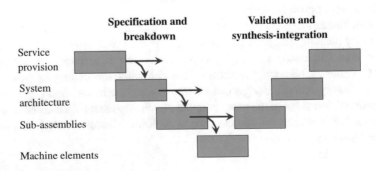

Fig. 2.13 General scheme of the "V" cycle

2.3.3   Project Leader's Management Tools: Planning, PERT Charts and Budget Reporting

2.3.3.1   Planning

An action can be planned provided it is divided into elementary tasks of known duration and dependency links. If the elementary tasks of the detailed design can be given a duration, we can construct a planning schedule for a systematic design project.

However, note that the elementary tasks are essentially limited to the detailed design phase and then validation. In practice, this means that the project generally includes a short, sparsely detailed "front-loaded" preliminary phase or pre-project with reasoning applied up to the detailed design. Strictly speaking, the planning, i.e. the part where tasks are organized such that they are "concurrent", relies solely on the validation and detailed design part.

2.3.3.2   PERT and Critical Path

PERT, which stands for "Program Evaluation and Review Technique", is a technique developed in the 1950s for overseeing large-scale American military projects (for an historical perspective on PERT and project management of these large military programs, see Lenfle and Loch 2010).

A PERT chart is constructed using a graph of tasks defined by their duration and inter-task succession constraints (see Fig. 2.14). The so-called "earliest dates" are then calculated, beginning from the project start date, propagating forwards and writing down for each task the "earliest date" (ED) and the "earliest end date" (see graph 2 on Fig. 2.14). The so-called "latest dates" (LD) are calculated starting from the project end date and propagating backwards through the preceding tasks. The LD is written down for the completion and launch of each task (see graph 3 on Fig. 2.14). The margin for each task is calculated, this being the difference between the "earliest" and the "latest" date. The path leading from the initial to the final task is critical if the margins for all the tasks on this path are zero (see graph 4 on Fig. 2.14).

Hence the PERT chart is able to organize the correspondence between tasks, handling a large number of tasks, but also continuously rescheduling and incorporating specific temporal constraints (mandatory start date, dates defined with respect to a benchmark, etc.). Note, however, that the PERT process takes no account of resources. This means managing production with no constraints on production resources (the capabilities of machines and personnel, etc.) (Fig. 2.14).

2.3.3.3   Management Tools: Budget Reporting

In particular, the PERT technique allows budget reporting tools to be developed. What is forecast is compared with what actually exists at some date t.

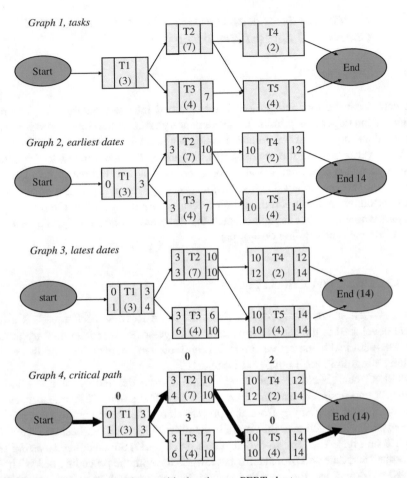

Fig. 2.14 Example of constructing a critical path on a PERT chart

Using the PERT chart we can construct a curve of forecast commitments including the forecast expenditure for each task (BCWS curve, Budgeted Cost of Work Scheduled). In addition, actual expenditure is monitored regularly (ACWP curve, Actual Cost of Work Performed). Finally the progress curve can be drawn, i.e. the BCWP curve, Budgeted Cost of Work Performed).

Timescales can be monitored using the cost difference between BCWP and BCWS (see graphs on Fig. 2.15)

Costs can be monitored by the difference between ACWP and BCWP (Fig. 2.15).

2.4 Conclusion

In systematic design, design follows a linear reasoning process as per the languages predetermined for the object. Design is project-based, the project having to hit a Quality-Cost-Time target with the resources already in place and maximizing the

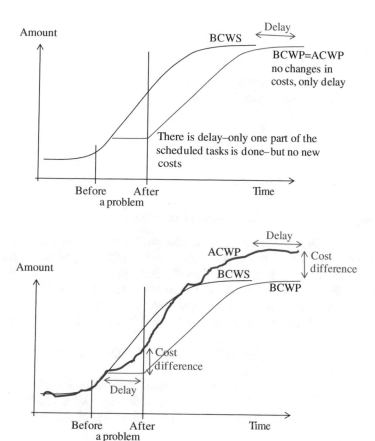

Fig. 2.15 Budgetary oversight tools

use of all available knowledge. We again emphasize that it is because the reasoning process is conducted in this way that the logic of performance and organization (especially organization in planned projects) can be deployed.

It is interesting to observe that, in the event of the failure (or success) of a project, there are always two possible causes: either the project team and project leader have been unable to make use of the systematic design tools (forgetting a function in the functional analysis, poor planning, forgetting a resource, etc.), or the conditions for applying systematic design failed to come together. In the rest of this book it is therefore necessary to set out these conditions. In particular, we contemplate how a knowledge base, i.e. expertise and skills, etc., can be built that is sufficiently well matched to the project that very little learning will be required.

With systematic design we have studied one of the most sophisticated and effective rule-based design regimes. However, what are the conditions laid on the rule base that allow this regime to function?

2.4.1   Main Ideas of the Chapter

- The stages and languages of systematic design
- The fundamental assumptions of systematic design
- The notion of function and functional analysis
- PERT charts and the critical path
- Project performance: CQT

2.4.2   Additional Reading

This chapter can be extended in several directions:

- on New Product Development Management, the reader can study the seminal reference works of (Clark and Fujimoto 1991; Wheelwright and Clark 1992), Front Loading (Fujimoto 1997; Thomke and Fujimoto 2000) and variants on Flexible Product Development (MacCormack et al. 2001) and Fuzzy Front End (Cooper 1997; Khurana and Rosenthal 1998; Reid and De Brentani 2004).
- On Integrated Product Development see (Olsson 1976; Andreasen 1987; Magrab 2010)
- On Project Management (Midler 1995; Lenfle and Midler 2009; Ben Mahmoud-Jouini et al. 2004)
- On the notion of function: the contradictions of functional analyses (Vermaas 2013); the house of quality (Hauser and Clausing 1988); see the notion of function in other disciplines—notably the famous "Form follows function" of the architect Louis Sullivan.
- On PERT and planning: (Moisdon and Nakhla 2010); for the logical processes of more advanced planning, see the work on interactive planning by (Hatchuel et al. 1997).
- On decision under uncertain conditions and project selection: see (Moisdon and Nakhla 2010) and classical courses on economic calculation. For more in-depth information, the reader can refer to (Hooge 2010) for the assessment of innovative projects.
- On systematic design and its variants: in engineering design (Pugh 1991; Cross 2000; Ulrich and Eppinger 2008); under more technical forms (Karniel and Reich 2011); in Stage Gate management (Cooper et al. 2001; Cooper and Kleinschmidt 1987, 1993) and its critique (Varnes 2005)
- In particular, see also the relationship between Systematic design and the Theory of Technical Systems (Hubka and Eder 1988; Eder and Hosnedl 2010)
- On recent development of evolutionary approaches on engineering design see (Vajna 2005; Vajna 2011)

2.5 Workshop 2.1: Functional Analysis

Exercise, first part: make a functional analysis of a bus station.

This exercise will highlight several critical elements in the practice of functional analysis. The main lessons are:

1. Functional statements should be neither too precise nor too general. If too general they may allow misguided interpretation; if too detailed they put the designers into a straightjacket of ready-made solutions or combining them leads to insurmountable contradictions. The functional statement is like a contractual clause binding the "requestor" to the "bidder" (the engineering department charged with carrying out the functions), and hence presupposes a validation procedure to check that the function has been properly fulfilled.

Example: "the bus station should allow users to wait in comfort" is too vague a statement to constitute a function; once the bus station has been designed, how would one check that it will be "sufficiently" comfortable?

2. A functional analysis might start by identifying the various constituent parts (clients at all levels—clients of clients of clients, etc.; users, the whole supply chain, maintenance, ecosystem, prescribers, institutions, etc.) and different environments (atmospheric conditions, day or night, type of country, place, situations encountered by the product within its life-cycle, etc.).

In the bus station example: don't forget the roadway, maintenance, residents, town services, etc.

3. Functional analysis depends on the business model associated with the object. On this point, note that business models do not correspond to a dimension "to be designed" in systematic design, and are in fact considered as data, especially in stakeholder models expressing what they expect from the object.

Example: in the case of Parisian bus stations, these are made available to the operator by the Decaux company which, in return, has the right to put up advertizing material in the station. This business model requires taking a new participant into account, namely the station's financing company (in this case Decaux): they will define the functions that are critical for them. Hence the station must possess suitable surfaces for posters which must be visible (also at night), clean, and easily accessible for changing the advertisements, etc.

4. Take care: functional analysis is *not* a user concept. On the one hand functional analysis incorporates dimensions other than those of usage (standards and norms including social norms not corresponding to a single particular user (see the functions associated with respect for the environment and sustainable development), the logic of risk demanding an awareness of very rare events for which there is no proven purpose). On the other hand, certain uses will not be included

in the functional analysis; for instance deviant uses that are not statistically significant will not be taken into account. In the general case, functional analysis does not have recourse to an analysis of usage, often relying as it does on too limited a number of uses and individuals to reach a statistically significant threshold. Rather, functional analysis depends on the modeling of standard users and stakeholders, characterized by a small number of attributes. There exists a conceptual model specific to functional analysis, a model which plays a critical role in formulating the functions (and associated tests).

Example: the observation that a writer has written his entire novel sitting under a bus shelter will be difficult to include within a functional analysis. On the other hand, the observation can be generalized and incorporated within the standard user model; for instance, a new function can be added such as "the user wishes to pursue some activities while waiting for the bus" (e.g. a bus station with internet access).

5. We can generally assume that there are three sets of specifications (i.e. three highly contrasting sources of functional specifications):

    a. The customer's specifications sheet, i.e. the list of specifications drawn up by the customer.
    b. The set of professional and industrial specifications, i.e. the list of specifications which are either not explicitly requested by the customer but essential nevertheless, or do not involve the customer directly but rather other stakeholders.
    c. The company's own requirements specification, i.e. the list of specifications associated with the company's strategy (their image of quality, robustness, high-tech, innovation, etc.).

*Exercise, second part: carry out a functional analysis of a **night** bus station.*

This question must be put back into context: when the Paris night bus service (Noctilien) was established, the transport authority, RATP, created new bus routes and new conditions for operating them. In particular, they had to design special night bus stations, requiring adaptation to the bus station's specifications.

This second part generally leads to an upgrade of certain parameters in the requirements specification ("better lit", "more comfortable", "better arrival announcements in real time", etc.).

One or two limits of the functional analysis tool can be highlighted:

1. Functional analysis ignores any exploratory efforts that might be required by new environments or stakeholders. Hence any analysis of the "night" bus station requires learning about what night is. We would discover: (a) that stations have already been operating "at night" (in winter night falls at 17.00 in Paris); (b) that there are several types of night in Paris: night with "day-like" activity (up to around

22.00 and then from 06.30 to 08.00), night with a drop in activity (from 22.00 to 01.00, the traditional closing time that bars and the transport system have in common; then from 05.00 to 06.30), and (c) the night when nothing happens, when the entire town is "closed" (from 01.00 to 05.00 in the morning). We can see that it is only this last version of night which poses problems, very new problems, in fact: for example, there is no simple emergency response in the event of illness or attack, such first response being provided typically by residents or passers-by in day time. New functions then appear: raising the alarm, giving first aid, having water available, etc.

2. Functional analysis also ignores the difficulties of these explorations: in certain situations it is impossible to know the stakeholders' expectations of the object, quite simply because the situation does not yet exist. Before the Noctilien service, it was difficult to anticipate the functional specifications related to the station's expectations. In cases of this type functional analysis might demand a far more complex prototyping and test protocol, assuming a design effort akin to innovative design (see Chaps. 4 and 5). With this kind of reasoning we can envisage, for example, "mobile stations" which can be imagined as fully equipped buses (heating, emergency response facilities, water, etc.) that might be deployed at a few critical points in the town to act as night bus stations in the depths of night.

2.6   Case Study 2.1: The Logic Underlying the Domestication of Design: The Origin and Success of Systematic Design

In this study of historical cases we shall present a few significant episodes in the history of the domestication of design and that of the invention of the "engineering department" by large companies. Within the scope of the canonical performance model (see the introductory chapter) we shall analyze the succession of rule-based design regimes and in each case set out the industrial context, the issues and underlying logic of performance, what had been designed and by which designers (for a more extended version see Le Masson and Weil 2008).

2.6.1   Wild Design: The Inventor-Entrepreneurs of the First Industrial Revolution in England

The first industrial revolution born in England took place in the absence of any engineering departments. However, it came about through the emergence of new industrial sectors (mechanical spinning machines, steam engines, machine tools, railways, steamships, etc.) and through the design and production of an extraordinary variety of new objects. These were made by highly inventive independent engineers with considerable business acumen making up a very active milieu. The exchange of information between them, including that with scientists and businessmen, were intense, particularly through the many learned societies such as the Lunar Society of Birmingham (Agogué 2012; Schofield 1957, 1963) or the Smeatonian Society of Civil Engineers. They were present at a burgeoning of ideas and experiments. The development of the railway was typical of the dynamic of the time. Essentially this was down to half a dozen or so "great" engineers such as Stephenson, Brunel, Locke, etc.

Take, for example, the case of locomotive design by the Stephenson father and son duo, often presented as the "fathers" of the English railway. This they illustrated with the opening of the first commercial line between Stockton and Darlington in 1825, going on to participate ceaselessly in the growth and rapid development of this new means of transport. However, they were not alone, and there was fierce competition to win bids for new lines. In this context they developed two companies: the first, dedicated to civil engineering, handled the design and construction of the lines, while the second was devoted to the design and manufacture of locomotives. Everything had to be designed: the characteristics of the line and civil engineering structures, as well as stations, the organizational principles of the various services, fee structures, etc. Locomotives were designed by quite a small number of engineers. The newly completed line served as a test-bed and prototype for refining the next generation. In this way the main difficulties they encountered were gradually resolved. Major enhancements were brought into improve the thermal or mechanical performance, without recourse to established scientific results or generating exhaustive scientific investigations. Drawings were extremely perfunctory and were little more than a diagram of some essential part or other and a list of components. Design relied on the manufacturing workshops where highly skilled workers, immensely capable of building the parts from the very basic sketches they were provided with, made the design workable by adjusting and modifying to obtain one of the most complex and sophisticated objects ever made.

We shall call this first design regime the "wild" design of the inventor-entrepreneur. How to characterize it? The resources were limited to a few engineers, their initial knowledge was poor, and the learning process, primarily through the trial and error of successive generations, served as a design space and hence learning for those that followed. In terms of expansion, there was still only a poor organisation for repeating or reproducing a particular design, and it was always just one product (and its associated process) that was designed: the designers attempted to discover the extent of its performance and gradually stabilize the object's identity. The customer at that time was not always well-informed and did not always know how to get what he wanted. Performance consisted of constructing a potential for a specific value/skill which would be brought to bear in winning bids and maintain progress in an ever-changing field (see the table and summarizing graph at the end of the case study).

2.6.2   Parametric Rule-Based Design: "Recipe-Based" Design or Pathways for Industrial "Catch-Up"

At a time when industrial development in England was in full swing, France and Germany were wondering about how to catch up industrially. One thing was certain: the process of building up a population of technical specialists and inventor-entrepreneur engineers similar to that which existed in England would take too long and seemed difficult to push forward. In France, scientists concentrated on conceptual developments with the aim of establishing a science of machines (mechanics, kinematics, strength of materials, heat, hydraulics, etc.) and scientific teaching at an advanced level (The Ecole Polytechnique, etc.).

In Germany, an original approach distanced itself from the tradition of applying the results of scientific investigation. Ferdinand Redtenbacher became the moving force behind this approach. He was initially professor of mathematics and geometry at the Ecole Polytechnique in Zurich, and then taught mechanics and mechanical engineering at Karlsruhe. He was also closely in touch with industrial machine manufacturers. In the preface to his 1852 book he stated his critique thus: "We don't invent machines using the principles of mechanics, for that would also require, besides an inventive talent, an exact knowledge of the mechanical process the machine was intended to fulfill. Using the principles of mechanics we cannot provide a sketch of the machine for that would also require a sense of composition, layout and shape. No machine can be built using the principles of mechanics for that would require practical knowledge of the materials with which we work and familiarity with tools and handling machines. No industrial project can be led using the principles of mechanics for that would require a particular personality and a knowledge of business affairs." (Redtenbacher 1852). His ambition was not to train designers capable of making use of all these sorts of knowledge and designing machines in their entirety. Rather, he imagined a two-stage design process. In his book he sets out a set of "recipes for design" which allowed the specialists under his tutelage to design all sorts of machines adapted to the varied situations they might encounter. It was sufficient for them to follow the stages and calculations defined by the "recipe" to be sure of obtaining a satisfactory result.

This method was much better than a mere catalog since each time it enabled a design suited to the situation to be obtained without having to design all possible machines beforehand. The procedure proposed by Redtenbacher displaced the effort of design: this was no longer a unique product to be designed but a recipe capable of generating an entire family of products. Moreover, the value of the recipe lay in its ability also to guarantee the performance of the products it generated. In conceiving his recipes for design, Redtenbacher came up against a dual problem, namely the development of a conceptual model linking performance with the design parameters of the device ($K(X)$), and that of a generator model (a previously defined series $d_i \in D$, where the d_i are parametric) ordering the design stages in a linear sequence (the idea of the generator model will be introduced and discussed in Chap. 3). The recipes given in his books covered a vast range of machines from water wheels (very widespread in Germany but often with mediocre performance) to the most modern machines such as locomotives. It was by following these famous recipes that the nascent Alstom company started to manufacture locomotives. Thus were forged the skills that would become essential for keeping up with the accelerated pace of change in the world of railway innovation.

We give an example of how the method works in a simple case, namely the design of water wheels.[1] In the first part of his book (Chaps. 1–3) Redtenbacher surveys the state of the art for water wheels and existing theories to gradually formulate a set of "equations of effects" covering the performance and dimensions of a water wheel.

Redtenbacher draws on the work of Poncelet (op. cit.), Navier, Morin and Smeaton, whose tests were already ancient (1759), and also provides results from his own tests. As Redtenbacher wrote: "one might think that water wheels were already widely understood... and that any practical or scientific treatment would be of no value today". Most of the work took account only

---

[1]Redtenbacher, 1858, op. cit.

Fig. 2.16 Smeaton's
experimental method (1759)

of the head, the volume of water, the speed of the water course, and the inlet
speed. For example, using an experimental method, Smeaton's investigation
sought the height of the water on entry to the wheel for optimizing the
transmission of motion (see Fig. 2.16).

However, these studies failed to deal with the particular arrangements of
the wheel or with the environment in which it was located. Also lacking were
the equations relating to the size of the wheel, its diameter and width, choice
of vanes or buckets, number of buckets, their shape, the depth to which the
wheel had to be immersed in the water, the constructional quality of the
mounting and control of leaks, etc. All these limits meant that the designers
were unable to use the scientific results hitherto obtained. Hence (and still in
the first part) Redtenbacher concluded his state of the art using complete
models of existing machines grouped together under major types.

Once these major descriptive models had been drawn up, Redtenbacher
moved on to the second, and most original, part of his book: the method of
ratios. Chap. 5 sets out the rules to be followed to assess "the specific forms
and dimensions on which the wheel preferentially depends under the conditions for a perfect realization of the structure". The method starts off by
following the major steps of a fictitious dialog between the designer-
entrepreneur and his client. According to Redtenbacher, the first question

concerns the budget that the client is prepared to spend on his machine since, depending on the response, the designer will tend towards either a metal or wooden wheel, wheels whose efficiencies and sizing equations are very different. Once the material has been chosen, two questions have to be asked: the head of the water course and the usable flowrate (or, which amounts to the same thing, the power expected on the shaft). The designer then has to use a chart (see diagram below) which, depending on the head and flowrate, enables him to choose the best type of wheel (e.g. a mountain stream of large head and low flowrate would use an overshot wheel, while a watercourse in the plains, of low head and high flowrate, would tend to favor a Poncelet type wheel). He then goes on to define the main dimensions (radius, fill rate, circumferential speed, bucket volume, depth of the wheel, number of buckets, number of arms, the clearance of the wheel in the race). At this stage the method allows the designer to choose a class of wheel, assessing the expected performance without yet stating all the dimensions (Fig. 2.17).

At that time this was for Redtenbacher the most critical part of the reasoning process, since (he observed) most wheels were ill suited to their context.

The second part of the design process consists of specifying, step by step, all the parts of the machine by following the methods of calculation or even the drawings (proposed in the engineering design) which correspond with the patterns (as seen in sewing patterns): the drawing is non-dimensional and also

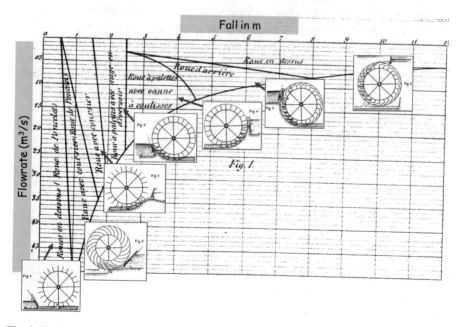

Fig. 2.17 Chart for choosing the type of water wheel depending on the conditions of use

provides the ratios between the parts as a function of some known fixed unit. It then sets out the modes of linkage and the level of precision with which the assembly must be put together. Finally, the last part deals with what we can call "tuning": Redtenbacher restates the theoretical efficiency formulae and the technique for measuring the actual efficiency. He invites the designer to compare the efficiency measured on the installation and indicates the means for improving the actual efficiency on an almost completed wheel.

In this second part of the calculations, Redtenbacher notes that industrial wheels are rather well designed, emphasizing the fact that his method allows performance to be checked, to "get rid of imperfections" and to "relate all uncertainties to solid rules".

We have gone from "wild" design to "rule-based" design, where the recipe allows parameterization. We now examine the various dimensions of this new regime (see Fig. 2.18 and Table 2.1):

Design rules figure at the very top of the resources, i.e. a generative model based on several conceptual models. Introduction of the recipe also leads to the need to distinguish between two types of designer: one makes the recipe and one uses the recipe; and there are two associated types of reasoning, different for each designer. Creating the rules demands a major exploratory effort, production of knowledge (experimental methods, tests, etc.), modeling and rare skills. The use of rules is compatible with limited and far more widespread skills.

The expansion made possible by these rules is no longer limited to a specific product, and leads to a diversified family of products. However, such variety is predetermined by the generative model contained within the recipe.

Finally, the logic underlying performance is that of industrial catch-up where the number of completed designs (conjunctions) has to be maximized while limiting, as

Fig. 2.18 Performance of the different recipe-based design regimes

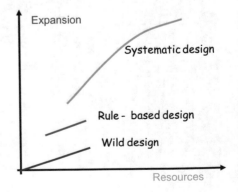

Table 2.1 comparison of rule-based design regimes

	Resources	Expansion	Performance
Wild design	A few engineers Initial knowledge poor Learning by trial and error P_i = learning space for P_{i+1}	Gradually work out the dimensions and stabilize the identity of the objects	Construct a potential skill/singular value
Rule-based design	Recipe (Generator model based on a conceptual model) Distinction between designing the recipe and designing the product	Product family but product determined by the generator model	Catch-up Maximize connectivity without producing additional knowledge
Systematic design	Design department 4 languages Division of labor and specialized skills	Product family (dominant design) = Variety + cone of innovation on known performance Set out in ranges and families	Expansion extended by controlling learning processes (δK) Still connective (robustness: minimize risk) Industrial system

far as possible, the additional effort in the production of knowledge. The recipe offers the ability to effectively exploit the product line, guaranteeing the variety and economies of knowledge. On the other hand, this regime is highly sensitive to any technical evolution that may require the recipe to be redesigned.

2.6.3   Systematic Rule-Based Design: The Invention of the Engineering Department

Under the joint thrust of some changes characteristic of the second industrial revolution, research and experimentation in other forms of rule-based design were not going to be held back. With the increase in production volumes, preoccupation with industrial efficiency became a priority. Customers (often in a B2B situation) became more competent and more demanding. Finally, the product dynamic made it essential to reconcile the new knowledge spaces: heat or electricity might be added to mechanics, for example.

2.6.3.1   The Baldwin Locomotive Works, or the Power of Expansion and Organization in Generator Models

One of the first and best models of this evolution was provided by the Baldwin Locomotive Works (BLW) of Philadelphia which, due to the methodology of its design organization, was to become the undisputed leader in this sector during the second half of the 19th century (for information on this company see Brown 1995). At its creation en 1831, the Baldwin company was similar to that of Stephenson. Very quickly, however, it found itself face to face with unprecedented problems: the vertiginous growth of the America railway market was punctuated by abrupt halts corresponding to the recurrent financial crises while the railway companies were engaged in a race for performance. Furthermore, these railway companies operated in varied contexts which led their "engineer" to make their orders very specific. How could this demand for variety, potential for evolution and the constraints of an industrial complex of then unknown size be reconciled (the factory employed about 10,000 staff in 1900)? Around 1900, BLW produced over 1200 locomotives per year, including nearly 120 different models, delivering their locomotives two months after being ordered (design included). How was such a feat possible?

Three elements played a significant part. First was the structuring into product families based on studiously analyzed reference architectures, covering all the requirements and supporting the constant improvement in performance. The directors of the company regularly took part themselves in this redesign effort, incorporating recent technical advances, including if possible any differentiating innovations but organizing variety as well (this meant having certain degrees of freedom) and authorizing enhancements to the primary dimensions of performance so that the locomotives designed on this basis would remain competitive. These families embodied the design strategy of the firm. Secondly, the commercial relationship with their customers was also handled directly by the directors. This relationship relied on a scalable description of the main elements of the locomotive. Hence the customer had complete freedom in defining the characteristics of his locomotive but in a language consistent with the product families mentioned above. Thirdly (and finally), on the basis of this requirements specification the designer had instructions to employ previously used standard parts for which pre-existing drawings would make the best use of the capabilities of the machine tools. Where the designer could not meet the requirements specifications with existing components, he was not permitted to design a special part, but had to design a new standard component which could be used in future projects and approved by the manufacturing department.

Resources: as in the case of the recipe, there were two types of designer: those who designed the generative model (how it would be possible to design a locomotive with the most standard components) and those who defined the bases on which the major conceptual models were used. In particular their effort was focused on the definition of several mutually articulated languages. These second designers formed increasingly numerous and specialized collectives, with the division of labor made possible by the effort in defining a dominant design. Rather, they designed the parts under a restrictive, but evolving, system of rules (dictated in particular by manufacturing and industrial capabilities).

Expansion: associated with the variety guaranteed by the product families were "cones of innovation" which enabled the primary dimensions of product performance to be improved from design to design.

A high level of design performance resulted from this strong ability to expand, obtained by limiting and controlling the effort in producing new knowledge. Risks were managed using an important property of the generator model, namely its connective power, by means of which the designer could predict whether or not he would succeed in designing a product meeting the initial specifications. Finally, the attention given to the industrial system drew out the best in it.

2.6.3.2   Germany and Systematic Design

Similar changes were occurring in Germany, where they were accompanied by important theoretical debate on the formalisation and development of a doctrine, namely systematic design. New industrial companies appeared such as AEG which, thanks to the organisation of their design department, were more successful than their older rivals (Siemens, for example) in designing a great variety of electrical machines for the mass market; these combined the use of standard components with steady progress in improving the performance of these machines.

Above all, however, there was a renewal in the debate around the teaching of design. Such debate emphasized the limitations of recipes and questioned how the production of knowledge could be integrated within the design process (König 1999).

The major advance was in the gradual and difficult distinction of four main languages used to describe the objects to be designed: functional language (expressing the needs of the customer in a language the designers could use), conceptual language (where the primary languages of the engineer are to be found: mechanics, strength of materials, kinematics, thermodynamics, hydraulics, electricity, etc.), morphological language (the assembly of machines) and the language of detailed design (in which the constraints of the manufacturing process might intervene in defining the smallest detail in the shapes of elementary parts).

Authors, teachers and consultants in regular contact with manufacturers suggested organizing design as a staged process articulating these different languages in sequence. Powerful engineering departments were not slow to organize themselves on this basis. Such was their performance that the two previous forms of design organization gradually disappeared or merged into this new model.

There thus appeared a new model of the industrial firm based on systematic design (see Fig. 2.18 and Table 2.1, Last line), and this was clearly the point at which the modern large enterprise was born (Segrestin and Hatchuel 2012). Drawing on the design office model, Taylor created the process and planning department for organizing the production system. R&D laboratories sprang up around the design offices (see also the case study on the history of the industrial research laboratory in Chap. 3). The rationalization of product development was accompanied by a rationalization of communication, commercialization, distribution and the emergence of powerful marketing departments (the start of the 20th century), and a rationalization of purchasing systems. The first design studios also appeared, sometimes as part of the company. These various entities were of greater or lesser importance in the big companies. However, behind the diversity of the organizations we should note the predominance of a systematic design model based on a stable dominant design, i.e. a design model in which products shared the same reference points in terms of performance, function, architecture, technology and skills. This is what we shall study in Chap. 4.

References

Agogué M (2012) L'émergence des collectifs de conception inter-industries. Le cas de la Lunar Society dans l'Angleterre du XVIII siècle. *Gérer et comprendre* 109:55–65.

Altshuller GS (1984) *Creativity as an exact science: the theory of the solution of inventive problems* (trans: Williams A). Studies in Cybernetics: 5. Gordon and Breach Science Publishers.

Andreasen MM, Hein L (1987) *Integrated product development*, IFS (Publications). p.

Ben Mahmoud-Jouini S, Midler C, Garel G (2004) "Time-to-market vs. time-to-delivery: Managing speed in Engineering, Procurement and Construction projects." *International Journal of Project Management* 22 (5), pp. 359–367.

Brown JK (1995) *The Baldwin Locomotive Works 1831–1915, A Study in American Industrial Practice*. Studies in Industry and Society. The Johns Hopkins University Press, Baltimore and London.

Clark KB, Fujimoto T (1991) *Product Development Performance: Strategy, Organization and Management in the World Auto Industry*. Harvard Business School Press, Boston.

Cooper RG (1990) Stage-Gate Systems: A New Tool for Managing New Products. *Business Horizons* May-June 1990:44–53.

Cooper RG (1997) Fixing the Fuzzy Front End of the New Product Process. Building the Business Case. *CMA Magazine* October 1997:pp. 21–23.

Cooper RG, Edgett S, Kleinschmidt E (2001) Portfolio Management for new product development: results of an industry practices study. *R&D Management* 31 (4):361–380.

Cooper RG, Kleinschmidt EJ (1987) New products: What separates winners from losers?. *Journal of product innovation management* 4:169–184.

Cooper RG, Kleinschmidt EJ (1993) Major new products: what distinguishes the winners in the chemical industry?. *Journal of product innovation management* 10:90–111.

Cross N (2000) *Engineering Design Methods. Strategies for Product Design*. 3rd edition edn. John Wiley & Sons Litd, Chichester, England.

Daloz P, Weil B, Fridenson P, Le Masson P (2010) Concevoir les outils du Bureau d'Etudes: Dassault Systèmes une firme innovante au service des concepteurs. *Entreprises et histoire* 58 (1):150–164.

Eder WE, Hosnedl S (2010) *Introduction to Design Engineering – Systematic Creativity and Management*. CRC Press/Balkema, Leiden (The Netherlands).

Ehrlenspiel K (1995) *Intégrierte Produktentwicklung. Methoden für Prozessorganisation, Produkterstellung und Konstruktion*. Carl Hanser Verlag, München, Wien.

Fridenson P (2015) Inventeur salarié, automatisation et entreprise: expériences internationales, tensions et dynamiques. In: Michel A, Benoit S (eds) *Le monde du génie industriel au XXe siècle: autour de Pierre Bézier et des machines-outils*. Pôle éditorial de l'UTBM, Belfort, pp 15–28.

Fujimoto T (1997) Shortening Lead Time through Early Problem Solving, a new Round of Capability Building Competition in the Auto Industry. In: Jürgens U (ed) *New Product Development and Production Networks, Global Industrial Experience*. Springer, Berlin, pp. 23–54.

Hatchuel A, Saïdi-Kabèche D, Sardas J-C (1997) Towards a new Planning and Scheduling Approach for Multistage Production Systems. *International Journal of Production Research* 35 (3):867–886.

Hauser JR, Clausing D (1988) The House of Quality. *Harvard Business Review* May-June 1988:63–73.

Heymann M (2005) *"Kunst" und Wissenchsaft in der Technik des 20. Jahrhunderts. Zur Geschichte der Konstruktionswissenschaft*. Chronos Verlag, Zürich.

Hooge S (2010) *Performance de la R&D en rupture et des stratégies d'innovation: Organisation, pilotage et modèle d'adhésion*. MINES ParisTech, Paris.

Hubka V, Eder WE (1988) *Theory of technical systems. A total Concept Theory for Engineering Design.* Springer.

Karniel A, Reich Y (2011) *Managing the dynamics of New Product Development Processes - A new Product Lifecycle Management Paradigm.* Springer, London.

Khurana A, Rosenthal SR (1998) Towards Holistic "Front Ends" In New Product Development. *Journal of Product Innovation Management* 15:pp. 57–74.

König W (1999) *Künstler und Strichezieher. Konstruktions- und Technikkulturen im deutschen, britischen, amerikanischen und französischen Maschinenbau zwischen 1850 und 1930,* vol 1287. Suhrkamp Taschenbuch Wissenschaft. Suhrkamp Verlag, Frankfurt am Main. 263 p.

Kroll E (2013) Design theory and conceptual design: contrasting functional decomposition and morphology with parameter analysis. *Research in Engineering Design* 24 (2):165–183.

Kroll E, Condoor S, Jansson DG (2001) *Innovative conceptual design: theory and application of parameter analysis.* Cambridge University Press, Cambridge.

Kroll E, Le Masson P, Weil B (2013) Modeling parameter analysis design moves with C-K theory. Paper presented at the *International Conference on Engineering Design, ICED'13,* Séoul, Korea.

Kroll E, Le Masson P, Weil B (2014) Steepest-first exploration with learning-based path evaluation: uncovering the design strategy of parameter analysis with C–K theory. *Research in Engineering Design* 25:351–373.

Le Masson P, Weil B (2008) La domestication de l'innovation par les entreprises industrielles: l'invention des bureaux d'études. In: Hatchuel A, Weil B (eds) *Les nouveaux régimes de la conception.* Vuibert-FNEGE, Paris, pp 51–67 (chapitre 3).

Le Masson P, Weil B (2010) Aux sources de la R&D: genèse des théories de la conception réglée en Allemagne (1840–1960). *Entreprises et histoire* 2010 (1):11–50.

Le Masson P, Weil B (2013) Design theories as languages for the unknown: insights from the German roots of systematic design (1840–1960). *Research in Engineering Design* 24 (2):105–126.

Lenfle S, Midler C (2009) "The launch of innovative product-related services: Lessons from automotive telematics." *Research Policy* 38 (1): pp. 156–169.

Lenfle S, Loch CH (2010) Lost Roots: How Project Management Came to Emphasize Control Over Flexibility and Novelty. *California Management Review* 53 (1):1–24.

MacCormack A, Verganti R, Iansiti M (2001) Developing Products on "Internet Time": The Anatomy of Flexible Development Process. *Management Science* 47 (1):133–150.

Magrab EB, Gupta SK., McCluskey FP, Sandborn P (2010) *Integrated Product and Process Design and Development: The Product Realization Process,* 2nd Edition, CRC Press, BocaRaton, FL. p.

Midler C (1995) ""Projectification" of the firm: The renault case." *Scandinavian Journal of Management* 11 (4): pp. 363–375.

Moisdon J-C, Nakhla M (2010) *Recherche opérationnelle - méthodes d'optimisation en gestion.* Presses de l'Ecole des Mines, Paris.

Olsson F (1976) "Systematisk konstruktion (Systematic Design)," (thèse), Lunds Tekniska Högskola, Department of Machine Design, Lund, Sweden. p.

Pahl G, Beitz W (1977) *Konstruktionslehre (English title: engineering design)* (trans: Arnold Pomerans KW). Springer Verlag, English edition: The Design Council, Heidelberg, English edition: London.

Pahl G, Beitz W, Feldhusen J, Grote K-H (2007) *Engineering design, a systematic approach* (trans: Wallace K, Blessing L, Bauert F). Third English edition edn. Springer, London.

Peirce CS (1879) Note on the theory of the economy of research. *United States Coast Survey.*

Pugh S (1991) *Total Design. Integrated Methods fo Successful Product Engineering.* Prentice Hall, Pearson Education., Harlow, England.

Raïffa H (1968) *Decision Analysis.* Addison-Wesley, Reading, MA.

Rasovska I, Dubois S, De Guio R (2009) Mechanisms of model change in optimization and inventive problem solving methods. In: *International Conference on Engineering Design, ICED'09,* 24–27 August 2009, Stanford CA, 2009.

Redtenbacher F (1852) *Prinzipien der Mechanik und des Maschinenbaus.* Bassermann, Mannheim.

Redtenbacher, F (1858) *Theorie und Bau der Wasser-Räder.* 2ème édition (1ère édition 1846), Bassermann, Mannheim. 264 p.

Reich Y, Hatchuel A, Shai O, Subrahmanian E (2010) A Theoretical Analysis of Creativity Methods in Engineering Design: Casting ASIT within C-K Theory *Journal of Engineering Design*:1–22.

Reid SE, De Brentani U (2004) The Fuzzy Front End of New Product Development for Discontinuous Innovations: A Theoretical Model. *Journal of product innovation management* 21 (3):170–184.

Savage LJ (1972) *The foundations of statistics.* 2nd edition (1st edition: 1954) edn. Dover, New York.

Schofield RE (1957) The Industrial Orientation of Science in the Lunar Society of Birmingham. *Isis* 48:408–415.

Schofield RE (1963) *The Lunar Society of Birmingham, A Social History of Provincial Science and Industry in Eithteenth-Century England.* Clarendon Press, Oxford.

Segrestin B, Hatchuel A (2012) *Refonder l'entreprise.* La République des idées. Seuil.

Thomke SH, Fujimoto T (2000) The Effect of "Front Loading" Problem-Solving on Product Development Performance. *Journal of Product Innovation Management* 17 (2):pp. 128–142.

Trigeorgis L (1996) *Real options: managerial flexibility and strategy in resource allocation.* MIT Press, Boston.

Trigeorgis L (2005) Making use of real options simple: an overview and applications in flexible/modular decision making. *The Engineering Economist* 50:25–53.

Ulrich KT, Eppinger SD (2008) *Product Design and Development.* 4th edn. Mc Graw Hill.

Vajna S, Clement S, Jordan A, Bercsey, T. (2005) "The Autogenetic Design Theory: An evolutionary view of the design process." *Journal of Engineering Design*, 16 (4): pp. 423–440.

Vajna S, Kittel K, Bercsey T (2011) "The Autogenetic Design Theory." The Future of Design Methodology, H. Birkhofer, ed., Springer London, London pp. 169–179.

Varnes CJ (2005) *Managing Product Innovation though Rules. The Role of Formal ans Structured Methods in Product Development.*, Copenhagen Business School, Copenhagen.

Vermaas PE (2013) On the Formal Impossibility of Analysing Subfunctions as Parts of Functions in Design Methodology. *Research in Engineering Design* 24:19–32.

Wald A (1950) *Statistical Decision Functions.* John Wiley & Sons, New York.

Wallace KM, Blessing LTM (2000) Observations on Some German Contributions to Engineering Design. In Memory of Professor Wolfgang Beitz. *Research in Engineering Design* 12:2–7.

Wheelwright SC, Clark KB (1992) *Revolutionizing Product Development, Quantum Leaps in Speed, Efficiency, and Quality.* The Free Press, Macmillan, Inc., New York.

Zwicky F (1969) *Discovery, Invention, Research - Through the Morphological Approach.* The Macmillian Company, Toronto.

Chapter 3
Designing the Rules for Rule-Based Design—Conceptual and Generative Models, Axiomatic Design Theory

What is the system of rules capable of supporting rule-based design projects? Are there "good" rules? Is it possible to characterize the criteria for a "good" system of rules? More prosaically, what is a "good" design department or effective R&D? In Chap. 2 we set out the action model for the project and project leader, but systematic design is not limited just to the project. On the contrary, we have seen (see the case history study, the birth of rule-based design) that rule-based design was born when a distinction could be made between the person who conceived the system of rules and the person who, given the system of rules, designed products using those same rules (see the Redtenbacher case (Le Masson and Weil 2010a); and earlier, see the "reduction in art" performed by engineers since the 18th century (Vérin 1998)). This second facet of systematic design, i.e. the conception of the system of rules and its assessment, is the focus of this chapter.

We shall analyze the conception of the rules for rule-based design within the general framework established in the introduction: reasoning, performance and organization. Only when we have successfully carried out this analysis will we have a satisfactory description of the regimes of rule-based design.

3.1 The Logic of Performance in Systematic Design—The Notion of Dominant Design

The performance we now seek to characterize is not that of a specific project but rather that of a system of rules supporting *several* projects. Given a set of designer participants (design department, enterprise, industrial sector), what is the performance of the system of design rules for this ensemble?

© Springer International Publishing AG 2017
P. Le Masson et al., *Design Theory*,
DOI 10.1007/978-3-319-50277-9_3

3.1.1 A Few Examples of Sector-Wise Performance

Let us take the case of the pharmaceutical industry. Using the data provided by the pharmaceutical companies worldwide we can plot the left-hand curve below (source PhARMA, Scherer 2011). If the newly launched molecules correspond with outputs and R&D expenditure with input, we obtain an R&D performance curve for the pharmaceutical sector shown schematically by the right-hand graph. This curve shows decreasing returns and negative returns for very large investments in R&D (See Fig. 3.1).

This type of curve shows performance shapes somewhat different from those of traditional productive performance: models of decreasing efficiency are generally adopted for these.

By contrast, the semiconductor industry defines its performance in terms of the number of transistors on a microprocessor (Moore's law) and for several years the companies in this sector have been seeking to limit their investment in R&D staff. The number of transistors per microprocessor may correspond to an output (a major shortcut, of course) and a company's R&D expenditure to inputs (here ST Microelectronics over the period 1994–2002).

Thus we find a curve with *increasing* outputs (see the graph below—note that the scale for the number of transistors is logarithmic) (Fig. 3.2).

These two examples illustrate the variety of performance encountered in systematic design. These examples also show that, corresponding to these contrasting levels of performance, the design capabilities are more or less appropriate to the intended expansion objectives—we shall study these design systems in the 2nd and 3rd parts of this chapter.

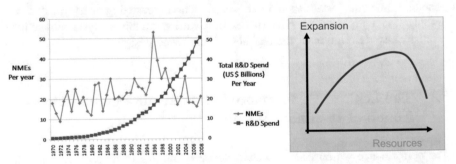

Fig. 3.1 Design performance of the pharmaceutical industry. New Medical Entities (NMEs) per annum as a function of R&D expenditure in billions of dollars (US$ 2008). *Source* U.S. trade association, Pharmaceutical Research and Manufacturers of America (PhRMA). *Right handside*: Schematic representation of the same data on a graph of expansion versus resources

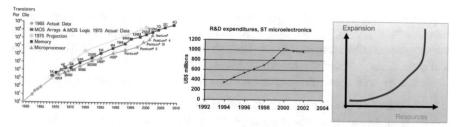

Fig. 3.2 Design performance in the semiconductor industry. *Left* microprocessors generations and the number of transistors per die (innovation intensity); *Center* R&D expenditures by TS-Microlectronics 1994–2002; *right* Schematic representation of these data on a graph of expansion versus resources

3.1.2 Characterizing the Performance of Systematic Design

More generally, to measure the performance of systematic design the inputs and outputs of a set of rule-based designers and their associated levels of performance need to be characterized.

3.1.2.1 Inputs and Outputs

Concerning *outputs,* the innovations of a sector must be measured. Simply counting "new products" gives rise to known problems (redundancy especially) and in particular, fails to take account of the logic itself of rule-based design: as we saw when studying the history of rule-based design, it is not exactly "innovation" which is being aimed for but the ability to repeat this innovation as a matter of course within the framework of known functional languages. From the point of view of systematic design, the design performance is good if it leads to regular suggestions for *new points in an algebra of fixed functions.* Within this algebra, innovation is either the ability to propose a new level on top of a known characteristic (higher calculating speed) or the ability to propose new combinations of characteristics at a known level (a chipset which may be slower but consumes little power).

We may note that this type of measurement is consistent with the economic theories of consumption developed in the 1960s, notably by Kevin Lancaster. In those days, development of an economy of variety (a consequence of the rule-based design we are currently describing) led economists to observe that "in the case of new commodities, the theory (of consumer in general equilibrium theory) is particularly helpless" (Lancaster 1966b, p. 133): indeed, the theory presupposed that all the utility functions (i.e. consumer preference) should be rewritten as soon as a new item appeared. From the 1960s onwards new items were appearing very regularly! Lancaster's proposal was to perform the utility calculations not so much for the goods but rather for the consumption *characteristics* associated with these goods, characteristics whose nature was considered fixed. These characteristics appeared exactly as functional specifications determining consumer preferences. In Lancaster's new theory of consumption, all goods can be described as a point in the vector space of characteristics (equivalent to functions) (Lancaster 1991, 1966a, b) and innovation corresponds to new points.

Hence we will assess the expansions of a sector according to two criteria:

(1) the variety of goods proposed and the ability of the sector to meet commercial demand, i.e. to make offers appropriate to different segments of the market;
(2) *improvement in performance* over known functions (reduction in vehicle fuel consumption, reduction of noise, reduction in the emission of pollutants, etc.). This second type of performance is generally made clear in an industry by using a small number of criteria (e.g. the speed of a microprocessor, the energy efficiency of a refrigerator, etc.): one might say that it is exploring a *cone of performance*.

As far as the *inputs* are concerned, the project is not the unit of measurement; we also need to state the reference points—these might be the design department, the company, the industrial sector, or a geographical area, etc. We shall assume that the inputs are the sum of committed resources for a design unit. For a company, we shall take account not only of the design department, but also research, marketing, etc. In some cases the indicator for the intensity of R&D can be an acceptable approximation.

3.1.2.2   Coupling Inputs with Outputs: Performance

Within this framework, performance will be the level of *expansion with given resources: for a given level of resources, obtain the broadest scope and the greatest level of functionality.*

However, performance can also be read off the ordinate axis for an expected level of expansion: *performance will then consist of knowing how to use the resources necessary for a given pace of innovation.* This means, for example: attracting talent, ensuring the training of experts, promoting coordination with outside partners both before and after the event or with outside research, developing internal learning systems for increasing the skills of experts, etc.

Performance can be deduced from the firm but also from the ecosystem: a sector will be considered to be performing well if its participants (companies, integrators, suppliers, specifiers, inspection and certification agencies, universities, public or sector-specific laboratories, etc.) are capable of occupying a variety of high-performance commercial positions; the industrial sector can also be assessed in terms of its ability to make the necessary resources available: the attractiveness of the sector (recruiting top-flight designers), technical schools for initial and continuing training, common research laboratories, standardization systems, standards, road-maps and sector-wise programming, etc. are all means for ensuring that design resources are available over a long period of time.

3.1.3   The Notion of Dominant Design

This system of measurement can be summarized in one indicator: the existence of a dominant dynamic design.[1] An industry is said to be engaged in dominant design if:

- the value of its products its business model and business relationships are stable
- the nature of the product performance is stable
- the skills necessary for their design make use of known professions or disciplines (the "nature" of the profession is known and the level within the profession can by dynamic)
- the design process can be split in a stable manner between companies and between professional divisions within the company.

The existence of a dominant design is a direct consequence of the existence of systematic design.

The effectiveness of systematic design is reflected in the *dynamic* of the dominant design: improvement in performance or in the performance mix, improvements in skills.

3.2   The Logic of Reasoning in Systematic Design: Conceptual and Generative Models—Axiomatic Design Theory

Intuitive motivation: We want to assess a set of rule-based design reasonings all using the same rule basis. Two difficulties are apparent: the quality of the basis itself and the underlying logic of its use must be assessed. We perceive that "skill", "expertise", level, quality, etc. all play a role; however, we sense that the capability for "using" this knowledge is also important: an "expert" or teams of experts actually mix these two dimensions. We shall clarify these intuitive ideas, and will see how systematic design provides highly structured answers to these questions of "knowledge management" in design.

3.2.1   Conceptual and Generative Models

The logic of the rule-based system is based on two essential ideas: conceptual models and generative models.

---

[1]The idea of Dominant Design was introduced by Utterback and Abernathy (1975).

3.2.1.1 The Idea of the Conceptual Model

We have seen several examples of conceptual models in Chap. 2 (see the historical case studies in particular). Thus the recipes for designing water-wheels rely on modeling the performance as a function of a number of design variables. In-depth research driven by Smeaton, the great English engineer, had been pursued in the mid-eighteenth century to define these relationships. He had developed an experimental methodology which allowed him to reconstruct the phenomenology and adjust the variables to measure their influence. Herein lies the essence of a conceptual model, for which we can give the following definition.

Definition of a conceptual model: a conceptual model is a modeling process (in K (X)) applied to relations between *known* objects, characterized by the limited number of parameters used (and consequently by the exclusion of many candidate parameters) and by the "actionable" or "useful" nature of these parameters (related to D(X) or P(X)). They constitute a "good summary" of the known, and may also be valuable for objects that are still unknown.

A few of its properties are indicated below:

- It has no specific link to a design language:

  - It can be mobilized at different levels of language: Ohm's law can be used in detailed design just as much for calculating the size of an electrical conductor as for choosing major technical alternatives at the conceptual level.
  - It can be oriented according to the criteria of value or design: in Ohm's law, U, R or I can also just as well be functions as design parameters (a particular voltage can be achieved by adjusting R and I, or a particular current can be obtained by adjusting U and R).

- This is not a "complete" modeling of the object.
- It can be easily activated or mobilized in the design reasoning process.

These conceptual models may be many and varied. They fulfill several roles: (a) to explore the possible effects, (b) analyze the causes leading to these effects, (c) use simplified relationships between cause and effect to avoid the laborious process of trial and error (as used to be the case with "wild design" (see the historical case study in Chap. 2)); and (d) enable a rapid assessment of products and hence their rationalization and improvement.

These models are very much linked to engineers. The engineering science which developed in the 19th century consisted precisely of generating and then organizing a rich range of conceptual models around varied phenomena. Teaching these conceptual models, classified by major disciplines, to engineers would occupy an expanding and favored position.

An example of a conceptual model: Carnot's principle (Carnot 1824)
In his 1824 work "réflexions sur la puissance motrice du feu et sur les machines propres à développer cette puissance" ["*Reflections upon the motive power of fire and on machines fitted to develop that power*"], Carnot proposed the development of a theory equivalent to that of mechanics to "obtain a prior understanding of all the effects of heat acting deterministically upon a body". He thus concluded with propositions constituting conceptual models. His fundamental proposition, for example: "the motive power of heat is independent of the agents implemented to create it; its quantity is fixed solely by the temperatures of the bodies between which heat is ultimately transferred" (p. 58). This proposition highlighted the decisive design parameters governing motive power (source temperatures) and also showed that certain parameters, frequently explored before then for improving the power, actually had no effect (e.g. the nature of the fluid, materials used, etc.).

However, Carnot was also very conscious of the fact that his conceptual model was not sufficient for designing with. He concluded his work with the words: "we must not fool ourselves into believing that all the motive power of the fuel can be made use of in practice". The designer must "in each case properly assess the considerations of suitability and economy which may present themselves".

However, knowledge of a conceptual model is insufficient for designing a product, and the information is insufficient to organize a process of design. This is what generative models will do.

3.2.1.2 The Idea of the Generative Model

We have already seen several examples of generative models: Redentbacher's design recipe for water-wheels, or for Baldwin, the hierarchy of languages such that the components and standard processes correspond to a customer's request formulated in the specifications via a reference architecture.

Definition of a generative model: Let there be a knowledge base $K(X)$ and a set of product concepts X with expected performance $P(X)$. A set of design rules (operators $d_i \in D$) associating K with X which, for any concept-product $\{X, P_j(X) \subset P(X)\}$, enables the construction of a sequence $d_i \in D$ such that, for minimal design effort, $d_1...d_n(X)$ satisfies $P_j(X)$, is known as a generative model. The d_i form an almost complete choice function.

A generative model defines a sequence of design sub-spaces $i*$ $\{X_{i*}, K_{i*}(X_{i*}), P_{i*}(X_{i*}), D_{i*}(x_{i*})\}$ such that:

(1) in each sub-space, the design efforts are minimal to conceive the sequence $d_{i*k} \in D_{i*}$ such that the $d_{i*k}(X)$ satisfy $P_{i*}(X)$

(2) the sequence of all the d_{i*k} leads to one or more X satisfying $P_j(X)$, i.e.
$P_j(\cap_k(d_{i*k})(X))$ is true.

In minimizing the expansions, the generative model therefore consists of making the most of the available knowledge. For example, it might involve injecting the maximum available knowledge into the object before starting to tackle the unknown parts. However, this is not the only possible generative model: systematic design, on the other hand, suggests a very gradual injection of the available knowledge.

Note that the idea of a generative model can be generalized. Let H be a criterion for the quality of design reasoning (reasoning, not X; this criterion is independent of X and hence distinct from P); in the case of systematic design, H is: "minimize the expansions". A generative model defines set of design rules for improving H. Later we will see cases where H could be: "review the identity of the objects".

A few properties of a generative model are given below:

- We are interested in the conjunctive power and the generative power of a generative model:

  - **Generative power** = measurement of the space P(X) covered by the generative model (i.e. for which there exists an almost complete choice function)
  - **Conjunctive power** = that part of P(X) which can be attained without δK, or more generally, conjunctive power can be defined for all fixed $\delta_{max}K$ as that part of P(X) which can be attained for $\delta K < \delta_{max}K$. Conjunctive power corresponds to the fairly substantial assurance ($\delta_{max}K$) offered by the generative model to designers that, once the design process has been committed to, it will conclude with a satisfactory result.

Systematic design is therefore a special generative model (design sub-spaces confined to each of the languages).

3.2.1.3 Analysis of Combinations of Conceptual Models and Generative Models in a Few Cases of Rule-Based Design Regimes

How should conceptual and generative models be combined? On the basis of these ideas we can return to the design regimes we have already studied and gain a better understanding of their relative levels of performance.

In the *wild design process*, innovator-entrepreneurs endeavor to solve by trial and error the problems encountered in the previous generation of the product, which therefore serves as a prototype. The conceptual models are contingent on the problems encountered, while generative models are limited to resolving the problem or to the choices in the catalogs of existing solutions.

In *rule-based design governed parametrically or by recipe*, conceptual models are well ordered and identified, and they were used to construct a recipe-based theory. For instance, in the case of Redtenbacher's water-wheels design, the recipe (the series of ratios) constitutes the generative model. This is strongly conjunctive: we can be sure to obtain a water-wheel with a minimal δK (dialog with the client, calculations, tuning). It is weakly generative, however: of course, the wheels will be suited to the situations codified by the recipe (flow-rate, head of water, etc.) but it will not be possible to attain certain levels of performance (the higher efficiencies might get with a turbine, for example); special situations: making the wheel operate in icy conditions, etc.). The recipe does, however, allow an effort at domestication which is not simply about looking in a pre-existing catalog nor about the effort in reconfiguring an "original" design for each new exercise in designing a water-wheel and it is very efficient.

In *rule-based systematic design*, conceptual models are many (at the level of conceptual design of course, but also at other languages such as functional, physico-morphological, etc.). The generative model relies on a hierarchy of languages. It assumes a backward compatibility to avoid retrograde steps that might ruin the performance of the design process. At each level of language designers can generate alternatives they will then have to choose between. This rapid "descent" guarantees a strong conjunctive power. The exploration of variants at each stage allows the generative power to grow (variety, increased performance, etc.). However, this expansion occurs within dominant design, i.e. based on stable functional, conceptual and physico-morphological languages. Systematic design allows a very effective domestication within the scope of this dominant design since it simultaneously enables great variety and regular innovation for certain dimensions of performance while maintaining maximum economy (to minimal design) of the resources committed to the design process. Putting the importance of conceptual and generative models to the forefront, systematic rule-based design also denotes the spaces where efforts must be made for the regeneration of conceptual and generative models.

3.2.1.4   Analogy with Statistical Regression—Regression in the Unknown

(The reader unfamiliar with the principles of statistical regression can skip this section).

We can illustrate conceptual and generative models using the image of "regression in the unknown": systematic design organizes the knowledge of known objects $K(X)$ and possible decisions D to make the as yet unknown object X_x exist, satisfying $P(X)$ in the manner of a regression.

Hence, X_x can be regarded as a (dependent) variable Y that we might seek to approximate in terms of an (independent) variable X.

This predicted Y relies, on the one hand, on $K(X)$, which is the knowledge of values already obtained in the past for the pairs (x_i, y_i);

On the other hand, it relies on choosing a model (design strategy D) which gives the best prediction of Y as a function of X (f(X)) for any value of x_0, regardless of the previous values drawn (K(X)).

The statistical regression proceeds in two operations which, by analogy, provide information on the nature of the conceptual and generative models thus:

- The model is obtained by minimizing $E(Y - f(X))^2$ i.e. minimizing the variance of the residual $Y - f(X)$. In design, this corresponds to the unknown residual part in Y once X is known. In statistics, we show that the residual obtained for the best f(X) is of zero expectation and that the covariance of the residual with X and with any function of X is zero (mathematically, $f(X)$ is the orthogonal projection of Y onto the sub-space L_X^2 of functions of X): the residual does not depend on the known X, only the unknown remains in the residual. *This minimization operation gives f(X), which is the best approximation of Y by X. For the design process, the function f is the combination of knowledge which is the nearest to the unknown that can be attained. This is therefore a generative model.*

  Note that we frequently make assumptions about the form of the regression: we assume, for example, that $f(X)$ is of the form $f(X) = \alpha + \beta X$, which makes it easier to predict Y from the known X; we shall also make an assumption (known as a homoscedastic assumption), which requires that the variance of the residual does not depend on X—another assumption which ensures that the residual does not depend on the unknown. All these assumptions are embodied in a *generative model*: they consist of *making assumptions on the approximation of the unknown by the known, such as to minimize the "distance" from the unknown to the known or to simplify the relationship between the unknown and the known.*

- The generative model leaves the parameters free—for example, the coefficients of linear regression or the constant variance of the residual in the case of homoscedasticity. These free parameters will acquire values as a result of the previously accumulated knowledge. Statistically speaking, we can *estimate* these parameters because of the known realizations (x_i, y_i). Finding the best estimator for these parameters consists of constructing a *conceptual model* of the pair (X, Y). *The conceptual model represents the best synthesis of the known to explore certain unknowns.*

  We recall that, given a cloud of points (x_i, y_i), the "best" straight line is that which minimizes the Euclidean distances between the y_i and their vertical projection onto the straight line; quite generally, this straight line is different from that which minimizes the squares of the distances from a point (x_i, y_i) to the straight line, or that which minimizes the distances between the x_i and their horizontal projection onto the straight line (the latter corresponds to a regression of x in y). In other words, the "best" summary depends on what regression is chosen; or rather, knowledge is synthesized as a function of the unknown to be addressed: *conceptual models are in fact constructed as functions of generative models.*

3.2.1.5   Application: Platform-Based Design

The logic of platform-based design is a simple case of the generative model, very similar to the regression logic seen above.

A platform is a generative model constructed on a simplified conceptual model of the object being designed; platform design consists of an almost complete choice function enabling (a) selection of all the necessary and available knowledge for designing the object, and then (b) reduction of the residual unknown without taking account of the known. Hence the Y to be designed is broken down into a part determined by the known $f(X)$ and a residual part which no longer depends on X.

Take the case of platform-based vehicle design: let $K(X)$ be the platform for vehicles of level M1 (low to mid range) comprising, for example, common parts, common principles of architecture (even a common assembly chart), common conceptual models (internal combustion engine, principles of deformation under crash conditions, etc.), common functions (perceived quality, etc.). Designing a new type A vehicle within the M1 range consists of reviewing all the knowledge inherent in the platform to design A, and then designing the residual part outside the platform (see illustration below of the PSA-Citroën platform for small vehicles and their derivatives in the mid-2000s) (Fig. 3.3).

We can then define:

- The generative power of the platform: all vehicles possible with the M_1 platform.
- The conjunctive power of the platform: all type A vehicles possible with zero or small δK.

The generative and conjunctive powers will differ depending on the platform adopted. A platform based on a commonality of parts (a platform whose aim is generally to lower the cost of components) will invariably have a high conjunctive power (there will be little in the way of additional detailed design if many parts are already available); on the other hand, the generative power will be limited. A platform based solely on sharing an assembly chart (and more generally on sharing the constraints of the process: after-sales care, distribution, etc.) will have a more limited conjunctive power (this time it will be necessary to redraw the parts, etc.) but also a greater generative power.

Fig. 3.3 PSA platform for entry-level vehicles in the mid 2000s

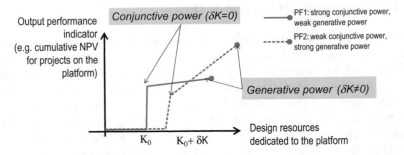

Fig. 3.4 Efficiency curves for two types of platform. A platform assumes an initial investment and is then able to make several products with virtually no new resources (the *vertical* part of the curve). The maximum platform-based output without new resources may be likened to the conjunctive power of the platform. With a few additional resources, the platform may then make new products up to a maximum point no profitable new design is possible on the platform): this maximum can be considered as the generative power of the platform. Two examples are given on the diagram: a platform of strong conjunctive power but weak generative power (a large number of common rules but few additional developments are possible—"lego" type); a platform of weak conjunctive power but strong generative power (there is a common core but new modules must always be developed on an ad hoc basis to have a product)

In the case of a platform we get efficiency curves such as those illustrated in Fig. 3.4.

See case study in this chapter for an exercise of application.

3.2.2   Assessing Systems of Rules: Axiomatic Design Theory

We have looked at the general structure of a system of design rules, and have seen that this structure determines the generative and conjunctive powers of the system of rules. Let us now assume a systematic design-type of generative model and a set of conceptual models. Can we assess the "quality" of these conceptual models (i.e. of these design rules)? It is to this question that the theory of axiomatic design due to Nam P. Suh endeavors to find an answer: in systematic design, how a set of rules enables a set of products sharing the same functional reference base to be designed in a reasonably easy and robust manner.

3.2.2.1   General Framework

Nam P. Suh, professor of engineering at MIT, was invited very early on by the NSF to assess some engineering projects. On that occasion he pondered the criteria for the "good result" in engineering. In his work of 1990, *Principles of design* (Oxford University Press) (this work would be followed by another published eleven years later: *Axiomatic Design advances and applications*, Oxford University Press.) (Suh 1990, 2001), he seeks to answer questions of design evaluation: "The fact that there may be good design solutions and unacceptable solutions shows that there are traits and attributes which distinguish between good and bad designs". Moreover, he is critical of systematic design and its so-called "morphological" part (Zwicky matrices): this can, of course, generate a very large number of solutions through combination, but does not allow choice between the solutions in the catalog.

Nam P. Suh models design from the definition of two "domains": the functional domain (objectives, functional requirements FRs) and the so-called "physical" domain or the domain of design parameters, DPs:

- The FRs are very similar to the functions of systematic design, and are required by all the participants concerned. According to Suh, "The FRs are defined to be the minimum set of *independent* specifications which completely characterize the design objective for some specific need". We note that it is not obvious how this independence requirement can be realized. Note also that the constraints are not FRs since they are not mutually independent nor are they independent of the FRs (e.g. "costs are only an upper limit").
- The DPs are defined by successive design decisions, and are the designer's levers for action: he sets the level of the DPs and thus realizes the FRs. This means that the designer draws on various technologies (which are particular relationships between the (levels of) DPs and the (levels of) FRs), and that he sets the actionable parameters for these technologies (the DPs) to achieve certain effects (the FRs).

In Suh's eyes, any successfully completed design is a mapping between the domain of DPs and the domain of FRs. A design is described by the matrix linking the FRs to the DPs (see Fig. 3.5 below). In terms of conceptual and generative models, the technologies that link DPs with FRs are conceptual models; the generative model is both the DP-FR distinction and the general structure of the matrix relating all the FRs of a family of products to all the DPs.

3.2.2.2 Axiom of Independence, Axiom of Minimum Information

Axiom of Independence

In this context, Suh proposed the first axiom of good design: a design satisfies the first axiom (or "is good in the sense of the first axiom") if it is decoupled, i.e. if the matrix relating the FRs to the DPs is diagonal.

Suh justified his first axiom thus: "In an acceptable design, the DPs and FRs are related in such a manner that a specific DP may be adjusted to satisfy its corresponding FR without affecting the other FRs". In other words, the design is easily adaptable to changes in FR (robust against variations in FR); and "A perturbation in the physical domain must affect just a single functional dimension". The design is

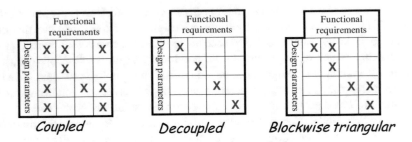

Fig. 3.5 Coupled, decoupled and blockwise triangular Suh matrices

robust against variations in the DPs. The first axiom therefore corresponds to a principle of robustness of architectures.

Suh showed that a design which is non-diagonal, but Jordan block-wise triangular (see Fig. 3.5) is still interesting. In such a case:

- it is still possible to adjust the DPs to the FRs by reasoning block by block
- in each block it is possible to find a fitting strategy: the first DP can be used to set the first FR (top left); then, depending on the value of this DP, a constraint on the second FR can be deduced, and the second DP can be used to set this second FR, and so on…[2]

The case of the mixer tap illustrates the first axiom (see Fig. 3.6). Let there be two sources of water, one hot, the other cold. We wish to design a system to provide a certain flow of water at a certain temperature. An initial design might use two taps, each on the hot and cold inlets respectively and such that the flow rates can be adjusted by amounts α and β. This matrix is coupled (adjusting an FR, e.g. the flow rate at constant temperature, assumes that both DPs are affected). A second design proposes a "moving mask" on the cross-sections of the two water inlets, the mask having two possible directions, θ and ψ: when the mask moves along θ the total flow changes but the proportion of hot and cold water in the total flow remains constant—hence the flow has been adjusted without changing the temperature. If the mask moves along ψ, the total area remains unchanged but the ratio of the areas changes, hence the total flow remains constant but the temperature has altered. This second design is decoupled.

A few points need to be emphasized:

1. the first axiom characterizes not one product (besides, this can never be completely specified) but a set of products making use of the same natures of DP and FR; all products from the same family can be designed using the same matrix.

---

[2]Readers with a penchant for linear algebra may observe that a triangular matrix can be diagonalized, meaning that after a few linear combinations a triangular matrix can be transformed into a design which satisfies the first axiom.

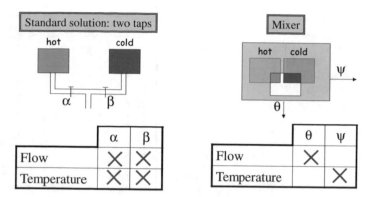

Fig. 3.6 Coupled and decoupled design for a device delivering a fixed water flow at a fixed temperature

Herein lies a good criterion for assessing the basis of the rules enabling a family of products to be designed.

2. The first axiom leads to the creation of "good" generative models: if the matrix is diagonal, it is then possible to identify the list of d_i enabling a new product to be designed quickly and efficiently (the d_i are independent and correspond to each of the DPs). If the matrix is triangular, the sequence of d_i is more complicated (successive d_i are interdependent) but it is still possible to have a sequence of d_i which converges to a product in a well-controlled period of time.

The first axiom has several operational consequences:

- As long as the list of FRs remains fixed, the matrix remains valid; on the other hand, Suh observes that just one new FR is sufficient to couple the matrix and that all the DPs need to be revised. The addition of a new FR (the customer may want a new "little" function) may mean revising the design in its entirety, and a new DP must be found to adjust the new FR; this new DP must not create any new interdependencies with the other FRs, and neither must the other DPs have any effect on the new FR.
- One simple consequence of Suh's axiom 1: once all the FRs have been chosen, we know that at least as many DPs will be needed. If the available rules do not offer as many as are required, the matrix will necessarily be coupled (and the design consequently less robust).
- The first axiom also helps to direct research work: among all the questions posed by a still-coupled system, axiom 1 tends to give priority to working on the couplings rather than to the uncoupled parts. In the diagram below, it is preferable to work on the coupling factor at the lower left than on the coupling factors in the triangular matrices.
- It is always possible to carry out linear combinations on the FRs which, as a consequence of the logic of linear algebra, will lead to a diagonal matrix. Aside from the mathematical aspect, this means that remodeling the FRs *taking*

Fig. 3.7 A coupled Suh matrix: how to manage the design of a product following such a rule base? Suh's first axiom leads to remove the coupling in the bottom left corner

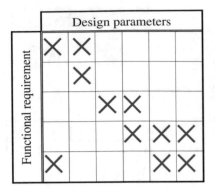

account of the technical constraints (DP) has the potential to satisfy Suh's first axiom. In practice, this means that it is the technical principles of the engineering department that can lead to revision of the functional language.

- Axiom 1 is directly related to the issues of division of labor:

  - If it is satisfied, design tasks can be decoupled so it is therefore an effective tool in organizing a V-type cycle (see Chap. 2).
  - If it is not satisfied, it nevertheless allows the difficulties encountered to be understood and leads to discussion of the questions of interdependency (Fig. 3.7). It orients design work on the most critical issue-such as removing the coupling in the bottom left corner in the system in Fig. 3.7.

Axiom of Minimum Information

Suh's second axiom stipulates that the best design is functionally decoupled (axiom 1) and minimizes the information the user must have to use the device.

Nam P. Suh says that design minimizes the "information content of the design" and defines this information content thus: this is a measure of the probability of obtaining an FR in a certain "design range" (the tolerance expected by the user) with a DP in accordance with a certain "system range" (all the values effectively achieved by the device) (see Fig. 3.8). This is calculated according to the formula: $I = \log_2$ (design range/common range). The ideal design is one in which the common range and the design range are the same, in other words the design range is "included" in the system range. Put another way, whatever the user's FR level in the design range, there exists a setting on the machine which allows this level to be attained; in the opposite case, there exist FRs that the user can only achieve by adapting them himself on the basis of the "nearest" value provided by the machine (hence the user must "import" supplementary information).

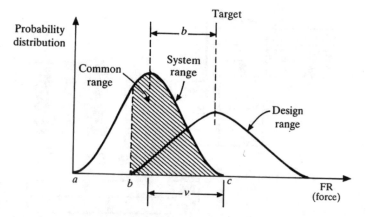

Fig. 3.8 Measure of the amount of information for a device (from Suh 1990)

3.2.2.3 Two Application Examples

The Bagless Vacuum Cleaner

In a normal vacuum cleaner, the bag is a DP which couples the separation of air and dust (FR1) (the bag is a filter) and retains the dust (FR2) (the bag is a container). The bagless vacuum cleaner (introduced by Dyson in the 1990s) uses a cyclone (DP1) for separation and a chamber (DP2) for retention: this is a better design in the sense of Suh's first axiom (see Fig. 3.9).

Several studies have shown the importance of the function: "easily empty the accumulated dust" (FR3). For this function, the consumable nature (DP) of the traditional vacuum-cleaner bag provides decoupling between FR3, FR2 and FR1. On the other hand, the "bagless" cleaner in fact couples FR 2 with FR3. Axiom 1 then invites the addition of a new DP which would decouple FR2 and FR3: why not add a disposable bag, which would then lead to a "bagless" vacuum cleaner "with a bag"? (Fig. 3.10).

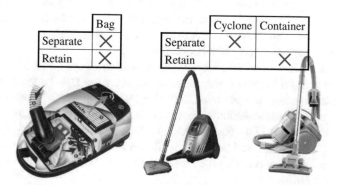

Fig. 3.9 Suh matrices for the vacuum cleaner with and without a bag

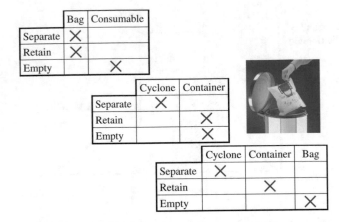

	Bag	Consumable
Separate	X	
Retain	X	
Empty		X

	Cyclone	Container
Separate	X	
Retain		X
Empty		X

	Cyclone	Container	Bag
Separate	X		
Retain		X	
Empty			X

Fig. 3.10 The bagless vacuum cleaner with a bag

The "Open Rotor" Engine

This case is one of the results of the work of Damien Brogard and Mathieu Joanny, who opted for the Design Engineering course at MINES ParisTech (Brogard and Joanny 2010).

Safran–SNECMA is a world leader in aircraft propulsion, Its best seller, the CFM 56, is fitted to the A320, A340 or B737; there are 20,000 examples in service in the world today, and an aircraft powered by them takes off every 2.5 s. In the prospect of more fuel-efficient aircraft, Safran is developing an engine project known as "open rotor" in which the external *casing* (the enclosure around the engine) is removed, reducing weight and drag and hence the hope of reducing fuel consumption in the order of 25%.

The main DPs and FRs for the standard engine can be represented as per the Fig. 3.11 (FR in rows, DP in columns). We see a triangular matrix, corresponding to the division of design labor at Safran: the "module" department deals with the first two FRs (torque and thrust, taken to be "primary functions") using the DPs "power body", "internal casing" and "ducted fan"; the "integration" department is involved after the "module" department and deals with the other FRs (called secondary functions, and functions interfacing with the aircraft system). The Suh matrix reveals the pertinence of this organization.

In this matrix it is possible to model the transition of the CFM 56 to an "Open Rotor" as envisaged by SNECMA: in an open rotor the DPs "ducted fan", thrust reverser" and "casing" are removed, and the DPs "gearbox", "propellor", "pitch controller" and "contra-rotation" are added to ensure the functions previously assigned to the removed DPs (see the two matrices below: on the left is the CFM 56 with the new virtual DPs still "empty", while on the right is the future Open Rotor for which the FRs addressed by the removed DPs are "redistributed" over the new DPs) (Fig. 3.12).

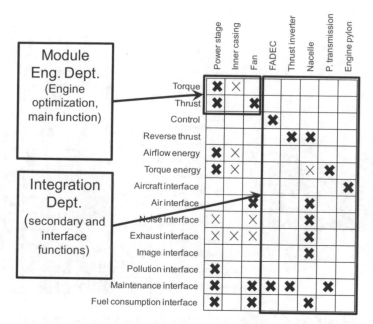

Fig. 3.11 Suh matrix for the CFM 56 (Brogard and Joanny 2010)

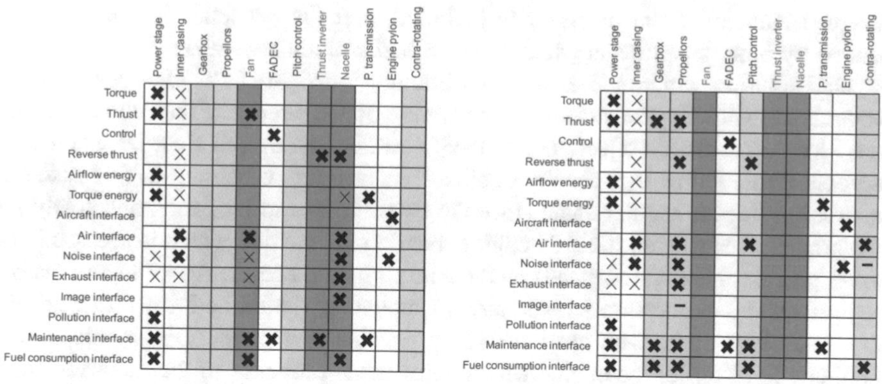

Fig. 3.12 Redistribution of the functions ensured by the removed DPs for the CFM56 over the new DPs for the future Open Rotor. The former DPs correspond to the *salmon-pink* (or *light gray*) columns and the new DPs to the *blue* (or *dark gray*) columns. On the *left* is the CFM with the new virtual DPs still "empty". On the *right* Open Rotor project

The new diagram anticipates several critical interdependencies: thus the propellor plays a critical role in the "module" department where it governs the thrust but also plays a critical part for the acoustic FR and for the "ejections interface" (in the event of a blade being torn out) with no other DP from the "integration" department allowing these FRs to be adjusted. In other words, the matrix is no

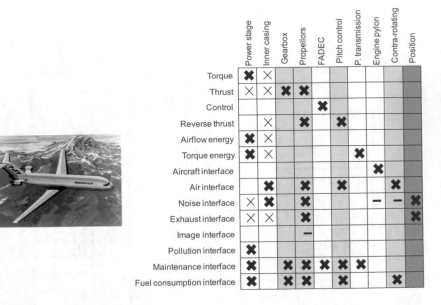

	Power stage	Inner casing	Gearbox	Propellors	FADEC	Pitch control	P. transmission	Engine pylon	Contra-rotating	Position
Torque	✘	✕								
Thrust	✕	✕	✘	✘						
Control					✘					
Reverse thrust		✕		✘		✘				
Airflow energy	✘	✕								
Torque energy	✘	✕						✘		
Aircraft interface									✘	
Air interface		✘		✘		✘			✘	
Noise interface	✕	✘		✘				—	—	✘
Exhaust interface	✕	✕		✘						✘
Image interface					—					
Pollution interface	✘									
Maintenance interface	✘			✘	✘	✘	✘	✘		
Fuel consumption interface	✘		✘	✘		✘			✘	

Fig. 3.13 Addition of the "location" DP to relieve the propellor of the "ejection interface" and "noise interface" FRs (return to a triangular matrix). The illustration shows an aircraft with engines in another position (but these engines are NOT open rotor engines)

longer triangular if the propellor is included, and the propellor is a critical DP shared by both departments which hitherto had worked separately.

One solution which addresses this difficulty is to add DPs. One possible DP consists of reviewing the location of the engine on the aircraft: if the engine is at the rear, loss of a blade ("ejection interface") and acoustics can then be taken into account by new drawings for the stabilizer fin (see figure below); in this case the matrix becomes triangular again since the two FRs are no longer carried solely by the propellor. However, the DP "engine location" is not one which is accessible to the engine designer, and depends on the aircraft manufacturer. Also, the Suh matrix suggests that a partnership with an aircraft integrator should be formed to work on the Open Rotor, and sets out the basis on which this partnership should focus: finding a location for the engine which relieves the propellor of the "acoustic" and "ejection interface" FRs (Fig. 3.13).

3.2.2.4   Design Logic Underlying Axiomatic Design. Analogy with Control Theory

Suh's axiomatic theory deals with the result of the design. Its aim is to assess a design (more precisely, certain elements of a design's rule system) and gives no indication as to the design process for the DPs enabling a decoupled design to be obtained.

Fig. 3.14 In control theory, it is possible for the transfer function to depend only on the evaluation function. *Source* Suh (1990)

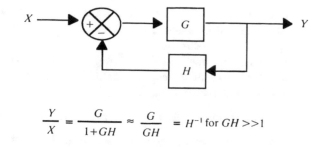

$$\frac{Y}{X} = \frac{G}{1+GH} \approx \frac{G}{GH} = H^{-1} \text{ for } GH \gg 1$$

Nam P. Suh is aware of this limit but provides an interesting argument: if the process of design is likened to a controlled system, designing consists of achieving setpoint Y via means X using two operations: an action (which transforms an X' into Y = G.X' which may be reasonably close to the setpoint Y_0) and feedback which evaluates G.X' (H.G.X') and compares this information with the value X (let X' = X − G.H.X', i.e. X' = X/(1 + GH)); we thus get the transfer function Y/X = G/(1 + GH) (see Fig. 3.14). Suh shows that for this type of transfer function, if the evaluation function is very efficient, we will have GH ≫ 1 and the transfer function is then Y/X = H^{-1}, which no longer depends on the generator function but solely on the evaluation function. In this way, we pass from evaluation to design!

3.3 The Organizations of Systematic Design

Having qualified the performance and rationality of the rule-based design (systematic) it is a fairly simple matter to characterize and position the primary skill-sets and major players in this regime.

3.3.1 Skill-Sets and Guarantors of the Company's Rule Base

3.3.1.1 The Primary Skill Sets

The organization of systematic design is based on those primary skill-sets that guarantee the rule base. We now define these skill-sets in the following manner:

- **development** is characterized by a controlled process making use of existing skills to address a set of predefined specifications.
- **research** is a controlled process for the production of knowledge (see the historical case study 3.1 on the origins of industrial research in this chapter).

- **marketing** is a controlled process for the production of knowledge about the customer and the market, responsible for writing and validating the functional specifications documents.
- Development draws on the skill-sets inherent in **engineering**. These skill-sets are the spaces of upkeep, accumulation and regeneration of the rules bases. These develop conceptual models.

The links between "R" and "D" are now easier to understand:

- matrix-based organization consists of enlisting the engineering skill-sets in product development; ideally, they should also allow the engineering professionals to renew their rule bases on these occasions.
- Research and development may be linked in many ways (see the historical case studies on the origins of industrial research): research can exist almost independently of engineering and design, e.g. when the research laboratory is responsible for measuring product quality (in the factory) or the quality of incoming materials (purchasing); when linked to design, research can, for example, be related to what has happened upstream (research helps engineering renew its system of rules).

3.3.1.2 A Few Guarantors of the Rule Base: Experts and Revisor

The experts are the guarantors of conceptual models and of their relationship with generative models (enlisted by the projects). Their training satisfies the "skill-set" logic, drawing on the universities as well as on the system of "mentoring" over the course of their careers (Weil 1999; Dalmasso 2008; Lefèbvre 2003). These experts have to manage a compromise between usage and updating, and renewing their expertise. Involvement in projects allows the expertise to be used and is accompanied by an important learning process, in particular when other skills come face to face with the product; however, the experts may also have an interest in developing their expertise by other means, less expensive, of course (since they may not lead immediately to products of added value) but also more effective than a learning process during product development: this might involve experiments and modeling and simulation work carried out alone or in partnership with outside laboratories (see case study 3.2, in this chapter, Renault acoustics).

Alongside the professional experts, one of the key figures of the organization around the rule base is *the revisor*, described in the historical studies of design departments (König 1999; Neuhaus 1904). The revisor is responsible for checking the drawings (detailed design) proposed by a project leader having concluded his work. He therefore appears as a form of "quality control". But what is he "controlling"? Paradoxically, the revisor does not check compliance with the specifications document—this assessment is down to the project leader himself (the Q in QCT). The reviser checks consistency with the company's rule base: even if a project fully satisfies the customer's specifications document, the reviewer can reject it on the grounds that the project designers have "over-designed" it and have made insufficient use of the knowledge within the company; a second possible reason for rejection may be this: naturally, certain new drawings may be necessary, but are made in such a way that they cannot be re-used by the company and cannot be incorporated within the company's rule base.

Although the term "revisor" may not have been adopted by industrial language, such a person still exists today: he might, for example, be the engineering department manager, business manager or platform director. Each of these players will, in effect, regularly ensure the "proper use" of the rule base (and of the "bearers" of these rules, the experts!) by the projects; this use will require that these two facets are continuously maintained: not only proper use of the existing rules, but also a proper "update", a regular enhancement as a result of successive projects.

3.3.2 Sector-Wise Industrial Organization—The Ecosystems of Rule-Based Design

Updating these logical processes of rule-based design explains the logic of subsidiaries and associated industrial sectors.

The logic of preserving improving design rules makes it possible to create stable ecosystems with the following characteristics:

- on the one hand the regular re-use of components, machinery and processes enables modular change in the production chains: the sectors are therefore structured by an integrator and by level 1, 2 and 3 suppliers. They are linked along a "value chain" and their integration may be supported by a set of norms and standards which codify the interfaces ensuring compatibility, meanwhile leaving the supplier free to propose products which are always of higher performance.
- on the other hand, the creation of stable skill-sets and scientific objects makes it possible to stabilize sectors of competence: technical schools, engineering schools, academic disciplines (engineering sciences), research laboratories, potentially common laboratories (historically, France has had Irsid for iron and steel, Cerchar for coal, and more recently, Renault and PSA's LAB for automobile safety, etc.).

Note that sectors of competence associated with product usage may also be created: specifiers, inspectors, assessors, repairers, user communities or associations, etc. grow up around the dominant product design. Hence, the automobile ecosystem cannot be described without mentioning the highway code, drivers' license and driving schools, auto journalists (who assess and classify the products), traffic police, etc.

Note that it is all these stabilizations (on the nature of goods, performance, technology, skills, etc.) which also allow the multiplicity of market-type relationships: the market at the supplier-customer interface, the market for work (or the skills market) for designers, etc.

These structures constitute the ecosystems of rule-based design.

3.3.3 The Logic of Generative Bureaucracies— Coordination and Cohesion

These professions and organizations behave like bureaucracies: they strongly impose a routine on processes and reach a very high level of division of labor, standardization of tasks, products and processes.

However, we have also been able to show that these bureaucracies are still generative (Le Masson and Weil 2010b, 2008)—this is the great paradox from the point of view of theories of organization. While relying on robust, stable routines they manage to undertake intense regular innovation, i.e. to propose products which do not exist, improve the performance of products and enhance their capabilities.

This is because "routinization" relies on very high level languages (functional, conceptual, embodiment) which are sufficiently general to tolerate expansion: these are neither the product parts list nor the list of expertises built into the organization but the languages capable of bringing as yet unknown objects into existence (generative models) and summary representations of the object capable of incorporating the unknown within actionable abstracts, and of remaining robust in the face of the emergence of new objects (conceptual models).

These very high level languages (conceptual and generative models) have two essential properties:

(1) they enable the local renewal of certain rules: better accuracy, increase in functional performance.
(2) they offer a stable framework: the list of functions (functional language) ensures coordination between the design department and marketing; the stability of the architectures also enables a skills reference to be stabilized, as well as the relationships between these skills, etc.

Hence these languages ensure coordination in innovation. They also ensure cohesion, though this second result is not obvious: do we not view innovation as "creative destruction"? In pushing to renew skills and products, does innovation not lead to an alteration of the balance between skills and the interests of each? Against these perceptions, the *organization of rule-based design is a cohesive system which is robust against innovation*. Once these languages have stabilized, i.e. a very high level framework, continuous evolutionary change becomes possible. If the languages themselves had to change or if it was the order of these languages of systematic design that had to be modified, then the logical processes of cumulative learning for the primary skill-sets could no longer function. However, stability at a high level allows constant evolution—within certain limits, naturally—of products and skills. Despite (or maybe because of) the evolution of products, markets, skills, etc. the company maintains a form of "common purpose". In this sense, the organization of rule-based design is a machine for innovating *without* the torments of Schumpeter's "creative destruction"; it is a fluid form of management of change without the organizational crises.

3.4   Conclusion: Routine/Generative Equilibrium

With systematic design comes a profound "routinization" of design that supports a form of generativity. Innovation has become domesticated, systematized, predictable, and within the grasp of many.

To be borne in mind:

1. We are no longer (and this has been true for a long time) in the era of the singular innovation, but rather it is the broad range of innovation which improves performance. A designed thing is not a one-off product but evolving families of products with multiple variants. With the invention of organization in rule-based design, for 150 years the big companies have surpassed a representation of processes of innovation that might consist of going "from idea to market".
2. Design organizations appear as generative bureaucracies, far from the image of the commando or solitary entrepreneur and far also from the image of specialist adhocracy, they involve structured ecosystems and coordinated factories with white-collar workers.
3. These organizations and this level of performance rely on structured systems of rules. these rule systems follow two precise and demanding logical processes: the logic of the conceptual model, which consists of a synthesis of effective knowns, and the logic of the generative model, which consists of knowing, for any new concept, how to limit that part of the unknown it contains, injecting pertinent knowledge wherever possible. These rule systems define any conceivable products and those that are out of reach; those organizations that can master them know how to innovate within a given cone of performance, and they also know how to keep at arm's length or retain development projects depending on the distance to the rule system, and any additional effort for producing the knowledge necessary to achieve the target.
4. This model has its own regenerative logic: successive projects, carried out under the logic of systematic design, will enhance the rule base for any projects which follow. This logic of regeneration is made possible and controlled by unique organizational figures such as the revisor; it is also based on how the reasoning is structured which facilitates the accumulation of knowledge while ensuring compatibility within the conceptual models, and the actionability due to generative models.
5. However, this logic of regeneration retains the principles of systematic design: expansion at least, preserving established knowledge. The more adventurous (more subversive, even) designers tend to appear on the margins of rule-based design: visionaries who create their own laboratories (Thomson-Houston, Baldwin, etc.) capable of orchestrating the genesis of totally new families of product lines.

We shall be examining the underlying logic of these apparently peripheral players in the next chapter.

3.4.1   The Main Ideas of this Chapter

- Cone of performance, meeting commercial demand, dominant design
- Conceptual model, generative model
- FR, DP, first and second axiom of axiomatic design
- Development, marketing, research
- Ecosystem of rule-based design
- Generative bureaucracy

3.4.2   Additional Reading

This chapter may be extended in several directions:

- On professional skills: see the models of professional career paths (Dalmasso 2008), work on professional identities (Sainsaulieu 1986), and on the new work crises (Minguet and Osty 2008)
- On axiomatic design: see Suh (1990, 1999a, b, 2001); on the question of robustness in design refer also to Taguchi methods (in particular, see Annex A of Suh) (Tsui 1992; Taguchi 1987); see also other instruments for the analysis of interactions in complex systems such as the DSM method (Ulrich and Eppinger 2008; Smith and Eppinger 1997; Whitney 1990)
- On the idea of dominant design: see the original work of Abernathy and Utterback (Abernathy and Clark 1985; Abernathy and Townsend 1975; Abernathy and Utterback 1978; Utterback 1994; Utterback and Abernathy 1975); see the more recent product or industrial life cycle S-curve models, either with offer models (Klepper 1996, 1997) or with demand models (Klepper 1996, 1997)
- On the primary skills and professions—on marketing: (Le Nagard-Assayag and Manceau 2011; Kotler et al. 2006); on research: see work on the history of research (Reich 1985; Fox and Guagnini 1999; Hounshell and Smith 1988; Little 1913; Meyer-Thurow 1982; Shinn 2001; Travis 1993) or a summary in (Le Masson et al. 2006; Le Masson 2001) (third part); on models illustrating the relationship between the engineering and research departments see the chain-linked model (Kline and Rosenberg 1986); on the value of research: see the idea of absorptive capacity (Cohen and Levinthal 1989, 1990)
- On platforms: a few evaluative indicators (Meyer and Dalal 2002; Meyer and Lehnerd 1997); the logic of modularity and its economic value (Baldwin and Clark 2000, 2006a, b)

3.5   Workshop 3.1: Calculation of Static and Dynamic Returns on Platform

3.5.1   Introduction: Design Function, Static and Dynamic Returns

Economic theories of firm have long been founded on the notion of production function (see Fig. 3.15). They consider that there is a finite, fixed list of goods and they construct profitability on a fall in the costs of production factors. Design opens up the prospect of a more general definition of the company, seen as a function of design, working not so much to "transform" quantity factors into goods factors but simultaneously to expand the space of goods and the space of skills (Hatchuel and Le Masson 2006; Le Masson et al. 2010).

Such a definition makes it possible to assess two types of *design* return. Given a design action, a project P_i giving rise to a new item of goods G_i and to new skills, we can define:

- a *static design return for project P_i*, defined as the ratio of the income generated by products from project P_i to the resources invested for project P_i. We get an estimate for this return by calculating the net updated value of the project (as we saw in Chap. 2).
- a *dynamic design return* which characterizes an improvement or deterioration in the company's design capabilities after the conception of project P_i. Estimating this dynamic return is not always obvious; it is, however, the assessment criterion that would be used by a "revisor" or platform manager in accepting the development of a new "module".

The purpose of this workshop is to harness these ideas for economically evaluating a project on a platform.

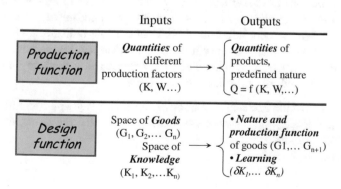

Fig. 3.15 Production function, design function

3.5.2 Platform-Based Project Evaluation

Let there be a company which designs and manufactures equipment for electrical networks (e.g. transformers, circuit-breakers, etc.) (e.g. Schneider Electric). This company is organized in terms of rule-based design and its products are designed on platforms (recall that the term "platform" means a set of common components, or more generally, common design rules that may be used for several products).

1. A project leader is nominated to develop a variant of a circuit-breaker for niche applications. The project leader is given several alternatives:

   - To develop the product independently for a total cost of 100 k€ over one year (year 0). Over years 1–5 the developed product will bring into the company 20 k€ per annum (alternative a).
   - To develop the product using an existing platform. The cost is then 40 k€ but the product is of lower performance and brings in only 15 k€ per annum (alternative b).

   Question: what advice should the project leader be given?
   Answer: the project leader legitimately reasons in terms of static returns. Hence the NPV of alternatives a *and* b *can be calculated. Calculating the NPV gives −24.18 for (a) and +16.86 for (b). The project leader must choose to implement his project using the platform.*

2. The manager responsible for the circuit-breaker platform suggests the following variant:

   - develop the product using the existing platform but at the same time develop a module which can be re-used for future products. This development might then cost 80 k€ (rather than the previous 40 k€) and the product would then bring in the expected 20 k€ (alternative c).

   Question: Does this alternative alter the decision of the project leader?
   Answer: the project leader still reasons in terms of static return. The NPV for this alternative is negative (−4.18); he therefore still chooses alternative (b).
   Comment: We see here how use of NPV as the sole indicator concentrates the whole point of the investment in refurbishing the platform on the "first" project of interest. The consequence is immediate: the project has a tendency not to allow refurbishment and all the rules tend to deteriorate.

3. As arbitrator of the exchanges of information between the platform manager and the project leader, the business unit manager asks that the dynamic returns be taken into account.

Questions: Calculate a dynamic return indicator for each of the three alternatives; under what conditions of adding value to the new module does alternative (c) become profitable?

Answer: The dynamic return indicator must take into account the acquired design capabilities. For example, this corresponds to what future users of the platform will not have to pay when developing their products. In alternatives (a) and (b) the dynamic returns are zero. In alternative (c) the platform manager proposes the development of a re-useable module which ought therefore to have a certain value later on. The dynamic return is therefore positive, and depends on the ability of the platform manager to add value to the module.

How to augment the dynamic return to make this scenario more profitable? For example, an internal transfer price associated with the use of a module can be introduced. This transfer price must

1. maintain the profitability of the primary project—the difference in profitability is an investment in the platform which must therefore pay not 80 but 80—(16. 86 + 4. 18) = 58.96, i.e. an investment of 21.04 on the part of the platform.
2. Allow the platform to recover its investment over a limited time. Other users will not have to pay the initial investment but they will have to ensure a return on the platform investment (which invested 21.04); a transfer cost bringing in 5 per annum over 5 years is sufficient.

The exercise thus shows that introducing the logic of dynamic returns restores an investment logic from the platform's point of view. This is an investment in revitalizing the rules.

Note that the practice suggested in the exercise is often difficult to implement in the companies of today.

3.6   Workshop 3.2: Design by Use—The Usefulness of Conceptual Models

The purpose of this workshop is to develop and use conceptual models, incorporating any innovation suggested by usage (hence extending beyond what is allowed by functional analysis alone).

The workshop is inspired by an actual case working with Jean-Pierre Tetaz, Vincent Chapel and the Archilab teams in 2004. Archilab provides a service helping business to innovate. Mob, a designer and seller of bolt-cutters (a site tool for cutting steelwork for concrete construction) asked Archilab to improve their product, but without touching the cutting head (the black part in the photo in front of the red handles) (Fig. 3.16).

Stage 1: propose alternatives for improving the bolt-cutter without touching the cutting head.

Over the course of this stage the functional space of the object has to be gradually explored. The functional analysis then concludes with an analysis of usage, which might, in particular, involve using the object.

Thus a typology of using a bolt-cutter appears: used on building sites (construction of buildings with shuttered walls and reinforced concrete slabs) for cutting round concrete reinforcing rods (6 m long straight metal rods), armatures (metal rods assembled in the form of 3D beams) and welded mesh (metal rods assembled in the form of a 2D lattice). Several standard cutting situations can be identified: cutting loose rods (cutting the metal rods to size is generally made on the ground (round, wall-ties, mesh)) on the ground or supporting against the body as per the diagram below and cutting fixed rods (the metal bars are cut to make an opening in the floor or shuttering). See the illustrations Fig. 3.17.

Fig. 3.16 Bolt-cutter

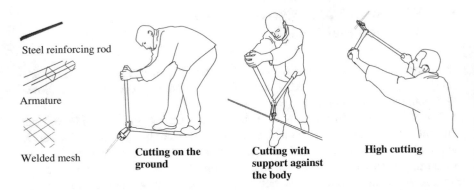

Steel reinforcing rod

Armature

Welded mesh

Cutting on the ground

Cutting with support against the body

High cutting

Fig. 3.17 Illustrations of ways in which the bolt-cutter can be used

Examination of the different stages involved with the bolt-cutter reveals a relatively long chain of events from the manufacturer to the end-user (artisan-builder, builder or scrap-metal merchant) via the distributor (hardware store or professional distributor) and the buyer (artisan-builder or site buyer). A study of typical sites (observation and tests) shows that usage demands considerable physical force and that positioning the tool is difficult (complex holding arrangements for cutting fixed rods, constantly looking for stable ways of holding so as to exert maximum force on the tool, etc.).

We are therefore led to suggest some functional improvements (better ergonomics, reduction of the force required, etc.), and solutions can then quickly be proposed: adjustable grips, telescopic handles to increase the torque, a ratchet-type "accumulator" system, motor assistance, etc.

At this stage we observe that while our reasoning has been made in terms of functional and embodiment languages, it has actually ignored (short-circuited) the conceptual language. Hence we come to a new stage in the workshop:

Stage 2: propose alternatives for improving the bolt-cutter, with two conditions: without modifying the cutting head AND drawing on conceptual models.

Three (at least) conceptual models can be developed:

1. A model based on the mechanics of materials: we know that the fracture of a metal rod can be modeled in terms of the curve of elementary strain (ε) under some elementary stress (σ), and we know that there are two phases of deformation: elastic strain followed by plastic deformation until fracture occurs. This curve enables the strain applied to the object to be calculated (and hence to consider the amount of energy required for external input); this is how a conceptual model is usually employed to produce a design. In particular, however, this leads to an examination of the parameters to which the curve is sensitive, i.e. the nature of the material, temperature, (hotter, but also colder), surface condition (cracks), the dynamic effects of the application of stress (shock, fatigue, etc.), surface oxidation or bulk contamination (e.g. hydrogen embrittlement

Fig. 3.18 Conceptual model
of fracture. The shape of the
curve changes according to
the material, temperature,
surface condition, etc.

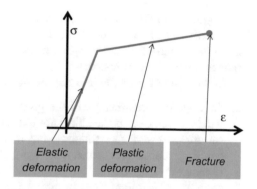

within the material). This additional information provided by the model represents innovative approaches for improving the bolt-cutter without modifying the
cutting head. All these innovative approaches were ignored during the first stage
(Fig. 3.18).

2. A kinematic model: using a kinematic approach we can model the number of
   degrees of freedom available to the user, in a situation involving both fixed and
   loose cutting. Hence on the one hand this approach enables us to see the issue of
   adding degrees of freedom for the fixed rod situation (we might consider
   articulated grips but observe that the hinge can be located somewhere else on the
   tool (pivoting head); on the other hand, we can see the issue of fixing the degrees
   of freedom in the situation where loose rod is being cut (ensuring the stability of
   the tool when vertical, also ensuring that the rod is orthogonal to the cutting
   plane of the tool). Here again the reasoning process opens up innovative
   approaches (a tripod arrangement for cutting loose rod, a system for attaching
   the rod in front of the cutting head jaws, etc.).
3. The third model focuses on situations of actual usage and is deduced from the
   kinematic modeling process. We have not three, but two contrasting situations
   which make use of opposing innovative approaches: cutting fixed and loose rod
   respectively. The first tends to transform the object into a manageable, flexible,
   light and adaptable *tool*; the second transforms it into a *machine* on a stand,
   minimizing effort and improving accuracy (possibly with external provision of
   power). This third model opens up two opposing views: either consider two
   tools which are more specialized, or work on the major difficulty with this tool,
   i.e. that of keeping the compatibility between the two situations and hence
   conceiving an object with the flexibility of a tool and the power of a machine.

We observe an overall magnifying effect of the conceptual models, opening up
further and more general alternatives (since being more abstract, they leave the
possibility of varied embodiments).

Stage 3: reactions to Archilab's proposed demonstrator.

Thirdly, we show the demonstrator built by Archilab to highlight the different innovative approaches the company proposed to Mob (see Fig. 3.19). We can then start to analyze the demonstrator.

In this type of situation it is important to distinguish two logical processes:

1. the logic of assessment: will the proposal be feasible at reasonable cost, what will be the market, is the object not too heavy, etc.? For the object concerned we can make several reservations—and if the object were made by a competitor, would the conclusion be that there would be few risks in seeing this object come to market?
2. the logic of learning: signs of interesting innovative approaches can be spotted on the demonstrator. In particular, the articulated support system shows that the demonstrator functions as both machine *and* tool; in this sense, and even if the suggestion of an articulated support is not the "best", the demonstrator shows an interesting innovative approach. And if this suggestion came from a competitor, the conclusion would be that the competitor was taking an original approach and that innovative and disruptive products could be appearing on the market in the short term.

Fig. 3.19 Archilab's innovative bolt-cutter demonstrator for Mob

3.7   Case Study 3.1: The Origins of Industrial Research

This case study deals with that "other" great part of R&D, research. We have studied the origins of the engineering design department (case study 2.1, Chap. 2) and have shown the gradual development of the increasingly generative doctrines and organizational methods of design. In research, the process is not the same—and the relationship between research and innovation is somewhat ambiguous.

As we shall see, (1) industrial research is historically not regarded as a profession for the designer; (2) methods of research are for rationalizing the production of knowledge—but not concepts; and (3) the ambiguous nature of the relationship between research and innovation is often due to ignorance of the complementary design processes that actually organize the relationship between the controlled production of knowledge and product renewal, and the associated skills.

3.7.1   A Brief History of the First Company Research Laboratories

Without wishing to embark on a complete investigation of the origins of industrial research and the creation of the first company laboratories (a complex task since the sources are very widely dispersed), we may recall one or two elements: from the mid 19th century the rise of such "science-based" sectors as chemistry or the evolution of metallurgy led to the creation of production control laboratories; in parallel, the improvement in products and processes was often entrusted to outside consultants (Caro for BASF before 1877, Le Châtelier in France in the 1880s) or directly from the founder of the enterprise (Eugen Lucius, the founder of Hoechst). From the 1870s (1877 at BASF, 1878 at Hoechst, 1876 at Pennsylvania Railroad, etc.) laboratories were created which did more than simple production control, and in fact incorporated within the company the services fulfilled by outside consultants. One of the most celebrated was the laboratory of Charles E. Dudley at Pennsylvania Railroad, whose first analyses would, for example, lead to new quality standards in the market for steel rails. These laboratories were often limited to one researcher and a few assistants, but they grew little by little. By the end of the 1890s, the laboratories of Pennsylvania Railroad occupied two floors of the Master Mechanic's building. The famous laboratories of General Electric and AT&T came into being a little later (1907 for AT&T with the arrival of JP Morgan as shareholders replacing the Bostonians with the firm ambition of developing a universal system: 1900 for GE); in both cases the research laboratory was an extension of the design engineering department (Carty directed the engineering department at AT&T, with new research directed by Jewett; at GE the research laboratory was created under the direction of Willis Whitney, in support of the engineering department directed by Steinmetz). In the 1900s the success of these laboratories would astonish the world: at GE, ductile tungsten (1900–1910) then X-rays, radio, electron emission from filaments, and plasma (1907–1914); at AT&T, work on repeaters for transcontinental telephone lines (1907–1914) and radio systems (1910–1917).

3.7.2   Questions that Would Stimulate the Emergence of Industrial Research

What were the reasons that pushed the public authorities or certain industrial concerns to take the risk of investing in industrial research laboratories? More precisely, how was the relationship between industrial research and design built up? Examining one or two of these reasons in a few particular cases, we shall see that industrial research was not responsible for design, but just certain phases of the wider processes of design.

The work of numerous historians gives a better insight today into the origins of "industrial research", for example the research carried out by the Académie des sciences in the 17th and 18th centuries (Hilaire-Pérez 2000), the Franklin Institute of Philadelphia in the 19th century (Sinclair 1974), the research undertaken by the Compagnie Parisienne du Gaz directed by such eminent scientists as Victor Régnault then Henry Sainte-Claire Deville (Williot 1988), the first laboratories of the telegraph or railroad companies in the 19th century (Israel 1998; Dennis 1987; Knoedler 1993), industrial chemical companies of the 19th century (Beer 1958; Meyer-Thurow 1982; Homburg 1992; Travis 1993) or metallurgical companies in the early 20th century (Charpy 1919; Chevenard 1923, 1933) or indeed the research departments of General Electric, AT&T (Reich 1985) or DuPont de Nemours in the 20th century (Hounshell and Smith 1988). Four main (sometimes mixed) reasons can be identified as to the origin of the creation of these different facilities: validating the invention, certifying the products (on purchase or sale), rationalizing production and nailing the competition.

In each of the following four cases, discuss the relationship between the research laboratory and the design process (the answer is given in the text below).

3.7.2.1   Validating the Invention

The 1699 ruling of the Académie des Sciences designated scientists as examiners in the service of the king (Hilaire-Pérez 2000) to confer privileges on inventors (against corporations). However, throughout the 17th century, the royal administration had already been calling on the services of the learned societies for judging inventions. The scientists therefore carried out their assessments, including the most thankless of tasks. Hilaire-Pérez writes: "Hellot tested the resistance of a dye with jam, Vandermonde tested a wax polish by wearing shoes (the inventor accused the academician of sweating)". The scientists were employed for their skills in providing evidence: chemical analysis, studies of drawings, descriptions and technical publications, and in situ validation.

As another example, in 1855 the Compagnie Parisienne du Gaz (the Paris Gas Company) was formed from the fusion of six small companies, with Victor Régnault as consultant engineer. He wrote: "every year new processes for the economic manufacture of gas for lighting are presented to the public; the inventors have every interest in exaggerating the advantages that the Compagnie Parisienne would enjoy from their use". He proposed the creation of a laboratory responsible for "the analysis of the proposed processes to verify the reliability of the submissions and flush out bad experimental practices". This goal would remain unchanged for thirty-odd years (Williot 1988).

In both these cases, "research" procedures were not put in place to ensure proper design, but to validate a design created by others.

3.7.2.2   Certifying the Product

As a further case, in 1875 Charles Dudley was employed by one of the great American railway companies, Pennsylvania Railroad, to set up a chemical laboratory within the recent "Department of Physical Testing" (Dennis 1987; Knoedler 1993). Dudley developed a three-part program: 1. to determine the most appropriate materials for each particular given task; 2. to prepare detailed specifications for purchasing; 3. to design methods for testing whether the acquired materials properly met the specifications. The first tests, for example, involved the talc used to lubricate the locomotive cylinders. They highlighted the presence of organic impurities and pressed the purchasing department to find a supplier of better quality talc. The laboratory also tested lubricating oils, steel plates and boiler water, etc. in each case determining the optimal quality for the operation of the machines. The company quickly demanded that, as a matter of course, its suppliers provide samples of their products to Dudley's laboratory to check that they were consistent with railway standards. The laboratory itself developed numerous machines for its test programs.

In 1878 one of Dudley's first reports, entitled "The Chemical Composition and Physical Properties of Steel Rails" analyzed the durability of several types of steel rail, the major steel-making innovation of the age. The report concluded that mild steels had a longer service life than hard steels, and recommended improvements to the mild steel purchased by Pennsylvania. The report provoked an outcry among the steel suppliers who were determined to keep control of the quality specifications of their products. They were critical of the results, claiming in particular that the proposals for improvement would lead to a considerable increase in the cost of the products. Dudley reported that, furthermore, they reminded the companies that they had no choice other than to accept the products on offer. This crisis was actually more general: with the coming of steel rail, the specifications became more and more varied, recommended by the companies or learned societies such as the American Society of Civil Engineers or the American Railway Engineering and Maintenance of Way Association. The tests were regularly contested by the steel-makers. Following his first report, Dudley gradually initiated a constructive dialog among the suppliers who (according to Dudley) "understand the problems of production and cost structures" and customers "who understand the long-term performance" to "develop specifications and tests which incorporate both an understanding of the behavior of the product during its manufacture and an understanding of its behavior in service" (according to Dudley). The American Society of Testing Materials was created in 1902 with the aim of "resolving the general conflict over the specifications between producers and consumers of various industrial materials". Dudley would be its president for the first 10 years.

Again we observe in these cases that "research" procedures are not set up to ensure the design but rather to validate designs already realized or, more interesting, to give a thrust to new product designs. Note that there is design, but it is a design for the certification procedure (machine, standard, etc.).

3.7.2.3 Rationalizing Production

The great German chemical companies Höchst, Bayer and BASF came into being in the 1860s. They would see an extraordinary growth over the following years, due in particular to the arrival on the market of new synthetic dyes: aniline red, which replaced purple, was introduced at the end of the 1850s; by the end of the 1860s it would be alizarin, a substitute for garancine (or Turkey red), that would ensure the fortunes of BASF; in the 1880s Hoechst profited from the "azo" dyes, and finally in the 1880s to 1890s a joint research program among the three companies concentrated on indigo. This industrial growth was accompanied by the creation of many well equipped and soundly based research laboratories (Beer 1958; Meyer-Thurow 1982; Marsch 1994; Travis 1993). Could it be that this "research" gave birth to an industry?

The above authors show that the establishment of an industrial research laboratory within the company was late in coming and actually accompanied the rationalization of production: at BASF the research laboratories were dedicated above all to production analysis, and it was only in 1877 that a distinct internal research laboratory began to form. Hoechst had to wait until 1878 until the director of azo dye manufacture initiated an extensive program of research for these products. The formation of a laboratory at Bayer was even later, but would also be particularly carefully considered: in 1891, after several years of research at the margins of analysis within the azo dye factory, Duisberg set up a complete research laboratory.

During the 1900s, the steel and glass-making industries provided themselves with internal laboratories. Their objective was to rationalize production under a Taylorian logic imported into France by Le Châtelier (Charpy 1919; Le Châtelier 1918). The laboratories analyzed production, potential causes of defects and the measures to be applied for their correction. From the 1870s and under the direction of Saint-Claire Deville, the experimental factory at the Compagnie Parisienne du Gaz (mentioned above), altered the direction of its activity towards the rationalization of processes to increase productivity.

In these cases the laboratories were able to incorporate a logic for the "resolution of problems", problems of quality or productivity.

3.7.2.4 Nailing the Competition

A further reason invoked for the creation of laboratories was to protect investments threatened by "disruptive" inventions. During the 1870s Western Union, the market leader in telegraphy, took on the young Edison to explore disruptive paths of innovation in telegraphy. The objective was not to follow these paths (costly, and above all destructive for the investments already made) but rather to file patents to avoid other entrepreneurs getting a foothold in the marketplace (Israel 1998). In the years that followed, William Orton, president of Western Union, continued to finance Edison's laboratories at Menlo Park for the same reasons.

Consider a further singular case: the central research laboratories of AT&T were created during the 1910s (they would then adopt the name under which they are known today throughout the world, Bell Labs). AT&T was subsequently the leader of the nascent telephone technology. One of the first lines of research at AT&T was radio (Brittain 1976, 1992; Reich 1985): Langmuir and Alexanderson would go on develop a complete radio communication system, whose primary objective was to hinder the development of forms of "mobile telephony" which would have rendered AT&T's infrastructure obsolete. Note, however, that the development of the radio industry is based on this work.

In these situations, research is responsible for the almost systematic exploration of possible disruptive paths. Herein lies a highly particular form of design, which we shall have occasion to study in greater detail in Chap. 5 (see the idea of concept-based research).

3.7.3   Rationalization of Industrial Research: Rationalizing the Process of Knowledge Production

We have seen that the motivation behind the creation of the first laboratories was not necessarily that of innovation. We shall now see how those in charge of these laboratories thought and organized their activities.

There are several descriptions of how to run a research laboratory well. We provide a few examples: during the 1830s, Liebig set up a private laboratory at Giessen for training experts in chemical analysis (as he says: "the university refused on the grounds that the job of the state, and hence the university, is to train civil servants and not doctors, soap manufacturers, brewers, dyers or distillers of vinegar"). There is a famous picture of the laboratory (see illustration Fig. 3.20). One of the essential features of the laboratory was the precision and rapidity of its analyses: for example, Liebig used the Kaliapparat developed by Berzelius for the rapid weighing of water and CO_2 in the assessment of matter in organic chemistry. Training was essentially in organic chemistry. One exercise is famous: the young chemist has in front of him a flask identified by a letter of the alphabet, and he has to identify its contents by conducting increasingly complex chemical analyses. At the start of the investigation, the student starts with the letter A; by the time he reaches flask Z, he is competent.

This example shows at what point the issue associated with the laboratory was the rigorous and efficient capacity for producing knowledge See the definition of research given in this chapter).

The development of research laboratories, particularly in the United States, was not without intense intellectual effort applied to the principles of organization. From

Fig. 3.20 Liebig's laboratory at Giessen around 1831

the 1880s onward, articles discussing the economic utility of research can be found. One of the most famous was written by C.S. Peirce in the United States Coast Survey (for whom Peirce had carried out statistical analyses since 1861) in 1879 (Peirce 1879). It is entitled "Note on the theory of the economy of research" and considers "the relations between the utility and cost of diminishing the probable error of our knowledge". The value of research is therefore in reducing uncertainty (see the logic of risk management in rule-based design, in this chapter and Chap. 2). Le Châtelier in France, Duisberg in Germany, Mees in the United States were researchers as well as organizers. Their communications on the organization of an industrial research laboratory were widely disseminated. In 1920 the National Research Council, created during the war under the direction of the National Academy of Science to make researchers (especially those in the industrial research laboratories) available to the government, continued its work and reflections on the organization of industrial research in the post-war period. In number 9 of the Reprint and Circular Series, Clarence J. West, of Arthur D. Little, draws up a reading list for the organization of industrial research, including no fewer than 19 pages of references covering the period 1908–1920. Among these can be found contributions from Whitney (GE), Steinmetz (GE), Carty (AT&T), Eastman and Mees (Kodak) or Le Châtelier. There are three types of source: reports from government or learned societies (including the American Society for Testing Materials, created by Dudley in 1902), general publications (many articles appeared in Science and Nature respectively) and especially many professional journals in the world of engineering (Transactions of the American Society of Mechanical Engineers, Engineer, Mechanical Engineering, Industrial Management, Fiber and Fabric, Textile World, Journal of Industrial and Engineering Chemistry, etc.).

What was to emerge from these different bodies of work was a very complete picture of the organization of a research laboratory.[3] Its justification for the company was an insurance against loss of control in its industry: "a final insurance against eventual loss of the control of its industry by any concern" (Mees, p. 764). Its objective was not to improve the process or reduce product costs but to undertake "fundamental developments in the whole subject in which the manufacturing firm is interested"; "the fundamental theory of the subject" was necessary. Its timeline was at least ten years. The laboratory had to set itself up, and then start by solving small problems in order to gradually have an effect on the industry as a whole (still Mees, p. 766). The critical point was training and the organization of work: the same specialists had always to be used, employing special instruments with special methods over long periods of time. It was necessary, moreover, to progress to large scale research, if possible at industrial scale. GRH: since we cannot always have "geniuses", we need "well-trained, average men, having a taste for research and a certain ability for investigation". Organization: "how do we get the greatest yield from a given number of men in a given time?". It was necessary to have high quality, constantly operational, machines, to increase the specialization of the workers, and increase cooperation among workers in different fields. Of course, one might hope to "increase the likelihood of the correct choice of a promising investigation but, unfortunately, very little can be done in this direction" (Mees, p. 768). What director for what laboratory? if possible, an industrial scientist; otherwise, a scientist who could be trained in the culture of industry. In practice, the organization of the Kodak laboratory (run by Mees) showed an additional trait not covered by the author: the laboratory was also responsible for producing special orders that the factory had difficulty in manufacturing. "The manufacturing section of the laboratory can carry out the work with a full understanding of the use to which the materials are to be put and can often materially assist the purchaser in working out his idea" (Mees p. 770). Outputs: "the best utilization of the results obtained in an industrial research laboratory is only second in importance to the organization required to obtain them". All results had to be published, in particular to maintain the interest of the laboratory scientists in pure science and not leave the research programs uncompleted. The logic of this scientific productivity, organized along the lines of a factory, is found in most of the work.

The research laboratory therefore appeared to be directly linked with engineering, responsible for producing the fundamental knowledge associated with the industrial domain of the company. It was necessary to reconcile its emergence with a profound renewal of the theories of design at the turn of the 19th century which left open the possibility of improving the parametric curves of rule-based design as a result of scientific progress (Le Masson and Weil 2010a, b, c).

---

[3]In the remainder of the section we draw on the definitive article of C.E. Kenneth Mees (director of research at Kodak), published in 1916, which would then be superseded by a book published in 1920 and reissued many times (again in 1951, where it was taken as the point of reference in organizing the central research laboratories of the great French companies).

We shall now examine in detail how one of the most innovative laboratories of the age functioned (for a more extensive analysis, see Le Masson and Weil 2016). At the start of the 20th century, one of the most celebrated of the industrial research laboratories was that of Chevenard at the Imphy factories. Chevenard carried on the work of Nobel laureate Charles-Edouard Guillaume (inventor of Invar, Nobel prize awarded in 1920), under the direction of Henry Fayol, director of the steel-making company and also the celebrated author of general and industrial administration (Fayol 1916; Fayol 1917). In 1933 Chevenard himself explained how he had organized the "experimental work" (Chevenard 1923, 1933). He emphasized the four following points, clearly analogous to the organization of a production system:

- One instrument per task: "to each measurement there corresponds a special instrument, more often than not constructed at Imphy itself", for reasons of economy. Construction in-house appeared "quicker, safer and less expensive".
- Quality instruments: Chevenard demanded "sensitive and reliable instruments" characterized by "their simplicity, robustness and ease of use". Hence it was the productivity of the instrument that contributed to its quality. "Automation is taken to the n'th degree to save on manpower and reduce the personal involvement of the users" to "ensure faithful reproducibility".
- Installation of the instruments: instruments must always be ready for action. They must be in "a clear and simple assembly, […] in working order and immaculately clean".
- Recruitment and training of users: "Use and handling is organized according to Taylor's principles: avoid pointless movements, make the best possible use of the staff's time without overworking them. The time taken for tests must be interwoven such that the same operator can handle several in parallel".

"The result was the rapid collection of abundant, accurate and perfectly uniform documentation" (Chevenard 1933, p. 22).

In 1947, Yvan Peychès revealed the organization of the central research laboratory of Saint-Gobain, of which he was director (Peychès 1947). He described in particular detail the organization of the workstation (see Fig. 3.21), the organization

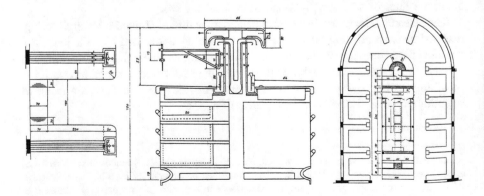

Fig. 3.21 The central research unit at Saint-Gobain (Peychès 1947). *Left* plan view and section of a bench in an analysis bay. *Right* plan view of analysis bays in the same room: the bays are on the periphery with common equipment at the center

of the different stations within the same room of the laboratory and the different rooms within the laboratory. The Taylorian influence is again very strong since the slightest movement of each experimenter had to be optimized, with the common instruments placed at the center of the room, and facilities common to several rooms placed at the centre of the building.

The organization of the laboratories therefore consisted mainly of optimizing a rigorous and controlled production of knowledge (see the definition of research, in this chapter, part III). Fundamentally, in the four motivational examples we saw previously, it was always the quality of the knowledge produced that mattered.

However, what are the conditions for this rationalization of the production of knowledge? The reader may ponder these conditions and analyze the consequences in terms of design. The answers are in the section below.

An analysis of the historical research laboratories shows that three conditions are required for rationalizing the production of knowledge in industrial research laboratories (for a detailed analysis, see Le Masson 2001):

1. the objects on which the research is directed must be already stable and well identified (organic chemical compounds, metals, electromagnetic waves, etc.).
2. It must have been possible for the associated scientific disciplines to enable certain fundamental principles to be posited (basic phenomena, properties, measurement principles, etc.).
3. The questions posed for research must be directly related to these objects, and it must be possible to answer them by making use of the available scientific knowledge.

These conditions are not strictly compatible with the creative process: does not the creative designer tend rather to conceive of new objects, explore new phenomena and roll back the frontiers of science? In practice, though, the entrepreneurial inventors of the 19th century sometimes had their own laboratories, but they were not organized along the same lines (we shall see the organization of Edison's laboratories in Chap. 5, for example). Hence there is something of a mismatch between this form of rationalizing the production of scientific knowledge and the process of disruptive innovation (we shall return to this point in Chap. 5).

On the other hand, rule-based design allows a very effective relationship to be organized between this type of knowledge production and the design process. The research laboratories would in effect appear as providers of design rules in the process of rule-based design, and the modes of research involvement would correspond to the phases of rule-based design:

- Industrial research is invoked for quality control, in production or at the end of the design process.
- Industrial research is invoked for the resolution of recurrent production problems, corresponding to the development phases in the design process (detailed design phase).
- Industrial research is invoked to provide what Reich called "technological theories", i.e. theories of designed products (and their defects). For example, the

AT&T laboratories studied the phenomenon of blackening of incandescent light bulbs in order to then reduce it (Reich 1985). These technological theories might then become useful conceptual models for designers.

- Industrial research is invoked to enlarge the scale of experimental plans. After the 1880s, the industrial chemical companies set up processes for the systematic screening of molecules that might be useful as colorants.

Hence we can conclude that, historically, research did not lead to innovation and that it was never set up for that purpose; it was gradually incorporated within the company precisely when the theories and methods of design were sufficiently well developed to take account of greater capabilities in the production of knowledge. Systematic design made it possible for the industrial research laboratory to be integrated within the company's processes of innovation.

3.7.4 The Origins of the Myth of Innovative Research: Nylon

One question remains to be answered: if it's not research that leads to innovation, where did this idea come from? We need to re-examine the origins of the myth of innovative research. The origins are, of course, varied and complex. However, one case remains the definitive reference: nylon. At the end of the second world war, the argument used to create a very large number of research laboratories would often be: by paying for research, you too will get your "nylon". The history of the discovery of nylon is one of the foundation-stones of the myth of "innovative research". Let us therefore examine the legend.

When he presented the nylon stocking to the press in October 1938, Stine, a member of the executive committee in charge of research at DuPont declared: "though wholly fabricated from such common raw materials as coal, water, and air, nylon can be fashioned into filaments as strong as steel, as fine as a spider's web, yet more elastic than any of the common natural fibers". In it, women immediately saw indestructible stockings, and nylon became associated with the magic of science. This was not just a marketing argument: according to the historians of R&D at DuPont, Hounshell and Smith (1988), nylon was by far "the biggest money-maker in the history of the DuPont company" (p. 273) and this success led the managers at DuPont to think that "further investment in high-caliber scientific research would lead to additional discoveries that DuPont's legions of chemists, engineers, and salesmen could develop into successful products. This became known as the "nylon model" of innovation, which dominated DuPont's research strategy for decades to come" (p. 274). The model of nylon inspired the organization of numerous research laboratories in the post-war period.

Is this therefore an exemplary case of research leading to innovation? Now that we have grasped the complexity of the processes of design, we can express some reservations about this claim. The historians who studied the processes of research and innovation at DuPont over that period have provided some remarkable material for discussion. Thus we can reconstruct the history of the discovery of nylon in three more or less overlapping stages of design.

3.7.4.1   Polymers in the Chemical Department

Phase 1: at the beginning of the 1930s, Carothers was leading a group of researchers working on a fundamental research program in the chemistry of polymers within DuPont's research department (the Chemical Department). This was to settle a controversy: do polymers have a special form of molecular bond, or are they ordinary bonds? Carothers wanted to prove the second assertion by synthesizing a polymer from initial components with known reactions (here we find the controlled process of knowledge production).

It was for this reason that Julian W. Hill, a researcher in Carothers' team, sought to construct longer chains by extending simple chemical reactions to doubly active basic molecules such as diacids and dialcohols to give esters and, possibly, polyesters. To this end, Carothers invented a molecular distillation device which gradually removed the water formed from the reaction and thus avoided hydrolysis of the synthesized molecules. The reaction [carboxylic acid + alcohol → ester] only occurred provided the water, which would otherwise hydrolyze the molecular chains, was removed. After several months of work, Hill obtained a polymer. He observed that this polymer could be formed into fibers which, after cooling, were very strong.

In terms of research, the result was very satisfying: the molecular weight of the polymer obtained was more than twice that of the longest chains produced until then, validating the assumption as to the nature of the chemical bonds within polymers. At this point, the research had certainly obtained a strong fiber—but still a long way from nylon stockings.

This fiber was then evaluated for textile applications and the "Rayon[4]" department would be given the results. The fibers turned out to be inappropriate for textiles since their melting point was less than 100 °C and the fibers were sensitive to water and certain dry solvents.

Let's stop there for a moment: that last step raises a question. The researchers drawing out the new molecule had emphasized the variety of possible applications for the new material (boots, photographic film, etc.). So why assess this component as a textile fiber? At this point we cannot exclude the intervention of a key person, the deputy director of the Chemical Department, Elmer K. Bolton, of whom we shall speak again. Another question is, what were the tests? Were these standard textile industry tests? Or polymer chemistry tests? That was not the case at that moment in time, for those tests were highly original. But who had designed them?

---

[4]Rayon is a cellulose textile fiber used for certain fabrics. It does not have the same strength and elastic properties as silk and consequently of the future nylon.

Hence we see the limits of an interpretation in which research might be responsible for innovation. Just the research on the polymers would be published with a small sample preserved, and the story would have finished there. Furthermore, by the start of 1933 the topic was dormant.

3.7.4.2 Designing Tests on a Fiber: A Request from the "Rayon" Department at the Central Laboratory

Phase 2: Charch's Pioneering Research Laboratory, attached to the Rayon laboratory, regularly tested samples from the factory. However, the test protocol required an already woven sample, which presupposed that large quantities of fiber were available. Such a system was appropriate for certifying a new product from a supplier. However, by the end of the 1920s the Rayon department was looking for new products with which to supply the textile industry. Creating a new fiber required that the textile quality of the fiber be rapidly understood without having to produce it on an industrial scale. The test protocol was therefore inappropriate for the latter situation. Hence Charch requested the *chemical department* (even before Hill's work on polyesters) to develop test protocols to assess the textile qualities of a fiber just on the fiber itself, without the need for a woven sample. John B. Miles, in charge of this work, was looking to assess certain fiber characteristics (such as their elasticity) based on special laboratory tests. These were the tests that were then used on Hill's polyester. For all that, was this merely a chance encounter?

At that point a major player appeared in the form of Bolton, who guided the strategy for innovative design. Bolton emphasized the economic issues around "artificial silk" and requested a "new textile fiber" from the *Chemical Department*. Bolton worked on the question as a designer. He had *two leads*, that of cellulose and that of Carothers' synthesized polymers. It was he, rather than Carothers, who had the essential parts of design reasoning. It was he, who in 1933, requested Carothers to put at least one person to work on synthetic fibers. In May 1934, Coffman, who was working for Carothers, drew a fiber that was already a nylon, a fiber whose value Bolton understood and which he had already earmarked for the textile industry.

3.7.4.3 From Polymer to Nylon 6-6

Phase 3: However, the story does not stop there. There was long way between the textile fiber of the laboratory and the nylon stocking. Again Bolton intervened. Where research promoted products that were easy to synthesize, Bolton guided work towards products which was more complex to synthesize, certainly, but which had a higher melting point and whose basic ingredients were inexpensive.

Another essential direction was that very soon, even though the researchers saw in nylon a universal product capable of replacing textiles as well as photographic film, and even though its development was only seen as a substitute for their own

Rayon, it was decided to develop "nylon" as a substitute for silk aimed at the lingerie and stocking market, a market whose value was beyond dispute.

Given these factors, development would mobilize multiple resources including research, but also those from the "Rayon" department and the chemical engineering group then emerging from within the *Chemical Department*.

Note that it was not the research laboratory that "made" the innovation. Nylon corresponds entirely to the logic of the controlled production of knowledge. Innovation depends on key players such as Bolton who guided the research (and development—see the interaction between Bolton and the Rayon department). It was Bolton who saw the value in the research results, and it was he who restarted research programs when necessary, and he again who then entrusted the fiber to be developed, all from a clearly identified value perspective and with skills that had already been developed.

Bolton therefore played an essential role in "activating" the research. However, his role was not that of the systematic design we studied in this chapter: Bolton explored new uses, and new types of product and processes. Bolton was in fact acting as an innovative creator of the type we shall be studying in Chaps. 4 and 5.

3.8   Case Study 3.2: Emergence and Structuring of the Acoustics Occupation in Automobile Engineering—Effectiveness of Conceptual Models (Jean-Hervé Poisson, Renault)

This case study was written by Jean-Hervé Poisson, who was himself an engineer in Renault's acoustics department between 1996 and 2002, and led the investigations to reconstruct the history of the department.

This case study analyzes the emergence and structuring of a skilled professional design occupation, namely that of acoustics at Renault. The study shows how the structure and performance of the occupation (known as a "métier" in French) depends on the conceptual models that are gradually being developed.

We shall start at the end, and for the reader unfamiliar with automobile engineering, we ask: what did an acoustics department look like at Renault in the middle of the 2000s? The department consisted of around 90 individuals (25 engineers, 50 technicians and 15 mechanics) with 10 test benches for physical prototypes, about 20 workstations (for signal processing), around 10 test benches and 300 km of test tracks at a secure center. The acoustics department provided a "service": at the interface between the customer and the engineering departments, a service provider transforms "customer expectations" into objectives and design recommendations for the engineering departments and in return, ensures that the finally designed parts meet certain expectations.

The reader should have little difficulty in making the link with the languages of systematic design. Note, however, that his position leads the service provider to supply the engineering departments with specifications (instructions for drawing the parts), but, from the point of view of the service provider, the language of this "specifications sheet" is a language of design parameters (DP) since the drawing instructions are in fact the department's modus operandi.

During the 2000s, the Renault range was distinguished by the quality of its acoustic performance: the Clio was the benchmark for its market segment in gasoline and diesel vehicles, the Megane also for diesels; the Laguna was considered to be the leader in terms of acoustics, while the Vel Satis was the first vehicle to be awarded the "Golden Decibel" for the quality of its acoustics.

How was the acoustics department able to attain this level? According to this chapter, this performance was certainly due to the quality of the conceptual and generative models implemented by the acoustics professionals. This case study focuses on these models and on the process of their genesis.

3.8.1   1964–1970: The Birth of the Acoustics Department— First Conceptual Models

Up until the 1960s, noise was associated with fault detection, with certain noises being due to particular mechanical components. For example, "whining" was a symptom of gearbox malfunction.

Nonetheless, various solutions had been introduced to reduce noise: exhaust silencers were widespread; in 1936 the Citroën Rosalie was fitted with a "floating" engine, i.e. mounted on rubber absorbers (Silentblocs).

Noise was then considered an annoyance, an unavoidable annoyance at that time. After the war four cylinder engines became increasingly common, but they were also the noisiest. The annoyance increased. In 1964, Renault produced the R4, the Dauphine, and the R8. Increasing numbers of customers complained of noise-related "fatigue". The company then decided to act, but at that stage the skills required for working on interior noise did not exist within the organization. The subject was entrusted to four "experts" in the "calculations" section, and thus the acoustics department was born.

What did they work on?

1. Methods for automatically quantifying noise: using a UV/paper galvanometer they could measure sound intensity (in volts converted to decibels, dB) over time. Vibrations and low frequency noise became quantifiable values.
2. Identification of noise classes: little by little they distinguished booming, harshness, intake and exhaust noise, whining and road noise. These new names were used to qualify the phenomenon and associate it either with some presumed cause (exhaust, etc.) or to characteristic noises such as booming or whining.
3. A boomimg model: it was shown that engine vibration obeys a relation of the form:

$$Vibration_{\text{humming}} = C \cdot \frac{mass_{piston}}{mass_{engine}} \cdot \frac{radius_{crank}}{length_{conrod}} \ (C \ is \ a \ constant)$$

This conceptual model would allow the design of new engines since it provided the new DPs (engine mass, piston mass, crankshaft radius, conrod length) required to act on an FR (booming).

This was a period in which the phenomenon was poorly measured, very little action was taken, and customer perception was negative. This period gave way to a situation in which names could be put to new potential solutions (reduction of booming, harshness, road noise); potential causes of noise could be identified (intake noise) and use could be made of new DPs (see Fig. 3.22).

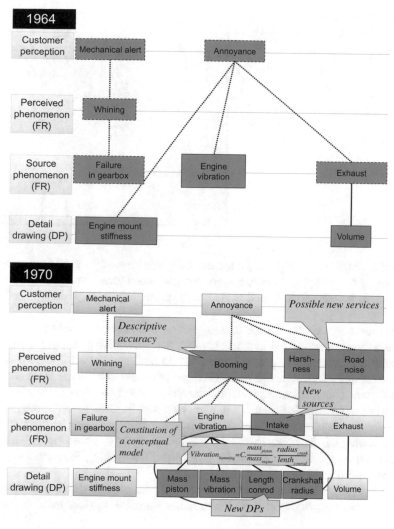

Fig. 3.22 Acoustic design rules in 1964 (*Top*) and in 1970 (*Bottom*). *Dark gray* new or modified boxes; *light gray* unaltered boxes. Main changes are the identification of possible new services with greater descriptive precision (humming, harshness, bearing noise), new sources (air inlet) and new DPs made possible by a conceptual model

3.8.2 1970–1979: Structure of the Profession—Complex Conceptual Models

The profession grew in the 1970s. Drawing on the rise of electronics, techniques were developed for the measurement and analysis of noise: analog spectrum analyzers provided the Fourier components of the noise, i.e. the noise intensity was known frequency by frequency. A "3D" representation of the noise was obtained

Fig. 3.23 Example of a
waterfall diagram

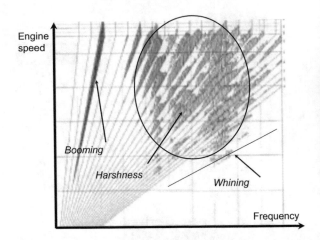

Fig. 3.23 Example of a
waterfall diagram

(waterfall diagram, see Fig. 3.23): the intensity was known for each frequency as a function of engine speed (ordinate axis). Noises "heard" were "visualized" on these 3D representations and had been named in the preceding years: whining, harshness, humming.

The advantage of the Fourier decomposition is that it establishes a relationship between "perceived" noise and mechanical parts resonating at neighboring frequencies. In a completely general sense, we do not know how to relate perceived noise to systems of assembled parts. However, the new calculation tools allow the use of more complex conceptual models: via relatively simple calculations (but somewhat more complex than the engine vibration formula given above) we can relate the geometric dimensions of certain assemblies of parts to their resonant modes such that, in order to reduce humming, we can predefine the resonant frequencies of the exhaust or intake pipes by dimensioning their cross-sections, volumes and diameters (see Fig. 3.24).

In the absence of detailed, tractable models, the experts developed simplified ones. Hence a simple model was able to relate humming to three main factors: the

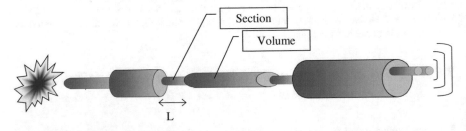

Fig. 3.24 Conceptual model relating noise transmission to the geometric dimensions of an assembly of parts (this is a simplified diagram of an exhaust system)

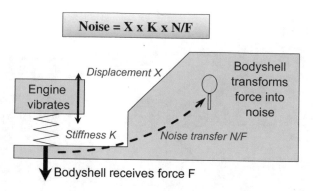

Fig. 3.25 Conceptual model relating humming to the three main design paths (engine vibration, stiffness of the engine mounts, and the body shell transfer function)

stiffness of the engine mounts (K in the diagram above), engine vibration (X) and the noise transfer of the body shell (N/F). Thus we have a high level conceptual model which distributes and coordinates the design effort among three design paths (Engine mounts, engine vibration, body shell transfer). Note, however, that at this stage the body shell noise transfer is not a very easy DP to activate (Fig. 3.25).

This work increased the extent of the design rules not only for the FRs but also the DPs (see the summarizing Fig. 3.26). Renault was thus able to address the growing demands of the market (see Fig. 3.26).

3.8.3   1979–1998: The Era of Fine Tuning

In 1979 the section consisted of about forty staff, organized along project lines: experts were grouped by project, with specialists in different types of noise working within the same project.

Acoustics became a unique performance service. The section was responsible for defining the noise target (FR) and provided the engineering department with the constraints to take into account for the engine, Engine mounts and, as far as possible, the body shell transfer characteristics.

What would be the effects of this system of rules on the projects?

The system of rules was sufficiently well developed to define the levels that had to be attained, speedy reaction on projects following the prototype phase, and good project oversight. That allowed the project logic to be strengthened and planning schedules to be shortened.

However, gaps in the body shell transfer characteristics raised recurring difficulties: given the lack of an accurate prescription (DP poorly understood upstream), any intervention could only be a corrective action applied too late: if prototypes were available, the acoustics experts would intervene to adjust the transfer characteristics of the body shell in order to meet acceptable levels of humming. These

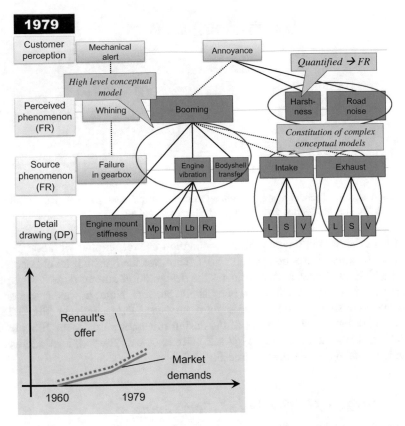

Fig. 3.26 Design rules in 1979 and associated performance. Three major transformations: we know how to quantify certain perceptions, opening the way to their transformation to FRs; complex conceptual models make new DPs possible (intake and exhaust), and high level conceptual models enable complementary DPs to be coordinated. The effects of these rules can be shown schematically by the figure on the *Bottom*: Renault is slightly ahead of the rise in customer demands

adjustments consisted of adding masses, dynamic dampers or tie-bars (which locally alter the resonant frequencies of the parts and hence the noise transfer) to achieve the required performance. This was time-consuming work, demanding an advanced level of expertise. Performance was not always guaranteed and vehicles became heavier.

This *modus operandi* had its limits: it was difficult to rank the performance benefits on account of a lack of coordination between experts working on different projects; knowledge was dispersed, capital spending was limited, there was no vigorous analysis of the phenomena and no upstream design anticipation. There was no improvement in the conceptual models, and the profession stagnated.

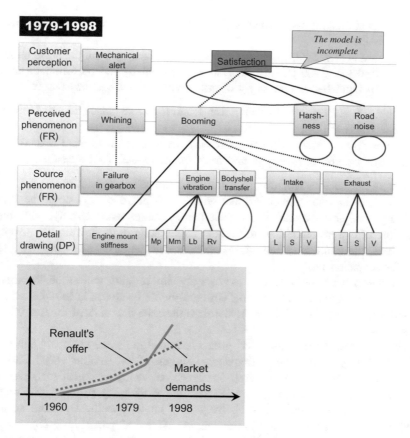

Fig. 3.27 The system of rules during the era of refinement (1979–1998)

During this period the section was in a phase in which the system of rules was being used, but very few new ones were being generated. The conceptual models, and hence the design rules, were soon inadequate for attaining the required performance (see Fig. 3.27): in the 1980s an increasingly demanding market led to new criticisms of Renault vehicles, in particular with regard to harshness.

3.8.4   1998–2005: Rejuvenating the Profession—New Conceptual Models

A new organization was set up at the end of the 1990s. Project grouping was abandoned, and teams were then built around the performance requirements (booming, harshness, road noise, etc.). These were focal points for the development of new conceptual models that would enable gradual control to be exercised over the new performance requirements. The development of rules for harshness is detailed below.

By the end of the 1990s, harshness was characterized by a unique noise indicator between 200 and 5000 Hz. An analysis carried out by the performance experts actually distinguished several types of noise within this frequency range:

- below 800 Hz: engine vibration noise, structure-borne noise and noise related to the stiffness of the engine mounts.
- Above 800 Hz: radiated engine noise (hard to measure), noise related to the airborne transfer of energy to and from the body shell (not measured).

The aggregate indicator was split into two indicators deemed to be more representative of acoustic perception and with which potential DPs could be associated. By studying the 200–800 Hz frequency range more closely, four major potential DPs were identified: driveshaft (eigenmode amplification, etc.), engine vibrations, the stiffness of the engine mounts and the structure-borne body shell transfer characteristics.

The first part of this work also led to a division of labor within the "harshness" group; two subgroups were formed, "structure-borne harshness" (200–800 Hz) and airborne harshness (800–5000 Hz). We shall describe the work of the first of these subgroups

The "structure-borne harshness" group followed the same logic of splitting up and identifying four distinct frequency ranges ("low rumble" [200–280 Hz], "sustained rumble" [280–400 Hz], "absence of damping" [400–550 Hz] and metallic tonals [550–800 Hz]). As before, this work on perception was able to associate particular DPs with each frequency range, specifically more accurate Design Parameters with regard to the "engine vibration" part (Fig. 3.28).

This model was satisfactory in terms of the accuracy of the modeling process, but was rejected by the organization; the section head anticipated difficulties in communicating four performance indicators for a single "structure-borne harshness" presentation and requested that the number of indicators be reduced to two in order to better highlight the performance. A new model was developed by combining the "low rumbling" and "sustained rumbling" classes into a single "rumbling" class, and the "absence of damping" and "metallic tonals" class into a single "engine running" class. In this new model, a "structure-borne harshness" (represented by about 20,000

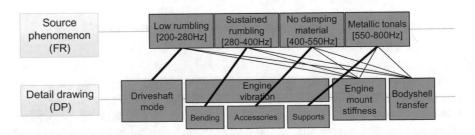

Fig. 3.28 Development of a conceptual model for "structure-borne harshness" (a test)

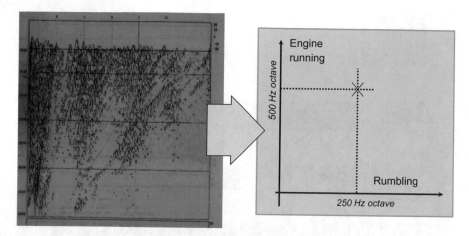

Fig. 3.29 A simplified performance model for better effectiveness

points on a waterfall diagram) was modeled by two parameters: octave emission around 250 Hz and octave emission around 500 Hz (see Fig. 3.29).

It remained for the relationships between these new performance requirements and the possible DPs to be set out. The experts started by checking that the booming formula (see above) is also true for rumbling with the engine running. They tested the formula on several models: the combination of frequencies for "engine mount stiffness", "engine vibration" and "body shell transfer characteristics" gave a theoretical value on the graph (250, 500 Hz octave) which was compared with the actual value. The difference was reckoned to be sufficiently small that the model was adopted. Use could then be made of the model for design purposes: for models in the process of development stiffness and mass figures could be recommended to ensure that the frequencies of the vibrational modes of the parts were optimal for minimum "rumbling".

This work led to a significant increase in conceptual models and the associated DPs and FRs (see Fig. 3.30). These conceptual models, built around the process of systematic design, simplified exchanges of information between projects and engineering departments. The numerical phases allowed performance to be predicted, with a possible positioning with respect to the competition, thereby strengthening the positions of the acoustic managers in the inevitable project negotiations. Design recommendations were made very early, and no longer on physical prototypes, providing a boost in effectiveness and cost. At the start of the 2000s, the Laguna, Megane, VelSatis and Scenic were classed among the best in their categories.

In conclusion: in 40 years the science of acoustics has become a well structured value space, going from the provision of a single performance service to around ten today (e.g. reduction of booming, engine rumble, high frequency noise, etc.), with these ten drawing on about a hundred design parameters recommended in the requirements specifications sent to the engineering departments.

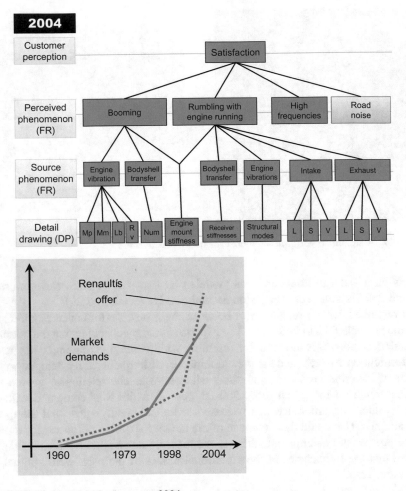

Fig. 3.30 The system of rules up to 2004

Historical work has revealed the gradual development of these FRs and DPs, and enabled discussion of the most favorable organizational forms for these developments.

The reader may ponder some contemporary or future forms:

- *concerning the FRs, how do we get from the idea of "satisfaction" to that of the quality of the sound space (this anticipates the subject matter of Chap. 4).*
- *what would this mean for the designers and engineering departments (e.g. consider the designers responsible for the vehicle radio, Hi-Fi systems, hands-free kits or the latest voice-controlled navigation systems)?*

References

Abernathy WJ, Clark KB (1985) Innovation: Mapping the winds of creative destruction. *Research Policy* 14 (1):3–22.

Abernathy WJ, Townsend PL (1975) Technology, productivity and process change. *Technological Forecasting and Social Change* 7 (4):379–396.

Abernathy WJ, Utterback J (1978) Patterns of Industrial Innovation. *Technology Review* 2:40–47.

Baldwin CY, Clark KB (2000) *Design Rules, volume 1: The power of modularity*. The MIT Press, Cambridge, MA, USA.

Baldwin CY, Clark KB (2006a) Between "Knowledge" and "the Economy": Notes on the Scientific Study of Designs. In: Kahin B, Foray D (eds) *Advancing Knowledge and the Knowledge Economy*. MIT Presss, Cambridge, MA, p chapter 18.

Baldwin CY, Clark KB (2006b) Modularity in the Design of Complex Engineering Systems. In: Braha D, Minai AA, Bar-Yam Y (eds) *Complex Engineered Systems: Science Meets Technology*. Springer, New York, NY, pp 175–205.

Beer JJ (1958) Coal Tar Dye Manufacture and the Origins of the Modern Industrial Research Laboratory. *Isis* 49 (2 (156), June):123–131.

Brittain JE (1976) C.P. Steinmetz and E.F. Alexanderson: Creative Engineering in a Corporate Setting. *Proceedings of the IEEE* 64 (9, September 1978):pp. 1413–1417.

Brittain JE (1992) *Alexanderson, Pioneer in American Electrical Engineering*. John Hopkins Studies in the History of Technology. The John Hopkins University Press, Baltimore.

Brogard C, Joanny D (2010) Stratégies d'innovation pour préparer les moteurs d'avion vert de 2025. Rapports de l'option Ingénierie de la Conception. MINES ParisTech, Paris.

Carnot S (1824) *Réflexions sur la puissance motrice du feu et sur les machines propres à développer cette puissance*. nouvelle édition de 1990 aux éditions Jacques Gabay, Sceaux. edn. Bachelier, Paris.

Charpy MG (1919) Essais d'organisation méthodique dans une usine métallurgique. *Bulletin de la Société d'Encouragement pour l'Industrie Nationale* Mai-Juin 1919:572–606.

Chevenard P (1923) Méthodes de recherche et de contrôle dans la métallurgie de précision. *Mémoires et comptes rendus des travaux de la société des ingénieurs civils de France* Juillet-septembre 1923.

Chevenard P (1933) L'installation et l'organisation d'un laboratoire sidérurgique moderne. *Mémoires de la société des ingénieurs civils de France* Septembre-octobre 1932.

Cohen WM, Levinthal DA (1989) Innovation and Learning: The Two Faces of R & D. *The Economic Journal* 99 (397):569–596.

Cohen WM, Levinthal DA (1990) Absorptive Capacity: A New Perspective on Learning and Innovation. *Administrative Science Quarterly* 35 (1):128–152.

Dalmasso C (2008) *Internationalisation des dynamiques de métier, dynamiques identitaires et organisation. Création d'une nouvelle entité d'ingénierie à l'international et mutation de l'ingénierie centrale dans l'industrie automobile*. Ecole des Mines de Paris, Paris.

Dennis MA (1987) Accounting for Research: New Histories of Corporate Laboratories and the Social History of American Science. *Industrial research and Innovation in Business, 1996, ed David E H Edgerton, Edward Elgar Publishing Limited, Cheltenham, UK* 17:479–518.

Fayol H (1916) Administration Industrielle et Générale. *Bulletin de la Société de l'Industrie Minérale* 10:5–164.

Fayol H (1917) De l'importance de la fonction administrative dans le gouvernement des affaires. *Bulletin de la société d'encouragement pour l'industrie nationale* novembre-décembre 1917: pp. 225–267.

Fox R, Guagnini A (1999) *Laboratories, workshops, and sites. Concepts and practices of Research in industrial Europe, 1800–1914*. Office for History of Science and Technology, University of California, Berkeley, Berkeley.

Hatchuel A, Le Masson P (2006) Growth of the firm by repeated innovation: towards a new microeconomics based on design functions. In: *11th International Schumpeter Society*, Nice-Sophia-Antipolis, France, 2006. p 18.

Hilaire-Pérez L (2000) *L'invention technique au siècle des Lumières*. L'évolution de l'Humanité. Albin Michel, Paris.

Homburg E (1992) The Emergence of Research Laboratories in the Dyestuffs Industry, 1870–1900. *in Industrial Research and Innovation in Business, David EH Edgerton, The International Library of Critical Writings in Business History, Vol 14* 25 (March):91–111.

Hounshell DA, Smith JK (1988) *Science and Corporate Strategy: Du Pont R&D, 1902–1980*. Studies in Economic History and Policy, the United States in the Twentieth Century. Cambridge University Press, Cambridge. L. Galambos et R. Gallman, Cambridge University Press, Cambridge. 756 p.

Israel P (1998) *Edison: A Life of Invention*. John Wiley and Sons, New York.

Klepper S (1996) Entry, Exit, Growth, and Innovation over the Product Life Cycle. *American Economic Review* 86:562–583.

Klepper S (1997) Industry Life Cycles. *Industrial and Corporate Change* 6 (1):119–143.

Kline SJ, Rosenberg N (1986) An Overview of Innovation. In: Landau R, Rosenberg N (eds) *The Positive Sum Strategy, Harnessing Technology for Economic Growth*. National Academy Press, Washington, pp. 275–305.

Knoedler JT (1993) Market Structure, Industrial Research, and Consumers of Innovation: Forging Backward Linkages to Research in the Turn-of-the-Century U.S. Steel Industry. *Business History Review* 67 (Spring 1993):pp. 98–139.

König W (1999) *Künstler und Strichezieher. Konstruktions- und Technikkulturen im deutschen, britischen, amerikanischen und französischen Maschinenbau zwischen 1850 und 1930*, vol 1287. Suhrkamp Taschenbuch Wissenschaft. Suhrkamp Verlag, Frankfurt am Main.

Kotler P, Keller KL, Dubois B, Manceau D (2006) *Marketing Management*. Pearson Education.

Lancaster KJ (1966a) Change and innovation in the technology of consumption. *American Economic Review* 56:14–23.

Lancaster KJ (1966b) A New Approach to Consumer Theory. *The Journal of Political Economy* 74 (2):132–157.

Lancaster K (1991) *Modern Consumer Theory*. Edward Elgar Publishing Limited, Brookfield Vermont, USA.

Le Châtelier H (1918) Du rôle de la science dans l'industrie. *La technique moderne* X (Sept 1918):3–14.

Le Masson P (2001) *De la R&D à la RID : modélisation des fonctions de conception et nouvelles organisations de la R&D*. Thèse de doctorat en Ingénierie et Gestion, sous la direction de Benoit Weil (thèse), Ecole des Mines, Thèse de doctorat en Ingénierie et Gestion, Paris. 469 p.

Le Masson P, Weil B (2008) La domestication de l'innovation par les entreprises industrielles: l'invention des bureaux d'études. In: Hatchuel A, Weil B (eds) *Les nouveaux régimes de la conception*. Vuibert-FNEGE, Paris, pp 51–67 (chapitre 53).

Le Masson P, Weil B (2010a) Aux sources de la R&D: genèse des théories de la conception réglée en Allemagne (1840–1960). *Entreprises et histoire* 2010 (1):11–50.

Le Masson P, Weil B (2010b) Les bureaucraties génératives: conception et action collective pour l'exploration de l'inconnu. Paper presented at the *Collège international de philosophie, colloque pragmatisme et bureaucratie, Paris*,

Le Masson P, Weil B (2010c) La conception innovante comme mode d'extension et de régénération de la conception réglée: les expériences oubliées aux origines des bureaux d'études. *Entreprises et histoire* 58 (1):51–73.

Le Masson P, Weil B (2016) Fayol, Guillaume, Chevenard - la Science, l'Industrie et l'exploration de l'inconnu: logique et gouvernance d'une recherche conceptive. *Entreprises et Histoire* 83:79–107.

Le Masson P, Hatchuel A, Weil B (2010) Modeling Novelty-Driven Industrial Dynamics with Design Functions: understanding the role of learning from the unknown. In: *13th International Schumpeter Society*, Aalborg, Denmark, 2010. p 28.

Le Masson P, Weil B, Hatchuel A (2006) *Les processus d'innovation. Conception innovante et croissance des entreprises*. Stratégie et management. Hermès, Paris.

Le Nagard-Assayag E, Manceau D (2011) *Le marketing de l'innovation: De la création au lancement de nouveaux produits*. Gestion Sup, Marketing-communication. Dunod, Paris.

Lefèbvre P (2003) *L'invention de la grande entreprise. Travail, hiérarchie, marché. France, fin XVIIIè-début XXè siècle*. Sociologies. Presses Universitaires de France, Paris.

Little AD (1913) Industrial Research in America. *Journal of Industrial and Engineering Chemistry* 5 (10, oct. 1913):793–801.

Marsch U (1994) Strategies for success: Research Organization in German Chemical Companies and IG Farben until 1936. *History and Technology* 12:23–77.

Meyer MH, Dalal D (2002) Managing platform architectures and manufacturing processes for nonassembled products. *Journal of product innovation management* 19:277–293.

Meyer MH, Lehnerd L (1997) *The power of product platforms*. The Free Press, New York.

Meyer-Thurow G (1982) The Industrialization of Invention: a Case Study from the German Chemical Industry. *Isis* 73 (268, September):363–381.

Minguet G, Osty F (2008) *En quête d'innovation. Du projet au produit de haute technologie*. Business, économie et société. Hermès-Lavoisier, Paris.

Neuhaus FA (1904) Der Einfluss des technischen Bureaus auf die Fabrikation. *Zeitschrift des Vereines deutscher Ingenieure* 48 (33):1221–1225.

Peirce CS (1879) Note on the theory of the economy of research. *United States Coast Survey*.

Peychès I (1947) Remarques sur l'organisation des laboratoires industriels. *chimie et industrie* 58 (6, décembre 1947):603–607.

Reich LS (1985) *The Making of American Industrial Research, Science and Business at GE and Bell, 1876–1926*. Study in economic history and policy, the United States in the twentieth century, R. G. Louis Galambos, Cambridge University Press, Cambridge. 309 p.

Sainsaulieu (1986) L'identité et les relations de travail. In: Privat (ed) *Identités collectives et changements sociaux*. sous la dirction de Pierre TAP, Toulouse.

Scherer FM (2011) R&D Costs and Productivity in Biopharmaceuticals. Mossavar-Rahmani Center for Business and Government, Harvard Kennedy School.

Shinn T (2001) The Research-Technology Matrix: German Origins, 1860–1900. In: Joerges B, Shinn T (eds) *Instrumentation Between Science, State and Industry*. Sociology of the Sciences. Kluwer Academic Publishers, Dordrecht, pp 29-48.

Sinclair B (1974) *Philadelphia's philosopher mechanics - a history of the Franklin Institute 1824–1865*. History of Technology. The Johns Hopkins University Press, Baltimore.

Smith RP, Eppinger SD (1997) Identifying Controlling Features of Engineering Design Iteration. *Management Science* 43 (3):276–293.

Suh NP (1990) *Principles of Design*. Oxford University Press, New York.

Suh NP (1999a) Applications of Axiomatic Design. In: Kals H, Houten Fv (eds) *Integration of Process Knowledge into Design Support Systems, 1999 CIRP Design Seminar*, University of Twente, Enschede, The Netherlands, 24–26 March 1999 1999a. Kluwer Academic Publishers, pp 1–46.

Suh NP (1999b) A Theory of Complexity, Periodicity and the Design Axioms. *Research in Engineering Design* 11:116–131.

Suh NP (2001) *Axiomatic Design: advances and applications*. Oxford University Press, Oxford.

Taguchi G (1987) *System of Experimental Design: Engineering Methods to Optimize Quality and Minimize Cost*. American Supply Institute.

Travis AS (1993) *The Rainbow Makers: The Origins of the Synthetic Dyestuffs Industry in Western Europe*. Lehigh University Press, Bethlehem. p 335.

Tsui K-L (1992) An Overview of Taguchi Method and newly developed statistical methods for robust design. *IIE Transactions* 24 (5):44–57.

Ulrich KT, Eppinger SD (2008) *Product Design and Development*. 4th edn. Mc Graw Hill.

Utterback J (1994) *Mastering the Dynamics of Innovation*. Harvard Business School Press, Boston.

Utterback JM, Abernathy WJ (1975) A dynamic model of process and product innovation. *Omega* 3 (6):639–656.

Vérin H (1998) La réduction en art et la science pratique au XVIème siècle. *Raisons pratiques* 9, institutions et conventions, la réflexivité dans l'action économique:28 p.

Weil B (1999) *Conception collective, coordination et savoirs, les rationalisations de la conception automobile*. Thèse de doctorat en Ingénierie et Gestion, Ecole Nationale Supérieure des Mines de Paris, Paris.

Whitney DE (1990) Designing the Design Process. *Research in Engineering Design* 2:3–13.

Williot J-P (1988) Un exemple de recherche industrielle au XIXème siècle: l'usine expérimentale de la Compagnie parisienne du gaz, 1860–1905. *Culture Technique* 18 (mars 1988):273–278.

Chapter 4
Designing in an Innovative Design Regime—Introduction to C-K Design Theory

Innovation in the 20th century was not just a singular event, but was continuous, incremental, robust—powerful. It was intentional, organized, manageable and controllable. The aim of innovation in the 21st century is to maintain the same constancy and the same power, while at the same time being radical, disruptive and creative. Stable dominant designs built the generative bureaucracies of the 20th century; in the 21st century, new design organizations are aiming to sweep aside, break and continuously regenerate the rules. The second industrial revolution invented the rule-based design regime, and by the same token it was this very regime that made this revolution possible. Following this logic, innovative design might be the heart of the revolution to come. What theories these days allow us to consider a continuous disruption? What methods and organizations today allow the implementation of these new innovative design regimes? The last few decades have seen the invention, construction and spread of theoretical frameworks and new practices. These will be studied in the next two chapters. Just as for rule-based design, we shall begin by studying the logical processes of innovative projects under innovative design (in this Chapter) before turning our attention to infrastructures and ecosystems in Chap. 5.

4.1 Reasoning in Innovative Design—C-K Theory

Design theories have enjoyed a revival over the last twenty years, centered about the theoretical schools in Japan (Tomiyama and Yoshikawa 1986; Yoshikawa 1981), America (axiomatic design (Suh 1990, 2001)—as seen in the previous chapter), Israel (Coupled Design process (Braha and Reich 2003) and Infused Design, (Shai and Reich 2004a, b)) and France especially. C-K theory appears not only as one of the most promising formalisms but also the most mature and, formally, one of the most generic and generative (see Hatchuel et al. 2011a and later

© Springer International Publishing AG 2017
P. Le Masson et al., *Design Theory*,
DOI 10.1007/978-3-319-50277-9_4

in this chapter). We shall therefore build an approach to innovative design regimes based on this formalism, and will then examine the relationship between C-K theory and other formal design theories.

4.1.1 Origins and Expectations of C-K Theory

C-K theory was introduced by Armand Hatchuel and Benoit Weil (Hatchuel and Weil 2003; Hatchuel and Weil 2009) and is today the subject of numerous articles in the literature (e.g. For a summary over 10 years of C-K theory, see (Benguigui 2012; Agogué and Kazakçi 2014); For practical applications in various contexts see (Elmquist and Segrestin 2007; Ben Mahmoud-Jouini et al. 2006; Hatchuel et al. 2004, 2006; Gillier et al. 2010; Elmquist and Le Masson 2009) recent work covers both its implications and its new developments, for example: (Kazakçi and Tsoukias 2005; Salustri 2005; Reich et al. 2010; Shai et al. 2009; Dym et al. 2005; Hendriks and Kazakçi 2010; Sharif Ullah et al. 2011)). In this chapter we make use of the most recent formulations (Hatchuel et al. 2013) but we provide the fundamental principles without necessarily giving the details of the formalisms.

The expectations of C-K theory are fourfold:

1. A "unified" Theory
2. A formalism for "Radical Creativity"
3. A method to extend the lists of DPs and FRs
4. A theory and method to overcome fixation

4.1.1.1 Expectations from the Point of View of the Professions: A "Unified" Theory

From the point of view of the *professions*, C-K theory proposes as unified a language as possible to facilitate dialog between the major design professions, namely designers, engineers and architects, independently of the specific nature of the objects they design and handle. The theory, ultimately known under the slightly enigmatic name "C-K", was initially presented as the "unified theory of design" (Hatchuel and Weil 1999).

In particular, C-K theory aims to combine the creative logic claimed by the artist with the logic of modeling and the creation of knowledge claimed by the engineer (or engineer-researcher). We might say that the theory seeks to combine two creative logics: that of the artist, who claims an ability to "see" new worlds, and that of the engineer, who claims an ability to create new knowledge. In practice we often find that these two approaches are far too simplistic, and that engineers can be visionary just as artists can be "savant"; C-K theory seeks precisely to formalize these logics, that of the unknown made thinkable (the logic of C-space, concept space) and that of the regeneration of knowledge (the logic of K-space, knowledge space) and especially their interactions (the operators linking C and K).

4.1.1.2   From the Point of View of Formalism: A Formalism for "Radical Creativity"

As with any theory of design, C-K theory tackles situations where $D(X_x)$ such that $P(X_x)$ is true is such that $D(X_x) \not\subset K(X)$ (see introductory chapter—this means that the initial knowledge *does not* include a set of decisions that enables X to have the property $P(X)$). But this time the aim of the theory is not to "minimize" the production of knowledge within the framework of a given dominant design. The theory must, on the contrary, reflect situations that show strong expansion of knowledge and reflect the design of objects deviating from hitherto known objects; furthermore, the theory should reflect the strongest forms of creativity, namely "radical originality" in the sense implied by Boden. As far as Boden is concerned, radically original ideas are those that cannot be produced by the set of generative rules whose purpose is to produce ordinary new ideas (Boden 1990, p. 40); hence this creativity explicitly assumes a revision of the rules, and the logic of this extension is not necessarily modular—they may lead to a radical questioning of the acquired knowledge and to a revision of definitions which hitherto seemed the most stable.

In this sense, C-K theory is a theory for the creation of new object definitions, a process consisting of two facets: first conceive the definition of hitherto unknown objects to bring them into existence, and then, on known objects, proceed to the propagation and re-organization required for the existence of the hitherto unknown new object while restoring or maintaining the conditions of existence of what had hitherto been known.

4.1.1.3   From the Methods Point of View: Consider the Extension of FRs and DPs

C-K theory will seek to extend and complete known theories and methods, in particular theories and methods of rule-based design. The limit of the theories and methods of rule-based design can be simply characterized: they work well while the nature of the functions and design parameters is known (to refresh your memory, see the functional analysis workshop in Chap. 2, especially the "night-time bus-station in workshop 2.1"). These days innovative design demands regular revision and extension of the FRs and DPs. The theories seen for rule-based design call for no formal framework to consider these extensions nor for any rigorous method of getting there.

4.1.1.4   From the Cognitive Point of View: Theories and Methods for Overcoming Fixation

For some time the cognitive sciences have shown the effects of fixation, where individuals in a creative situation that is both and individual and collective are victims (see (Jansson and Smith 1991; Ward et al. 1999; Mullen et al. 1991);

see (Hatchuel et al. 2011b) for a summary). This is associated in particular with a "fixed" representation of certain objects. For example, it is the effect of "fixation" that makes the puzzle below difficult to solve (see Fig. 4.1): how do you form a square by moving just one of the four matches arranged as in the figure? The solution is given on the right. We are conditioned to represent a square as a geometric form, and we fail to consider the "square" as in the sense of a mathematical operation.

Moreover, we can show that often the objective of training in industrial design these days is to overcome the effects of fixation. In this respect, they are inheriting the traditions of the Bauhaus: a study of the courses at the Bauhaus, in particular the introductory courses given by Itten, Klee and Kandinsky, showed the sophistication of the means used in training the young artists to overcome their fixations (Le Masson et al. 2013b). One of the expected results of C-K theory is in allowing the development of such methods—and (more modestly) in understanding the logic of existing methods.

More generally, and historically, the aim of the effort put into developing theories and methods of design was to correct any cognitive bias identified by the teachers and professionals of design. In the 1840s, Redtenbacher himself sought a method to prevent the designer of water wheels from always re-using the same wheel model without taking account of the context; the invention of systematic design also corresponded to a willingness to explore as much as possible, rather than be content with using only the available rules (see (Le Masson et al. 2011), also the historical case study in Chap. 2).

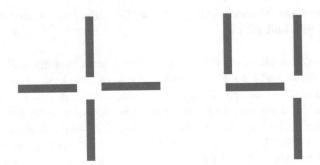

Fig. 4.1 An example of fixation. Form a square by moving just one of the matches in the left-hand figure. The problem seems insoluble as long as we think of the square as a geometric shape. The problem is solved by recalling that a square may also be the result of the mathematical operation of raising to the power of two. Four is a square, whence the solution given on the right. Note that this example illustrates fixation, but is still hardly generative: of course, we are playing on the two definitions of a square, but these definitions do not have to be revised!

4.1.2   Main Notions: Concepts, Knowledge and Operators

4.1.2.1   Intuitive Motivation Behind the C-K Theory: What is a Design Task?

C-K theory focuses on one of the most troubling aspects of the theoretical approaches to design, namely the difficulty of defining the starting point of a design task, i.e. what professionals describe as "specifications", "programs" and "briefs". This involves describing an object by giving it only certain desirable properties without the ability to give a constructive definition of the object and without being able to guarantee its existence on the basis of pre-existing knowledge. While mapping type theories of design tend to equate design with research in a space that is indeed complex, not to say uncertain (but known), C-K theory tries to preserve the fact that it is the ambiguous, equivocal, incomplete or vague character of the starting point that will allow the dimensions of the mapping to be regenerated. C-K theory therefore suggests a model that allows the design of a desirable but unknown object whose construction cannot be decided using the available knowledge.

This intuition raises a number of problems: how to reason about an object whose existence is a priori undecidable? and how to model the changes in the knowledge base that the initial "brief" sometimes tries to revise? In a rigorous sense, the object exists only at the end of the design process; at the start it is hoped that this future object might have certain properties and it will then be necessary to "gradually construct the new, as yet unknown object whose existence is undecidable".

4.1.2.2   The Space of Concepts and the Space of Knowledge

The underlying principle of C-K is to model design as an interaction between two "spaces",[1] the space of concepts (C) and the space of knowledge (K), which does not have the same structure or the same logic. These two spaces (or more precisely, the logical status associated with the propositions which make them up) determine the fundamental propositions of the theory.

Definitions of C and K

Definition of K space: the propositions of K space are characterized by the fact that they *all have a logical status* (true or false).

Definition of C space: C space is the space in which as yet unknown objects are developed. The propositions of C space focus on objects whose existence is still undecidable on the basis of the propositions available in K. We say that *the*

---

[1]In theory, a "space" is a collection of propositions; spaces are characterized by the nature of the logical status of their propositions and by the nature of their mutual relationships.

propositions of C are undecidable with respect to the propositions in K space. These propositions are known as *concepts*. Propositions such as "there are boats that fly", "there are mobile bus stations" (see workshop 2.1 Chap. 2), " there are smiling forks", "there are effortless bolt croppers" (see workshop 3.2 Chap. 3) are concepts. A concept is an *interpretable* proposition (all the terms used are referred to in K space) that is undecidable with knowledge in K space: the proposition is neither true nor false. It is not possible to say that there exists a boat that flies (otherwise the design would cease), but neither is it possible to say that no boat that flies can exist (otherwise the design would also cease).

Example: Let us give a mathematical example: suppose that the knowledge space of a young mathematician includes only reals as knowledge about numbers. If one assumes that this young mathematician is not a designer, he will assume that it is impossible to take the square root of a negative number since the numbers available to him all have positive square roots. This means that, in K space, he actually accepts a proposition of the form "all numbers are real" (*sub specie aeternitatis*). Suppose now that this mathematician becomes a designer. Hence when he says: "there exist real numbers whose square is negative", for him, this proposition is an undecidable concept with respect to his knowledge space. Actually, it means that his knowledge space contains the proposition that "all numbers *known to me* are real" (and not the proposition "all numbers are real"). We shall return later to this example when dealing with the design of complex numbers.

Note that concepts are not necessarily "surprising"; designing a camping chair that is cheaper and lighter than all other known chairs is also a concept. This means that, excluding special cases, a functional set of specifications such as those used in systematic design, is a concept.

Structures of C Space and K Space

Structure of C: concepts are of the form *"there exists a (non-empty) class of objects X for which a group of properties p_1, p_2, p_k is true in K"*.

In C space, since the proposition is undecidable, the proposition can only be worked on by *comprehension* (addition of properties) and not by extension (working directly on one or more elements in the class).

The structure of C is therefore constrained by the fact that the concept is an undecidable proposition. The most recent work proposes two approaches for the structure of C:

1. A set-wise approach: a concept can be considered as a particular kind of set, known as a C-set, for which the existence of an element is undecidable. This is the essential idea behind C-K theory and indeed the most critical aspect of its modeling. It is obvious that assuming the existence of an element in the C-set contradicts its status of concept (since we would then have to talk of elements with no possibility of defining or constructing them, contradicting the standard

elementary approaches of set theory (Jech 2002; Dehornoy 2010)). Also, the propositions that "a C-set is empty" or "a C-set is non-empty" cannot be decided with K. Only when the design has been completed can this question be answered. Technically speaking, Hatchuel and Weil suggest the C-set be governed by axioms using the axioms from set theory, rejecting those axioms which presuppose the existence of elements, namely the axiom of choice and the axiom of regularity. More generally, it is not possible in C space to have an inclusion relation, this relation having meaning only from the instant at which the existence of elements is proven. Rather, we shall speak of partial order (see below).

2. A logical approach: Hendricks and Kazakçi (2011, 2010) studied an alternative formulation of the C-K theory based only on first order logic, and which does not refer to C-sets. They obtained similar results on the structure of design reasoning.

In the remainder of this book we shall generally be using the set-wise approach, likening a concept to a set and the structure of C space to a set-wise structure without the axiom of choice.

Structure of K: the structure of K is a free parameter of the theory. This corresponds to the fact that design can use any type of knowledge, but also all types of logic, true or false; K can be modeled using simple graph structures, rigid taxonomies, flexible object structures or specific topologies (Braha and Reich 2003) or Hilbert spaces if there are stochastic propositions in K. The only constraint, from the point of view of C-K theory, is that propositions with a logical status (decidable) might be distinguishable from those that are not decidable.

Hence the K spaces of an engineer and a designer might be very different, with that of the designer containing, for example, knowledge about emotions, perception, theories of color or materials, etc., Such knowledge will clearly influence the way the (industrial) designer or engineer designs things. However, from the point of view of design, the models of reasoning are the same.

4.1.2.3 The Design Process: C-K Partitions and Operators

Design starts with a concept C_0, an undecidable proposition with knowledge in K space. The issue with the theory is that of formalizing the manner in which this undecidable proposition becomes a decidable proposition. This can come about through two processes: a transformation of the concept, and a transformation of the knowledge space to be used to decide on the concept. Transformations continue until they come up against a proposition derived from C_0 that becomes decidable in K' (i.e. K as it was at the instant the decidability of the concept was studied, i.e. when proof of existence is obtained). The concept then becomes a true proposition in K, and is no longer a concept.

During the process, the spaces evolve via expansions in K and partitions (or departitions) in C.

Expansion of K, Partitions of C

Expansions in K: it is possible to expand the K space (by learning, experimentation, remodeling, etc.); this expansion can continue until a decidable definition for the initial concept is obtained in K.

Partitions in C: it is possible to add attributes to the concept to promote its decidability. This operation is known as *partition* (see below). In C-K theory, the partitions of a concept C_0 are the classes obtained by adding properties (from K space) to the concept C_0.

If C_k is: *"there exists a (non-empty) class of objects X for which a group of properties p_1, p_2, p_k is true in K"*, then a partition consists of adding to property p_{k+1} to obtain the concept C_{k+1}: *"there exists a (non-empty) class of objects X for which a group of properties p_1, p_2, p_k, p_{k+1} is true in K"*. If C_{k+1} is the result of a partition of C_k, we say that $C_{k+1} > C_k$. Hence we have a partial order between the successive partitions of a concept (note that in a set-wise approach without the axiom of choice, we might speak of in inclusion relation $C_{k+1} \subset C_k$, though this relation should be constructed in accordance with the above principle and not according to an element-based logic).

Partition presents a rather specific problem: what is the status of the new C_{k+1}? This status must be "tested", i.e. its decidability with respect to the K space must be studied. This corresponds to making prototypes, mock-ups and experimentation plans. In turn, these operations can lead to expansions of the K space that are not necessarily related to the concept being tested (surprise, discovery, serendipity, etc.). The test has two possible results for C_{k+1}: (1) either C_{k+1} turns out to be undecidable with respect to K and the proposition therefore becomes a K space proposition, and the design ends in success; or (2) C_{k+1} remains undecidable in terms of K and the proposition is in C space.

Example: let the concept be "a boat that flies"; the designer is aware of flying fish and obtains, via partition, the concept of "a boat that flies like a flying fish". This concept must be tested in K (the test may consist of answering the question: do there exist boats that fly like flying fish?). The test will (probably) have two results:

- to proceed to the test, exploration in K will demand reflection on the flight of flying fish and hence will lead to an expansion of knowledge on this topic (e.g. modeling the flight of a flying fish).
- once this knowledge has been acquired, it will be possible to proceed to the corresponding test. Exploration in K may turn up boats that fly "like flying fish" (cf. Tabarly's hydrofoil) or otherwise (e.g. if one does not think that the hydrofoil flies exactly like a flying fish).

We may observe that the C-K partition does not exactly correspond to the definition of partition in mathematics: the status of undecidability does not allow the construction of a complete family of disjoint propositions whose "union" might reflect the previous concept.[2] Hence the C_{k+1} stated previously will correspond to the concept C_k, but also the concept: "*there exists a (non empty) class of objects X for which a group of properties p_1, p_2, p_k, **but not-p_{k+1}**, is true in K*". However, another concept cannot be excluded, that might be: "*there exists a (non empty) class of objects X for which a group of properties p_1, p_2, p_k, p_{k+1}, AND not-p_{k+1} is true in K*". We cannot have the law of the excluded third (*principium tertii exclusi*) in C space. However, the dichotomous logic (p_{k+1} on the one hand, non-p_{k+1} on the other) is often effective in C-K (see the workshop in this chapter).

Operators

All the operations described in C-K theory are obtained via four elementary operators representing the internal changes within the spaces ($K \rightarrow K$ and $C \rightarrow C$) and the action of one space on another ($K \rightarrow C$ and $C \rightarrow K$) (see Fig. 4.2 below for the four operators).

1. In C-K theory, the classical operations of inference, deduction, decision, optimization, etc. are operations of K in K.
2. The operator K to C is known as the *disjunction* operator, and consists of creating a new undecidable proposition on the basis of decidable propositions in K. The formulation of an initial C_0 is thus the result of a disjunction. In the same way, a partition ending up with a proposition C_{k+1} that, once tested, is a concept and also a disjunction.
3. The operator C to K is known as the *conjunction* operator, and consists of creating decidable propositions on the basis of undecidable propositions. For example, we have seen that a test might lead to the creation of new knowledge. In particular, a conjunction is a concept that has been partitioned to the point that it has become decidable. This conjunction corresponds to a "design path" that goes from the initial concept C_0 to a proposition C_k such that C_k is decidable in K. Note that if C_k is of the form "*there exists a (non empty) class of objects X for which a group of properties p_1, p_2,... p_k is true in K*" is decidable, then all C_i such that $C_k > C_i$ (in the sense of the order relation defined above, hence $i < k$) are also decidable and hence are in K.
4. The operator C in C is an operator that generates undecidable propositions on the basis of other undecidable propositions, using only C propositions; this is

---

[2]It is possible to retrieve, in design theory, the usual idea of partition in mathematics, we always need to introduce an "other" category and check that the intersections between the various alternatives are indeed empty.

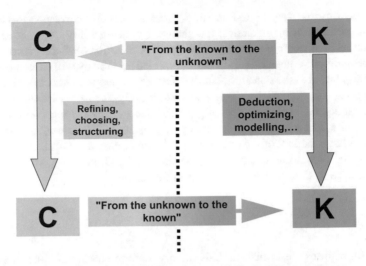

Fig. 4.2 The four operators in C-K theory: C → K, K → C, K → K, C → C

used, for example, if we seek to obtain as complete a partition as possible. If we have the concept "*there exists a (non empty) class of objects X for which a group of properties p_1, p_2, p_k is true in K*", the operator C → C will enable the concept "*there exists a (non empty) class of objects X for which a group of properties p_1, p_2, non-p_k is true in K*".

The main ideas of the theory are summarized in the Fig. 4.3.

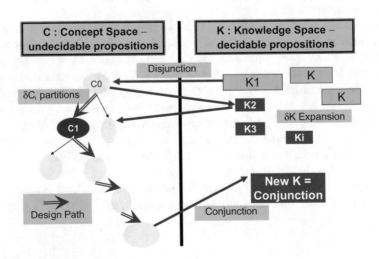

Fig. 4.3 The main ideas of C-K theory

4.1.3 Main Properties

4.1.3.1 Tree-Structure of a Concept C_0

One of the immediate results from C-K theory is that of showing that, for a given C_0, the C space necessarily has a tree-structure (associated with the order relation created by successive partitions).

This result is not trivial: it shows that the structure of the unknown (more precisely, the unknown thinkable with the propositions) is very particular. This means, for example, that if a brainstorming session is held on boats that fly, the set of ideas (each idea being likened to a concept) might be ordered as a tree structure based on the concept C_0.

4.1.3.2 Restrictive and Expansive Partitions

C-K theory allows us to distinguish between two types of partition: restrictive partitions and expansive partitions.

Properties of Known Objects

To this end an additional structure has to be introduced into K: properties common to the known objects. Given a family of objects X we can consider properties common to all objects X. This is what gives them their "identity" at a given instant (see the idea of the revision of identity of objects).

Note that we have avoided using the idea of "definition" here: these common, identifying properties do not constitute a general (fixed) definition of the objects. On the contrary (as we shall see) the identifying properties considered here can be "captured" from the perspective of their revision, rather than from their stabilization.

Examples:

- Hence in the case of complex numbers, we can say that, for the young mathematician, "all known numbers (the real numbers) have magnitude, namely their position on the real line".
- Similarly, for the designers of the boat that flies, we can say that "all known boats have a hull", and can even say that all boat hulls are of type A or of type B (wood, metal, etc.).
- For the designer of the camping chair (cheaper and lighter), all camping chairs have legs.

Restrictive Partition

A restrictive partition is a partition that makes use of these "identifying" properties of the known object or is compatible with them. Thus, in the design of a boat that flies, this can be partitioned into "a flying boat with a hull" (then to "a flying boat with a hull of type A" and "a flying boat with a hull of type B"). This operation is restrictive in the sense that it functions as a gradual selection in a set of known properties of the object "boat"—however, the concept thus formed remains a concept (of course we recognize that it is not enough to say that "the flying boat has a hull" to make it exist, to create a conjunction: undecidability still remains). The restrictive partition functions as a constraint: it obliges the flying boat to share an additional property with some of the known objects (namely the objects in the selection). Similarly, we can design a "two-legged cheaper and lighter camping chair", etc.

Expansive Partition

By contrast, an expansive partition is a partition that makes use of attributes that are not compatible with the identifying properties of the known objects (a flying boat without a hull or a flying boat with a hull that is neither of type A nor of type B; a number that might not be defined by its magnitude on the real line, etc.; a legless cheaper and lighter camping chair). Expansive partitions have two roles:

- they lead to *revision of the definition of objects*: if the "flying boat without a hull" ends up with a conjunction then there will exist in the new K space boats with and without hulls, so requiring the definition of a boat to be revised. In the case of complex numbers, we know that the conception of a number with a negative square leads to the creation of complex numbers that are not defined by their magnitude on the real line. Complex numbers require the previous definition of numbers to be revised.
- They steer the exploration towards new knowledge that is no longer deduced from the available knowledge. Hence working on the design of a "cheaper and lighter legless camping chair" can lead to experimentation: take a chair, cut off its legs and study the situation thereby created (See Fig. 4.4). We might discover that being seated on the ground raises new problems of balance-problems that were unknown with chairs with legs (whatever their number). It might lead to establishing a model of seated equilibrium in which balance might be ensured by the chair but also by the person on it, or by the interaction between the chair and the seated person. Hence we will have an operation in which new knowledge is created, driven by the expansive concept (see the chair example illustrated below). Thus is modeled a process by which the desirable unknown pushes to create knowledge, i.e. the imaginary stimulates research.

The generative power of C-K theory (discussed more formally further on) relies on this combination of the two effects of expansive partitions. Causing disruption with the definition of objects allows the potential emergence of new objects and the promise of new definitions; however, since their existence in K must still be brought about, expansive partitions lead to the creation of new knowledge steered by the disruptive concept (Fig. 4.4).

Rule-based designed chairs   Innovative-design chair

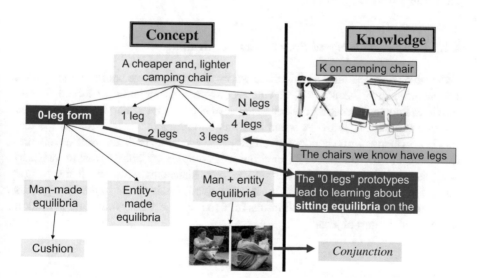

Fig. 4.4 Designing a cheaper and lighter camping chair. C-K theory allows a rigorous process of reasoning resulting in the so-called "Sangloyan" of Le Vieux Campeur or the Chairless of Vitra design; it also enables the systematic design of other "neighboring" objects sharing the definition of a legless, cheaper and lighter camping chair

Crazy Concepts—Chimera

The idea of the expansive partition thus captures what we normally call imagination, inspiration, analogies or metaphors. These ingredients of creativity are well known, but their impact on design is not easy to assess and seems to verge on the irrational. C-K theory models their effect as expansive partitions and reveals a double effect, namely the possibility of new object definitions, and giving rise to the creation of new knowledge. By distinguishing between these two roles and the value of their interaction and superposition, C-K theory explains the design rationality of "crazy concepts" and "chimera".

In particular, we may observe that only the second effect can be preserved: the attempt at a new definition comes up against a dead end; even so, the explorations made will have created interesting knowledge for future exploration even though they may not be aiming for such a radical revision as the definition of the object. This expansive partitioning is not the same as a standard trial and error process since, in contrast to standard trial and error tests, "crazy concepts" are not selected from a previously known list but are generated by expansion. The knowledge acquired is not related to an "error" but rather to an exploration down a deliberately original path, a path for which a realistic or possible solution could not have been known in advance.

4.1.3.3 New Objects and Preservation of Meaning

Expansive partitions raise a difficult question: if the expansive partition ends with a conjunction, then the new object will require that the definition adopted for the previous known objects be revised. The design of complex numbers requires the revision of what we know as a number: this is no longer a magnitude on the real line but an element in a commutative field. However, this revision itself means that others must be revisited as well (functions of a complex variable, new approaches to analysis, etc.). In revising the definitions, inconsistencies between all the former objects in K and the new objects must be avoided. Design thus implies a rigorous re-ordering of the names and definitions in K to preserve the meaning and definition of new and former objects.

Main definitions and first results in C-K theory (See also Fig. 4.5)

1. A set of propositions having a logical status is known as **K space**.
2. The addition of a proposition in K is known as an **expansion of K space**.

   By definition this proposition has a logical status

3. Given a K space, a proposition of the form $\{x, P(x)\}$, interpretable in the base K (P is in K) and undecidable in base K (P is in K), is known as a **concept** (the proposition $\{x, P(x)\}$ is neither true nor false in K).
4. The addition of some supplementary property to the concept (which becomes $\{x, P(x), p_k(x)\}$) is known as a **partition**.

   Remark: C is K-relative.
   In a set-wise approach, a concept is a set from which no element can be extracted
   Theorem: a concept space has a tree-structure.

5. Given a concept and its associated base K, an **operator** is an operation (using K or C) consisting of transforming a concept (partition) or of transforming the K space (expansion).

   Primary operators: $C \rightarrow C$, $C \rightarrow K$, $K \rightarrow C$, $K \rightarrow K$.

6. A **disjunction** is an operator $K \rightarrow C$: passing from decidable propositions to an undecidable proposition (using the known to work in the unknown).
7. A **conjunction** is an operator $C \rightarrow K$: passing from an undecidable proposition to a decidable proposition (using the unknown to expand the known)
8. Given a space K and C ($\{x, P_1 P_2 \ldots P_n(x)\}$ on this space K, an **expansive partition** (conversely **restrictive**) is a partition of C making use of property P_{n+1} which, in K, is not considered to be a known property associated with X (nor with any of the P_i, $i \leq n$) (conversely a property P_{n+1} such that P_{n+1} is associated with X in K or there exists an i, $i \leq n$ such that P_i and P_{n+1} are associated in K).

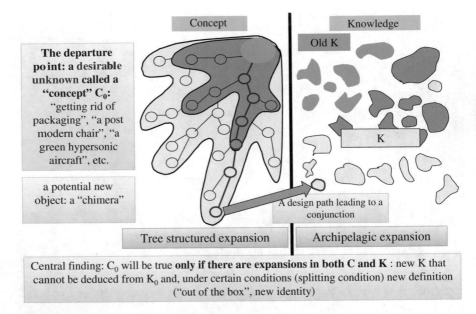

Fig. 4.5 A Synthesis of main notions of C-K theory

4.1.4 C-K Theory and Other Theories of Design

4.1.4.1 C-K Theory and Systematic Design

It can easily be verified that systematic design can be represented in C-K theory (see Fig. 4.6). We observe that systematic design consists of the a priori definition of partitions (partitions for functional, conceptual, embodiment and detailed design) and the types of knowledge to be invoked at each level, in addition to the nature of the knowledge to be produced at each stage.

In other words, in C-K the generative model appears as sequence of operators and the conceptual model as a set of items of knowledge—the theory allows the profound difference between these two ideas to be understood.

Recent work has analyzed several theories of rule-based design using C-K theory (Le Masson and Weil 2013) and has shown that, historically speaking, theories of rule-based design have always sought to preserve a strong conjunctive power while increasing generative power.

The representation of systematic design in C-K also emphasizes C-K's contributions with respect to systematic design:

1. *In C-K theory, design does not necessarily begin with functional language.*
   Hence the design of the cheaper and lighter camping chair starts with the number of legs, which pertains to the language of embodiment in systematic design.
2. In C-K theory it is possible to *revise the definitions of objects* in K. Hence the design of the legless chair is not constrained by the definition of a chair (chairs have legs).

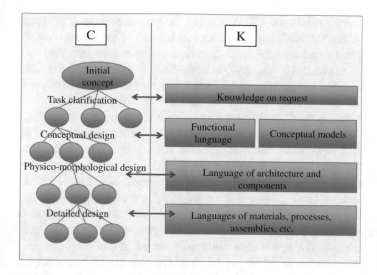

Fig. 4.6 Systematic design represented in C-K formalism

3. This revision of definitions may focus in particular on the languages of systematic design themselves and hence lead to their revision. This is one of the expected consequences of C-K theory: *revising the list of known functions and the list of known DPs*. This revision might take the form of a (modular) add-on. However, in directing the logic of the revision of definitions towards the languages of objects appearing at each level (functional, conceptual, embodiment, etc.), C-K theory offers a rigorous method for redefining entire segments of these languages. For example, if the purpose of a chair is to be "comfortable", it is possible to work on the concept of an "uncomfortable cheaper and lighter camping chair" that would certainly lead to a revision of the functions of a chair; similarly, if the basic technology of a refrigerator is a two-phase thermodynamic cycle, C-K theory allows for working on "a refrigerator concept which does not operate according to a two-phase thermodynamic cycle".

4.1.4.2   C-K Theory and Other Formal Theories: Generativeness and Robustness

While C-K formalism allows the extension of FRs and DPs to be considered, other theories of contemporary design obtain a similar result via different processes. It is instructive to reposition C-K theory in what appears today as a *continuum of formalisms as a function of their generativeness*. We shall provide a brief presentation only—for a more complete treatments, see Hatchuel et al. (2011a).

We start by one of the most sophisticated formalisms that appeared in the 1980s, the "General Design Theory" (GDT) of Yoshikawa (Reich 1995; Takeda et al.

1990; Tomiyama and Yoshikawa 1986). Design is represented as a mapping between FR and DP (as for Suh's axiomatic approach); one of the major inputs is that of formalizing the structure of the relationships between DPs and FRs as a function of knowledge about the "entities", already known objects from the same family (or even, from the perspective of an "ideal knowledge", all objects yet to come): these entities are the resources used to generate the DPs and FRs and the relational systems between them. Designing something is therefore that of making a selection from a subset of DPs and FRs on the basis of known structures; one of the major results of GDT is showing that the space of entities is a Hausdorff space, though for any set of specifications expressed by the FRs in this space it would be possible to "design" (i.e. extract) a mapping using DPs corresponding to these FRs. The generative power of GDT is thus that of its initial set of entities—this is a combinatorial, rather than expansive, generativeness. If we take the example of designing a camping chair, GDT enables cheaper and lighter chairs to be designed by combining the elements of knowledge obtained from all past chairs.

Suh's axiomatic system (see Chap. 3) is also concerned with the mapping between FRs and DPs, but rather than following the structures in a Hausdorff entity space, it suggests the construction of an ideal mapping with a one-to-one correspondence linking FRs and DPs. As we saw in Chap. 3, the axiomatic theory is one of evaluation and not of process. Hence it does not provide a generative power higher than the initial FRs and DPs, although it can occasionally lead to the development of specific DPs to "diagonalize" certain excessively coupled situations. In the case of the chairs, one might be driven to design modular chairs separating, for example, the structure of the seating part for greater comfort and less weight.

Using GDT, CDP theory (Coupled Design Process) (Braha and Reich 2003) still operates on the FR-DP mapping but on this occasion introduces phenomenological relations linking certain FRs to certain DPs, but (and herein lies the originality of their contribution) potentially by way of parameters that were never at the underlying origin of the process. These new parameters will therefore become new FRs or DPs. These "closure" operations mark the transition from a set of initial FRs to a set of extended FRs, similarly for the DPs. Thus we have a process of possible extension, associated with the closure structures known to the designers. In the chair example, CDP can lead to a functional extension: the chair is also a table, a traveling case, etc. and the constraints associated with the chair's environment (chair and table, chair and transportation, etc.) are amalgamated by "closure" and become new FRs for the chair (see Fig. 4.7).

The logics of "closure" are extended by the theory of Infused Design (ID) (Shai and Reich 2004a, b: Shai et al. 2009): the theory makes use of duality theorems and correspondence between systemic models which detect local "holes" (voids, see also the relation between C-K and forcing). These voids tend to create new relations and define new objects, and are therefore powerful levers in the creation of new DPs and FRs. In the case of a chair, for example, when applied to the question they will enable very different structural principles to be explored (rigidity of inflatable structures, tensile structures, etc.) and thus also deduce new associated FRs.

Finally, C-K theory allows extensions via expansive partition, i.e. via partitions making use of properties that the new object does not have in its usual definitions. Whence the legless chair.

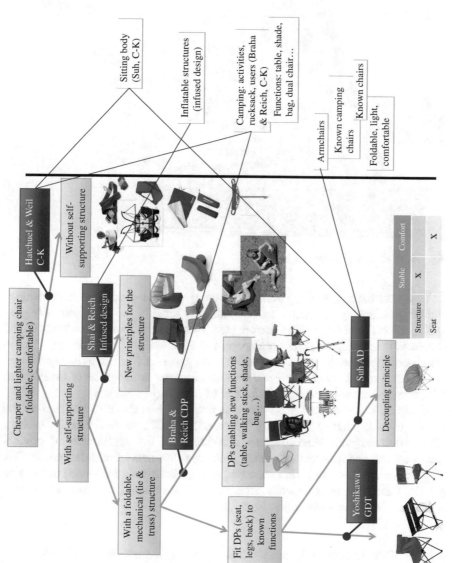

Fig. 4.7 A continuum of theories of design for a variety of generative forms

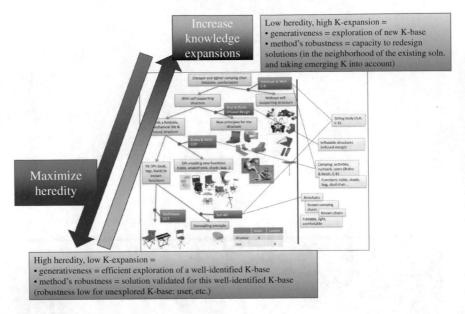

Fig. 4.8 Heredity and generative power

Today we have an ecology of mutually complementary and reinforcing theories allowing reasonably powerful forms of extension for FRs and DPs (and more generally, object definitions). We therefore pass from theories that rely on well-formed structures in entity space (Hausdorff space, DP-FR relationship according to Suh) to theories of dynamic structures (extensions). We also pass from *generative power by combination* of known elements to a *generative power by extension* of the FRs and DPs, or even by extension of the definition of objects (Fig. 4.7).

It will be observed that these different strategies are also characterized by the weight given to what we might call "*heredity*": in GDT, we design on the basis of known objects, with generativeness depending on the exploration of original combinations, and robustness depending on the robustness of past designs. In C-K on the contrary, heredity is limited, not to say systematically reassessed (expansive partition) and robustness depends rather on the ability rapidly to create knowledge as a result of new questions (see Fig. 4.8).

4.1.4.3   C-K Theory and Forcing: Theory of Design on Models of Sets in Mathematics

Armand Hatchuel has shown that, for objects, C-K theory is equivalent to the theory of forcing for models of sets (Hatchuel 2008, Hatchuel et al. 2013). In this more technical part (the reader less interested in formalism may skip this part), the study of forcing, i.e. a mathematically high level of design, leads us to emphasize some of the properties of C-K theory.

A method, forcing, has been developed in (mathematical) set theory which creates (or designs) new set models responding to certain "desired" properties. This technique was developed by Paul Cohen in the 1960s to prove certain important theorems of independence, in particular the independence of the Continuum Hypothesis (CH) from the Zermelo-Fraenkel (ZF) axioms of set theory. Gödel had proved in the 1930s that ZF was compatible with CH by constructing a ZF model that satisfied CH. It was therefore necessary to conceive a ZF model that did not satisfy CH. Using forcing, Cohen constructed just such a model, and showed that he could construct as many reals as parts of $\mathbb{R}$ (which is a non-CH ZF model).

The design of these models with the aid of forcing is based on the logic of extension (see forcing discussions in (Hatchuel 2008; Dehornoy 2010; Jech 2002)): using an initial model M, a new model N is constructed containing M, and for which certain properties can be controlled. The construction of the field of complex numbers we covered in previous sections follows precisely a logic of extension (Cohen refers to this in his "intuitive motivations" (Cohen 1966)): starting with the field of real numbers $\mathbb{R}$ we construct an extension $\mathbb{R}[\alpha]$ stipulating that α is the root of the polynomial $X^2 + 1$ (in other words, α satisfies $\alpha^2 = -1$). The extension $\mathbb{R}[\alpha]$ contains all possible "numbers" constructed by addition and multiplication on the basis of the field $\mathbb{R}$ and α, i.e. all "numbers" of type $a_n\alpha^n + \ldots a_1\alpha + a_0$. Put another way, the new numbers are described by polynomials with coefficients in $\mathbb{R}$. Indeed, α satisfies $\alpha^2 + 1 = 0$, hence some of these numbers are mutually equivalent (e.g. $\alpha^2 + 2 = (\alpha^2 + 1) + 1 = 1$ and similarly $(\alpha^2 + 1) \cdot (\alpha^2 + 1) + 1 = 1$, etc.) and it can therefore be shown that any new number is in fact equivalent to a number of type $a + b\cdot\alpha$ where a and b are in $\mathbb{R}$ and α satisfies $\alpha^2 + 1 = 0$ (we recognize the form of complex numbers where the common usage is to write α as i).

In Cohen's method, we no longer wish to construct an extension to a field (a very sophisticated set of mathematical objects) but rather an extension to models of sets (these are mathematical objects that are far more generic than a field). Cohen constructs this extension M[G] by adding to a model M a unique (generic) set G whose properties are specified by a partially ordered set P. The elements of P, called conditions, provide fragments of information about the set G whose addition has been proposed (just as we knew for α, that $\alpha^2 + 1 = 0$). Typically, should it be proposed that a new subset G of N be added to M, one condition might be a piece of information of the type "3 is in G and 5 is not". Cohen showed *how to organize these fragments of information* to obtain new ZF models: in other words, forcing creates new sets but the properties of former sets are preserved, what might be called their "meaning". Even if forcing does not form part of basic engineering knowledge and is taught only in advanced set theory courses, it is such a general technique that it is possible to understand the basic elements, elements that will emphasize some important properties of C-K theory.

Let us see how to construct a new set G from M, but outside M such that M[G] preserves the "meaning" of M. Five elements are required:

1. a basic ground model M, a collection of sets, ZF model (equivalent to a K space in C-K)
2. a set Q of conditions defined on M. Each condition extracts a subset of M. A partial order, noted <, can be constructed on these conditions: if we let q_1 and q_2 be in Q we say that $q_2 < q_1$ if the subset extracted by q_2 is included in that extracted by q_1. Hence we can have in Q a series of compatible conditions of increasing refinement: $q_0, q_1, q_2 \ldots q_i$ such that for all i we have $q_i < q_{i-1}$. Such a series is known as a filter.[3] We may observe that a filter can be regarded as the gradual definition of an object by "constraints" q where each constraint refines the previous one—a definition close to the successive partitions in C-K theory. We would imagine that the successive nesting of subsets of M could result in a set that is in M; surprisingly, as we shall see, certain nestings lead precisely to sets that are not in M.
3. The third elements: dense subsets. Given the set of conditions Q and the partial order <, we have (Q, <). We define a *dense* subset of Q, as a set D of conditions of Q such that any condition of Q is refined by at least one condition belonging to D. Put another way, even very long series of constraints (hence constraints associated with very "refined" subsets) are further refined by the constraints of D. Let $D_f = \{$the set of constraints satisfying a property $f\}$, and assume that D_f is dense. Whatever subset of M may be described by a condition q, this constraint is refined by q' satisfying f. This means that in any subset of M defined by the constraint q there exists at least one included subset, defined by q' that refines q and that satisfies f (Any subset defined by a constraint such as q at least "slightly satisfies" f; however, this does not mean that the whole set associated with q has the constraint f), hence f is a kind of "general property", "common" to any constraint q, even if this constraint q is not itself in D_f.
4. The fourth element is fundamental: let G be a generic filter, i.e. a filter that intersects all dense parts. In the general case (and this is an essential property), *G is not in M*.[4] We take things "out of the box", as it were, creating an object that has a property constructed on the basis of the properties of objects in the box, but which no object can actually possess. Things are taken "out of the box" "from the inside". This is very close to an expansive partition: the property is constructed on the basis of the known (all the constraints of the filter G are known) yet it creates an unknown object. Why is G generally outside the box? Let us take an arbitrary object O in M, the part D_O being defined by "the set of constraints that

---

[3]Filters are standard structures in set theory. A filter F is a set of conditions Q satisfying the following properties: it is non-empty, it is "upward-closed" (if $p < q$ and p is in F then q is in F) and it is consistent (if p, q are in F, then there exists an s in F such that $s < p$ and $s < q$).

[4]Actually, G is not in M the moment Q satisfies the "splitting condition": for any constraint p, there are always two conditions q and q' which refine it and which are incompatible (incompatible means that there will be no condition s that will refine q and q' "further on"). Proof: (see (Jech 2002, Exercise 14.6, p. 223): suppose that G is in M and assume D = Q\G. For any p in Q, the splitting condition means that there exist q and q' that refine p and which are incompatible; hence one at least is not in G and therefore is in D. Hence any condition in Q is refined by a constraint on D, and so D is dense. So G est generic and must therefore intersect D. Whence the contradiction. (see also Le Masson et al. 2016). For longer and more detailed explanations see Sect. 5.2.2.1, 199

are not included in this object O" is dense (for any subset—an arbitrary constraint q of Q—even very near to the object in question, always contains objects that are different from the object O; in other words, q can be refined by some q' in D_O). Indeed, G intersects it hence there exists at least one constraint of G that distinguishes it from the object in question. This argument is the same as that of Cantor's diagonal. G differs from all sets M but at the same time G intersects all the "general" properties in M (i.e. all the properties valid for the constraints of Q, i.e. of subsets of M), G collects all information available on the subsets of M.

5. Finally, the new G is used to construct M[G], the extended model. This requires an operation known as "naming" that allows all new objects in M[G] to be described uniquely on the basis of the elements of M and G (all just as the complex numbers described above).

Example: the generation of new real numbers Cohen gives a simple application of the Forcing method: the generation of new real numbers from integers (see Fig. 4.9).

The ground model is the set of parts of $\mathbb{N}$

Forcing conditions: these are functions that, with any ordered finite series of integers $(1, 2, 3,\ldots k)$ associate with each integer a value 0 or 1, and hence associates the k-list with 0 and 1, e.g. $(1, 0, 0\ldots 1)$. This condition is defined on the first k integers and extracts among these first k integers the subset of integers taking the value 1 via this constraint. We may also suppose that such a constraint corresponds to the set of reals written in base 2 and starting with the first k terms $(1, 0, 0\ldots 1)$. Given a constraint of length k, it is possible to create a constraint of rank $k + 1$ which refines the preceding constraint while keeping the first k terms unchanged and assigning the value 0 or 1 to the $k + 1$'th term. We thus obtain Q and the order relation $<$. Note that this order relation satisfies the splitting condition: for any condition: for any condition q_k, $(q(0), q(1), \ldots q(k))$, there are always two conditions that refine q_k and are inconsistent $(q(0), q(1),\ldots q(k), 0)$ and $(q(0), q(1), \ldots q(k), 1)$.

A generic filter is formed by an infinite series of conditions which intersects all the dense parts. The filter G contains an infinite list of "selected" integers and *is not in M*. We can prove this latter property by observing that Q satisfies the splitting condition; we can also present a detailed proof: let there be a function g in M (a function that associates a value 0 or 1 with any integer, i.e. a real number written in base 2) and let $D_g := \{q \in Q, q \not\subset g\}$, D_g is dense in Q hence G intersects D_g so G forms a new "real" number different from all the reals written in base 2!

The parallels between C-K theory and forcing are particularly valuable in that they allow certain characteristic features of design formalisms (for a more complete treatment and in-depth discussion of the relationships between C-K theory and forcing, see (Hatchuel et al. 2013)). Hence with forcing we find some aspects already highlighted with C-K theory:

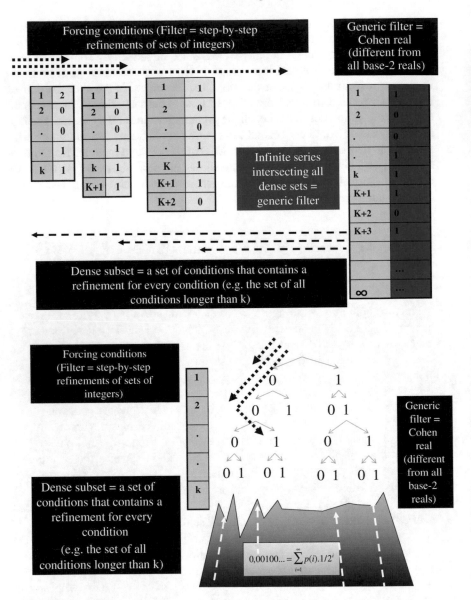

Fig. 4.9 Two representations of the creation of new real numbers by Paul Cohen

1. Expansion processes: in C-K theory as in forcing, a new object is constructed via progressive refinements. Moreover, we can show that a "design path" (C_0, … C_k) in C-K corresponds to a generic filter.[5] For all that, the generation of new

---

[5]For the entire dense subset D in C space, there is a refinement of C_k that is in D. C_k is also in K (the first conjunction) hence any refinement of C_k is in K and not in C, hence the refinement of C_k is C_k itself. Hence C_k is in D. Hence C_k does indeed intersect all the dense parts.

objects in C-K does not rely on an infinite number of conditions as in forcing, but on the existence of an expansion in K (introduction of a new proposition having logical status), even the revision of a definition in K. The two approaches differ in technique, but both depend on a logic of generic expansion.

2. Processes for preservation of meaning: the new objects created must remain consistent with past objects. Forcing imposes a "naming" phase on the process of generic expansion; C-K theory operates by "conjunction" of the progressive development of new propositions that are true in K space and by K-reordering.

The relationship between C-K and forcing also enables several other critical properties of a theory of design to be highlighted:

1. Invariant ontologies and designed ontologies. Forms of expansion are found in Forcing just as in C-K theory; however, forcing also tends to put the emphasis on structures *conserved* by forcing, hence the ZF axiomatic system is conserved from M to M[G}. In design, we will thus have an invariant ontology, a set of rules that remain unchanged over the course of the design; this ontology defines the conceived ontology by complementarity, i.e. the set of rules that can be changed by design (and there are a lot of them! We might imagine that a large part of human knowledge is constructed on such conceived ontologies); intuitively, we might think that the more general invariant ontology is, the more design would be generative—however, we might also think that a lack of stable rules would undermine the creative power of design.

2. Knowledge voids—independence and undecidability. In set theory, forcing allows the construction of set models that are ZF and satisfy a property P, and others that are also ZF but which do not satisfy P. We therefore show that P is undecidable in ZF or independent of ZF. P can be considered as a "void" in the knowledge over the sets; this void is in fact the condition for which forcing can be applied. In C-K theory, concepts are also undecidable propositions that can be viewed as "voids". The undecidability of concepts is assumed, and is necessary to start the design process. These "voids" are therefore common to both approaches, i.e. C-K theory and Forcing. Design "fills" the voids; forcing shows that "filling a void" is the same as showing the existence of independence structures in knowledge.

This idea of "void" also emphasizes the fact that *design is not based on the accumulation of knowledge, but on the existence of independence structures ("voids") in knowledge.*

4.1.5   Why C-K Theory Meets Our Initial Expectations

While the presentation of C-K theory here is still relatively succinct, the reader can be assured that, using the elements given above, the theory meets the initial expectations:

- "Professional expectations": the theory enables the relationship between the K-oriented professions (engineering) and the C-oriented professions (design) to be considered; it also reveals that there is K in design (the designer's K spaces—but see also the most recent work on K structures in design (Hatchuel 2005b, 2013; Le Masson et al. 2013b) and C in engineering (see below the interpretation of systematic design in C-K).
- Formal expectations: taking note of the creative act: see the notion of expansive partition, heredity, conceived ontology, invariant ontology, etc.
- Methodological expectations: the theory allows the revision of object definitions, and hence the extension of FRs and DPs (see C-K theory and systematic theory, C-K theory and other theories of innovative design).
- Cognitive expectations: C-K theory enables the effects of fixation to be overcome: fixation will arise from the definition of certain objects; indeed, the theory allows these definitions to figure in K space, then to be rigorously and systematically rediscussed via expansive partitions in C (see also the C-K exercise in the remainder of this chapter workshop 4.2).

4.2 Performance of the Innovative Design Project

In this chapter we study the performance indicators of a project team responsible for an exploration in innovative design. We shall be following the logic of the canonical model (applied to a single project): we give the inputs and outputs of the innovative design and the associated measurement methods.

4.2.1 Fundamental Principle of Performance in Innovative Design: Giving Value to Expansions

While systematic design gives value to minimizing expansions in order to attain a known objective, innovative design provides value to expansions. From a concept and a knowledge base we know that a concept tree and new propositions in K will necessarily be deployed; the concept structure is tree-like (see Sect. 4.1 of this chapter); In K space, the structure will generally be archipelagic in the sense that certain propositions will have no links with others (see Fig. 4.10).

In the exploration of "crazy concepts", this might give rise to new knowledge (expansions in K) which could be of value in the creation of a less original design path. Hence value must be given to the set of expansions in K and partitions in C.

In C-K, a rule-based design project minimizing the production of new knowledge will have the profile below. A "good" C-K exploration should rather tend to create "balanced" trees (exploration in "all" directions) and create new knowledge (see Fig. 4.11).

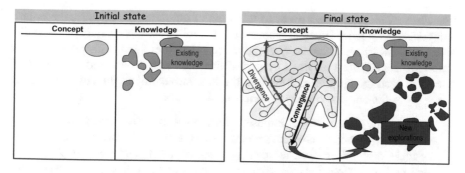

Fig. 4.10 Inputs and outputs for innovative design reasoning according to C-K theory

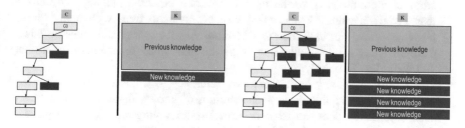

Fig. 4.11 Schematic representation in C-K of a "good" rule-based design exploration (*left*) and a "good" innovative design exploration (*right*)

4.2.2   Outputs: V2OR Assessment

How do we qualify a "good" tree and "good knowledge" in practice? C-K theory provides criteria for assessing outputs that allow an exit from an assessment restricted to the singular product without being confronted by the logical contradictions of knowledge for knowledge. Two families of criteria can be identified: those associated with C space and those associated with K space.

4.2.2.1   Criteria Associated with the Structure of the C Tree

For the C space, we draw inspiration from the assessments used for tests of creativity. One of the great contributions to psychological work on creativity (Guilford 1950, 1959; Torrance 1988) was the very early proposal for measures of creativity that would measure this form of intelligence differently from the traditional measure of IQ, but with the same rigor. For these authors, creativity is thus the ability to answer questions along the lines of "what can you do with a meter of cotton thread?"—questions for which there is no single good answer (as in IQ tests) but several possible answers. Measuring creativity is therefore that of *characterizing the distribution of answers* given to this type of question. Historically, the criteria suggested are: fluency (number of answers), flexibility-variety (variety of categories used to answer) and originality (originality being measured with respect to the reference distributions obtained by giving the same test to other individuals). C-K theory is used to apply these criteria to the innovative project. Just two criteria are normally sufficient (the fluency criterion is not used):

- **Variety**: the variety of the proposed solutions is assessed. In tests of creativity we refer to previously constructed categories (for 1 m of cotton thread there will be ideas centered on measurement (meter), on the thread (flexibility, tension, etc.) and on cotton, for example). In the case of the innovative project, the a priori distribution is generally not simple; hence the assessment is constructed on the basis of the proposed tree (a posteriori): variety is therefore measured in terms of the number of partitions but also their potential ranking (long chains may be given value). Thus, value will be given to trees with many "long" branches spread out in numerous directions. On the other hand, trees on which there are many ideas but all going along the same lines (technical or functional) will score low in variety.
- **Originality**: creativity is measured by reference to a known distribution (the yardstick given by the average distribution of known distributions); actually, such a yardstick does not exist in situations involving an innovative project! Another known alternative consists of evaluating the solutions suggested by experts (see the CAT method, Consensual Assessment Technique, developed by Amabile 1996; Amabile et al. (1996)); however, quite apart from the process being rather expensive and difficult to implement for innovative projects, it is intrinsically limited since these experts themselves may be victims of fixation, leading them to fail to recognize what is in fact original (Agogué 2012; Agogué et al. 2012) or to consider paths to be original when they may not be. C-K theory enables a more endogenous measure to be constructed: it is sufficient to count the expansive partitions, i.e. the cases in which the project managers will consider that they themselves add attributes to the concept that are not standard attributes in the knowledge base. The assessment protocol therefore enhances the process since it forces these project leaders to clarify the redefinitions they have used.

Examples (for the reader to discuss) (these examples are taken from Gardey de Soos 2007): taking the case of the night bus station, a *collapsible* bus station is more original than a *comfortable* bus station; a *summer* metro station is more original than a *well-lit* metro station.

4.2.2.2   Criteria Associated with K Space

It is not obvious how to assess the knowledge acquired: any project (especially a failed project) can show that it has created knowledge. The argument of knowledge creation is generally insufficient for a positive assessment of a project. Contenting oneself with an assessment of the concepts and ideas would hardly reflect the value of the expansions that had been made (see Elmquist and Le Masson 2009 for more on this debate). To assess the knowledge produced, one criterion is to evaluate it *according to its contribution to some future rule-based design*. To a first approximation, we consider a piece of knowledge to be useful in a design if it satisfies one of the following conditions: either it is a proposition that enhances the functional language, or it is a proposition that enhances the design parameters, whence two criteria: one "value" criterion and one "robustness" criterion:

- **Value**: in rule-based design, value is normally obtained by validating the functional criteria previously set out in a requirements specification. In innovative design, the value of an exploration is the ability to create new knowledge about the stakeholders and their many and sometimes unanticipated expectations (opinion, leaders, specifiers, customers, residents, third parties). In other words, the value assessed here is not the value of an object that has validated a criterion but is simply the ability to identify a new assessment criterion, whether that criterion has been validated by a product of the project or not.

For example (still with the bus station, same source (Gardey de Soos 2007)): in a base K where the functional criteria of the bus station focus generally on the problems of transport, the proposition that "certain residents (associations, shopkeepers, municipal authority, etc.) have certain expectations of the bus station" is a proposition that represents an increment of value, hence it is a new piece of knowledge that increases the value of the innovative project.

- **Robustness**: in rule-based design, robustness is often seen as equivalent to the validation of a functional criterion as a result of some well-mastered technical solution. In innovative design, "robustness" increases when new technical principles are identified, i.e. the list of potential solutions is enhanced. Included here are the new conceptual models accumulated by the explorations.

Variety, Value, Originality, Robustness (V2OR) constitute alternative criteria to the CQT criteria.

4.2.2.3 An Example: The RATP Microbus Project

In the 2000s RATP (*Régie Autonome des Transports Parisiens*) launched a new type of bus route, covering local routes and requiring buses that took up little space, known as microbuses. The first microbus project was considered a failure according to standard project management criteria—the project was delivered late, the new hybrid microbus was not ready when the line was inaugurated by the mayor of Paris, etc. However, an analysis based on C-K formalism and the V2OR criteria confirmed the intuition of the teams working on micromobility: the exploration brought by the first project was very rich in terms of V2OR and the outputs gathered at that time gave rise to many products and services that appeared later in the field of micromobility (see (Elmquist and Le Masson 2009) and see the Fig. 4.12).

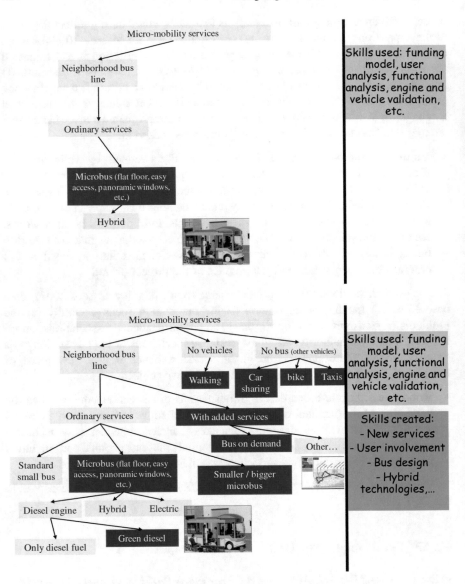

Fig. 4.12 Assessment of an innovative project: keeping only the main path/keeping all learned items. Within the standard CQT context (inherited from rule-based design projects) the project is perceived as a failure: it consumes many resources for a limited result (the first microbus was delivered late and was not a hybrid). From a V2OR perspective, it turns out that the microbus project was able to make a very broad exploration of the field of micromobility and build resources into the ecosystem—resources that would later allow an entire range of micromobility products and services to take off. The microbus itself would evolve into a whole range of vehicles

4.2.3 Inputs: Estimation of the Resources Consumed in the Case of an Isolated Innovative Project

Formally, the primary input of design is the *initial knowledge* (the skill of the designers). Hence we can estimate these resources by their "cost of use", i.e. the designers' salaries. We are also familiar with the strategies for reducing the cost of these resources (externalization, open innovation, etc.), and we can envisage a certain input "quality" (level of skill, ease of coordination, activation, etc.).

Another less obvious input is the *initial concept*. It is hard to put a figure on this input but it can play a major role. One might be tempted to liken the concept to a "good idea"; however, a "good idea" is a rather ambiguous notion (Is this a feasible idea? Is there a market for the idea? Or is it an original idea?) while a "good concept" is simply a well formed concept (the lack of logical status is obvious); on the other hand, a "bad concept" is a poorly formed concept, equivalent to a piece of knowledge ("services for the elderly" is a bad concept: such services already exist; implicitly it almost certainly means "cheaper services for the elderly, 'better' services focusing on life at home, independence, etc.").

Finally, the last critical input: the *expansion procedures* necessary to operate between C and K. In innovative design, the production of knowledge is not marginal; the tools for producing knowledge are therefore a critical input. Essential resources also include the quality of browsers, scientific equipment, relationships with research laboratories, the design studio, and other knowledge and concept producers; the capacity for making prototypes and demonstrations, validation procedures and tests, etc.

4.2.4 The Logic of Input/Output Coupling

4.2.4.1 Returns from Expansion and Returns from K-Reordering

Formally, input/output coupling can be complex. We recall that in the case of rule-based design, this coupling held to being the miracle of having "the competence of its products, and the products of its competence" (see Sects. 2.2 and 3.1). The "closer" the initial requirements specification (concept) was to the available knowledge, the better the performance (in a broad sense: not just conceptual models but generative models as well)—meanwhile allowing a marginal renewal of the rules, under the logic of dynamic efficiency.

In the case of innovative design, the logic of renewal becomes the most critical. A concept may be "far" from the knowledge base, but above all this "distance", this tension, must give rise to expansions and to a V2OR performance—at minimal cost. This efficiency is constructed in two parts:

- on the one hand, an efficiency in the phases of disjunctions and partitions in C (including the production of associated knowledge)—this is the most obvious efficiency.
- however, on the other hand performance is involved in the operations of conjunction and K-reordering: the "K-reordering" phase, i.e. the reordering of the knowledge base, may be fairly costly and reasonably "profitable" depending on the initial quality of the knowledge and successive partition strategies. This K-reordering phase is often critical for the efficiency of innovative design.

  - Examples of cost: certain disruptions can force an in-depth review of the skill necessary not just for the new product but also for all the preceding products (not just technical skills but also skills in production, distribution, commercialization, certification, branding, etc.). Hence, a new hypoallergenic filter system for the passenger compartment of automobiles may oblige all the pre-existing vehicles in the range to be revisited, or develop solutions for bringing previous vehicles up to date, etc.
  - Also an example of profitability: putting knowledge in order can "adorn" the value of previous products (Le Masson and Weil 2010): the Eiffel tower brought about an "adornment" of all existing iron architecture) (for the idea of "adornment" in design, see (Hatchuel 2006))

4.2.4.2   Towards a Logic of the Constitution of Resources

We observe that outputs introduce a feedback loop on the inputs: acquired knowledge and stated concepts constitute resources for later designs. This leads to two remarks:

- *pending concepts are also resources*; the ability to draw on already "designed" imaginary items is a priceless resource. These "imaginary" items are sometimes part of the knowledge of experts (who not only understand the solutions that have been developed effectively but also all the dreams of some technical domain that have already been tested without success, or those that have simply been thought about) in the manner of mathematical "conjectures", "utopias" or "great technical challenges" (e.g. see the work on imaginary space ideas) (Cabanes 2013).
- if the innovative project creates resources, then we can take account of this future "revenue" in the initial allocation for the innovative project. A limited initial budget can be a wise and effective solution, provided the project is left to benefit from its own dynamic returns.

We see the *logic of repeated innovation* allowing teams to gradually build up their resources. We also understand that these logical processes exceed the "singular project", and we shall discuss them in greater detail in this Chapter.

4.3   Organization of an Innovative Design Project

First of all we shall examine aspects of coordination (processes, etc.) and then questions of cohesion.

4.3.1   Design Space and Value Management

In rule-based design, linear reasoning made the process predictable and allowed it to be split into phases. Hence it supported stage-gate and planning. In innovative design, difficulties mount after the announcement of an initial concept C_0:

- The value associated with the concept may be poorly identified: "find a response to Toyota hybrid vehicles", "find applications for fuel cells", "find applications for natural fibers in construction" are possible concepts but their associated value remains to be explored (in contrast to the purpose of a normal requirements specification, which is to start with a "customer request").
- How to start the design process when the knowledge base is absent or obsolete? Expansions in K space are necessary, but where to begin? Even worse, sometimes the missing knowledge itself is not obvious, and it is the role of innovative design to reveal it. For example, the world specialist for petroleum drill pipes works on pipes "without lubricant": it would appear that it is simply a matter of finding a substitute for the contaminating lubricants used to facilitate screwing up drill pipes on offshore platforms—surely just chemistry of some sort? In fact, the project would reveal the necessity of working on the entire logistics of the pipe, on machining tolerances, the tools used by the fitter, the software used on the drilling rig, etc.
- How to avoid the premature death of the concept, surrounded as it is by obvious and apparently unsurmountable obstacles? How many innovative projects have ground to a halt simply because they were unable, right from the start, to demonstrate that they were satisfying some essential technical specification? In this case, the K base seems rich but a strong negative conjunction seems to have to come into play, linked for example to cost or draconian certification imperatives (e.g. demonstrating the airworthiness of an innovative drone).

Suppose reasoning gets under way and that the process starts, how do we explore without losing our way? How, during the exploration, do we avoid fixation or being attracted to "good ideas"? Reasoning does not occur in just one step. However, how do we define such steps, given that the definition of the steps results from successive learning processes?

C-K theory gives us the opportunity to identify the major difficulty: given an initial knowledge base K and a concept, the organization can only focus on the (mathematical) operators.[6] Previous difficulties are all related to questioning the operators to be used. *The creation of knowledge (ΔK) and its use in reasoning in fact represent organization of the exploration of a field of innovation.*

Formally, the elementary design operators ($C \rightarrow K$, $K \rightarrow K$, $K \rightarrow C$, $C \rightarrow C$) need to be managed; the combination can be sophisticated, thereby corresponding to such design actions as simulating, modeling, testing, validating, discovering, building prototypes, calculating, optimizing, selecting, organizing a focus group, observe uses, etc. Organizing the process of exploration in a field of innovation consists of *making these elementary actions possible.*

This essential management purpose—the possibility of partitioning to explore a concept—is a design space. We shall define a design space as working space in which the learning processes necessary for design reasoning are possible (Hatchuel et al. 2005, 2006). Formally, it is a subset of the initial set $\{C_0, K_0\}$ *in which designers can learn what needs to be learnt for exploring the concept.*

Design spaces in C-K formalism: The definition of a design space can be set out within the framework of C-K formalism. A design space can be defined as a configuration C_0^* - K_0^* with a *clear link* to the initial C_0-K_0 configuration:

- C_0^* *is linked to C_0 by changing the attributes of the same entity*: Given that C_0 is of the form "entity x with properties $P_1...P_n(x)$", C_0^* can be "entity x with properties $P_i...P_j \cdot P_1^*...P_m^*(x)$" where $P_i...P_j$ are properties chosen from among $P_1...P_n$ and $P_1^*...P_m^*$ are new attributes, chosen to support the learning process.
- K_0^* *is a set of knowledge items which can be activated specifically within a design space* (pending expansion). Hence $K_0 - K_0^*$ is the knowledge base that may *not* be used by the designers working in the design space. It may seem strange that the design space *restricts* the K space to be explored. However, K_0^* may also force knowledge to be implicated that might not be immediately activated in K_0.

The design process in C_0^* - K_0^* is always a double expansion δC_0^* (new attributes added to C_0^*) and δK_0^* (new propositions added to K_0^*). In other words, C-K formalism is still useful within a design space.

The link between the global C_0-K_0 and the design space is modeled by two types of transition operators. The first are operators going from C_0-K_0 to C_0^* - K_0^*, known as *designation operators*; the others are the *extractions* made on the δC_0^* and δK_0^* to bring what is extracted into the C_0-K_0 context. The

---

[6]The temptation might be to "select" the favorable C_0-K_0 configurations. However, what would be the criteria for such a selection, to the extent that the value is precisely an expected result of the process? This is why the issue is rather, to control the exploration.

designation operators may consist of adding a few attributes to C_0 or adding knowledge to K_0.

C-K formalism is therefore useful in describing expansion processes not only at the global level (value management space working on C_0-K_0) but also at the level of each of the particular design spaces (C_i^* - K_i^*).

An example of design space: designing an innovative drone without studying any flight certification (Taken from SAAB Aerospace)

The initial concept is C_0: "an innovative pilotless aircraft". However, the first design space is constructed on "an autonomous helicopter for the surveillance of automobile traffic" with research focusing on artificial intelligence and image analysis:

- C_0: "x = a flying vehicle", P_1 = "flight certified", P_2 = "pilotless", P_3 = "innovative".
- K_0: all knowledge is available or can be produced.
- C_0^*: remove P_1 and add P_4 = "being a helicopter" and P_5 = "for traffic surveillance duties".
- K_0^*: all knowledge about aircraft, military missions or automated flight is deliberately avoided. Why? Because normal drones are built on the principle of automated flight, which immediately determines the modes of reasoning. The design space explicitly excludes anything automatic in order to explicitly steer the learning process towards those disciplines that are underestimated in the world of drones: Artificial Intelligence (IA) (how an object can "decide" when faced with an original situation) and image analysis (what are the tools that can scan and analyze the environment)
- Validation in C_0^* - K_0^*: validation is linked to the disciplines concerned, and air certification is not considered.

The design space "emerges" from a more global exploration process, and feeds this process in return. We shall call this space that initiates the design spaces and summarizes the learning processes the "value management" space. The relationships between the design and value management spaces are modeled by designation operators—constitution of the design space (and extraction)—and integration of the learning processes in the design space within the overall reasoning. These various ideas enable the process of exploration of a field of innovation to be represented as per the diagram below (Fig. 4.13).

This modeling process describes the actions to be taken when faced with any difficulties encountered in exploring the fields of innovation:

- The initial concept can be poorly stated, the disjunction is barely visible and the unknown is hardly desirable. This is a poor point of departure for design reasoning. It is then possible to launch an exploration of a concept derived from the

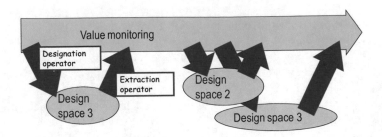

Fig. 4.13 Design Space and Value Management

initial concept. "A hybrid other than a Prius" might become, for example, "A hybrid with a French touch".

- When knowledge is lacking, the logic of the design space allows it to be created and to be created in a managed way. In contrast, the design space allows a knowledge overflow situation to be managed by arbitrarily limiting the exploration to a small number of K bases.
- When a killer criterion seems inescapable, it is possible to focus the exploration by explicitly rejecting this criterion: "We will do the study first without calculating the costs". For drones: "We will restrict the exploration to drones in simulated flight"; or "we will limit the exploration to a small number of flights in a secure airspace".

As the process gradually progresses, the double expansion occurs not only at the value management level but also at each of the particular design spaces.

New tools for the creative innovative project:
These days the designers of tools for creative designers are developing software suites enabling "design workshops" to go from the most exploratory phases to development phases that are not far from rule-based design. For a long time these workshops and software suites have been considered as constrained by the tension between generativeness and robustness: upstream, the possibilities for generation are very open, but explorations are fragile and not robust against standard assessment criteria; downstream, products become robust but the creative possibilities become very limited. Hence we had software suites and workshops which, taking this constraint on board, tended to augment the initial originality so as to better resist the feasibility constraints that would inevitably reduce the initial creativity.

However, recent work (Arrighi et al. 2012) demonstrates software that overcomes the "generativeness-robustness" conflict, simultaneously providing an improvement in robustness and generativeness. Given an initial sketch (let us say a concept state) for a pocket torch in the form of an eye (say), a designer using a standard tool would tend to increase robustness (see below: the object designed from the sketch follows certain constraints on the surface

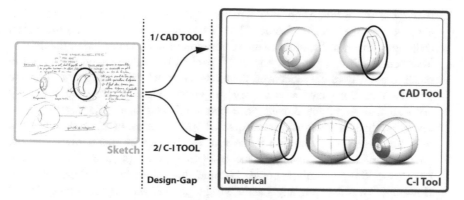

Fig. 4.14 New tools for the creative designer: a logical process of acquired originality (using this tool the designer can overcome the constraint (acquired robustness) while still being creative in how the constraint is satisfied—whence acquired originality

optical quality, here a sphere); the designer using creative design tools obtains good robustness (better, even: the shapes drawn using the software automatically satisfy a level 2 optical quality) but also achieves greater originality since he is exploring the space of allowable shapes and thus invents a surface that is "more original" than the sketch, but still of level 2 optical quality (namely, a "faceted" sphere"). Hence these software tools can provide a form of "*acquired originality*".

If such tools can be generalized, it becomes possible to envisage design paths richer than the traditional creativity-feasibility compromise (Arrighi et al. 2015) (Fig. 4.14).

4.3.2 New Principles of Cohesion: Strategy and Commitment

In rule-based design, it was possible to study just coordination. In innovative design, cohesion also plays an important role.

In the case of rule-based design, the value and legitimacy of the project were defined at the start. The project's relationship with the company strategy is ensured by agreement on the CQT objective, thus allowing services to be committed to the project. These conditions are not met by the innovative design project (for a detailed discussion of these questions, see (Hooge 2010)). The project organization not only has to manage the coordination (see above) but also the cohesion of the project.

1. *Managing the relationship with strategy:* the strategic nature of exploration evolves over the course of time. Thus Vallourec, a world leader in threaded drill-pipes for oil wells, initiated an exploration of the concept "after the threads have been cut": initially, this was about prudent risk management with not too much in the way of consequences, the expected conclusion being that "after the threads have been cut" was a very long term view; exploration gradually revealed that "after the threads have been cut" was in the dangerously near future—or had the potential for unexpected opportunities. In this case, it was not only the position of the project in the strategic framework that evolved, but the project itself led to a review of the company's strategic line of action. The innovative design project could become the strategy development tool. However, it was the company's underlying logic that was brought into question: this was the *common purpose* so dear to Barnard that could be invoked for the project, whence the management of innovative projects at the highest strategic level in the company, involving all stakeholders.

2. *Managing the commitments:* since the value and character of the innovative project were not well assured, the allocation of resources also became questionable, whence the internal "sponsoring" and the constant necessity for the project manager to secure the commitment of the stakeholders both within and without the project. Note that we are talking about design resources in the broad sense (skills, concepts, etc.) and not necessarily about financial resources. We shall see in Sect. 5.10 that the allocation of financial resources can have counter-intuitive effects (speculative bubble for some technologies) and pre-supposes particular forms of management.

4.4 Conclusion

In innovative design, reasoning follows a double expansion process: expansion of knowledge and new definitions of objects (no longer minimizing the production of new knowledge as in rule-based design). The performance of an exploration project consists of measuring these expansions in accordance with V2OR criteria (and no longer a convergence with respect to some CQT target). The organization rests on managing the learning processes describing the spaces where learning is possible (and often focuses only on certain facets of the concept), taking advantage of local expansions for the gradual structuring of all the alternatives (this is no longer a classic stage-gate where the phases can be predefined). This work demands a constant exchange between design and strategy, and between design and the stakeholders, whose commitment may change over the course of the process and due to the process itself (in contrast with the rule-based design project, whose legitimacy is guaranteed when it is first launched).

In our study of rule-based design, we saw that the success of the "rule-based" project did not rely solely on the management of the project but also, broadly speaking, on the set of rules on which the project was based. What is the equivalent for the innovative project? The innovative project itself also rests on an innovative design "infrastructure" which ensures the conditions for its success. It is clearly not the rule base itself that plays the most critical role (we have seen on several occasions, as much from the formal as from the managerial point of view, that this rule base is not the most critical element in innovative design): the innovative design infrastructure relies much more on the metabolism of knowledge and concepts, and on the ability to re-use and recycle the expansions produced over the course of time.

4.4.1 Main Ideas of the Chapter

- Concept, and knowledge in C-K theory
- Expansion of the K space, partition of the C space
- Operators (conjunction, disjunction)
- Expansive partition
- Design space, value management

4.4.2 Additional Reading

This chapter can be extended in several directions:

- On the "ecology" of theories of design:

  - see the following theories:

    General Design Theory {Tomiyama and Yoshikawa 1986 #2425; Yoshikawa, 1981 #882
    Axiomatic Design {Suh, 1990 #635; Suh, 2001 #2732},
    Coupled Design Process (Braha and Reich 2003)
    Infused Design, (Shai and Reich 2004a, b)

  - See also models supporting design processes: SAPPhIRE (Chakrabarti et al. 2005), N-Dim (Subrahmanian et al. 1997)
  - See papers comparing theories: ASIT and C-K (Reich et al. 2010); Parameter Analysis & Systematic Design (Kroll 2013); Parameter Analysis and C-K (Kroll et al. 2013);
  - See papers summarizing generativeness and robustness (Hatchuel et al. 2011a):
  - See the special edition of Research in Engineering Design in Design Theory (April 2013). Contributions from (Taura and Nagai 2013; Shai et al. 2013; Le Masson and Weil 2013; Le Masson et al. 2013a; Kazakçi 2013; Hatchuel et al. 2013; Kroll 2013):

- On C-K theory: a few "historical" papers (Hatchuel et al. 2011b; Kazakçi et al. 2010; Hatchuel and Weil, 2003, 2002a, 2009; Kazakçi et al. 2008; Hatchuel and Weil 2007; Kazakçi and Tsoukias 2005; Hatchuel 2005a; Hatchuel et al. 2004):
- "10 years of C-K theory" (Agogué and Kazakçi 2014; Benguigui 2012):
- On applications of C-K theory numerous publications—for an extensive review see (Agogué and Kazakçi 2014; Benguigui 2012); for applications see this and the next chapter.
- On the assessment of innovative projects:

  - on creativity and how to measure it (Csikszentmihalyi 1999; Boden 1999; Weisberg 1992; Torrance 1988; Guilford 1950, 1959):
  - on V2OR and its practice: (Le Masson and Gardey de Soos 2007)
  - on managerial questions associated with assessment: (Elmquist and Le Masson 2009).

- On value management and design space: in management (Hatchuel et al. 2005); Model in engineering design (Hatchuel et al. 2006); examples of such processes: see (Le Masson et al. 2010, Chaps. 11–13) or (Arrighi et al. 2013).

4.5   Case Study 4.1: Mg-CO$_2$ Motor

We give below a detailed example of C-K reasoning (see (Shafirovich et al. 2003; Hatchuel et al. 2004)).

4.5.1   Before C-K Work

First of all we give an account of work done before using C-K.

The reader can try to identify the concepts—sometimes implicit.

"How to design an Mg-CO$_2$ motor for Mars exploration"? This was the question to which the laboratory for Combustion and Reactive Systems (*Combustion et Systèmes Réactifs*) at CNRS, working notably for ESA (European Space Agency) endeavored to reply at the start of the 2000s.

What was the origin of such a proposal? Let us reconstitute a few elements of the initial knowledge base. While a vehicle engine burns fuel using an oxidant provided by the air (oxygen), a rocket has to carry both fuel and oxidant. For a mission intended to return samples from Mars, the initial mass rapidly becomes considerable: a mission of 500 kg must carry more than 10 tonnes of fuel and oxidant on launch. Several individuals have sought to use an energy source available on Mars, which would mean that the propellant otherwise required for a two-way trip would only have to be sufficient for one way. Given that the Martian atmosphere is 95% CO$_2$, could one use this CO$_2$ as an oxidant? Although the CO$_2$ molecule is quite stable, it can nevertheless support the combustion of metals under particular conditions of temperature and pressure. All that remained was to identify the metal fuel. One of the world's leading combustion specialists, Evgeny Shafirovich, was working at the CNRS laboratory. Along with other investigators, they showed in the 1990s that it was possible to generate a "specific impulse" using magnesium (Mg) particles in an atmosphere of CO$_2$. Carried from Earth, this result made magnesium a serious candidate for a motor capable of returning the mission to Earth.

The reader can check that the (implicit) concept "Mg-CO$_2$ motor for a mission to return samples from Mars" is a starting point from which the above reasoning can be reconstituted (check that this concept is consistent with the knowledge available; check that this concept lies at the origin of the new created knowledge).

Since the first test of the concept was a success, it was tempting to carry out a second, the criterion being the mass landed on Mars. Using Mg-CO$_2$, is the mass landed on Mars less than that which the same mission would require with classical propulsion? Work on this question showed that the answer was negative, and hence the proposal failed the second test. Did they have to give up on this Mg-CO$_2$ motor?

How should the project be relaunched?

Show that the initial concept should actually be written differently; show that the initial concept "Mg-CO2 motor for a mission to Mars" takes account of all the phases seen above and that it allows design work to be continued.

One route involved seeking mission scenarios where an Mg-CO$_2$ motor might provide advantages over classical propulsion. All mission scenarios using Mg-CO$_2$ propellant were analyzed systematically. A team was specially entrusted with this work, and each scenario was assessed according to the criterion *mass landed on Mars*. However, the failure was again unambiguous: for all scenarios, Mg-CO$_2$ is not as good as classical propellant.

The story might have ended there, with the research falling victim to the constraints of development or its own inability to better account for these constraints. However, the director of the laboratory, Iskender Gökalp, suggested to one of the students on the design course at the Ecole des Mines in Paris, Mikael Salomon, that he should make use of the C-K formalism to revisit the previous results. This involved seeing whether the design reasoning had been sufficiently rigorous and whether or not it was possible to identify new leads that had remained hidden in the shadows and that might be able to breathe new life into the project. As a result of this work carried out in 2003, an article was published that same year entitled "Mars Rover vs. Mars Hopper" (Shafirovich et al. 2003) demonstrating new avenues for Mg-CO$_2$ combustion in the mission to Mars.

4.5.2 C-K Reasoning in the Endeavor

The rest of the work made use of C-K reasoning in the endeavor.

A. First of all, C-K formalism took account of the first stages of reasoning. The initial question was a concept in the theoretical sense since the proposition "an Mg-CO$_2$ motor for Martian exploration" had no logical status but could nonetheless be interpreted in the K base ("motor", "Mg-CO$_2$", "mission to Mars" were known terms). This disconnect was written as a concept in C-space. The two successive partitions linked to research then featured in this space (sufficient thrust, then mission with return of samples or not). The new pieces of knowledge produced by research on that occasion were written under K (see Fig. 4.15).

Let us now examine the research stage of the mission. The concept became "an Mg-CO$_2$ motor for a mission not requiring return of samples"; mission scenarios were generated in K-space. The concept was partitioned with each of the n scenarios generated and scenarios were assessed one after the other (in K). Each scenario ended with a negative conjunction.

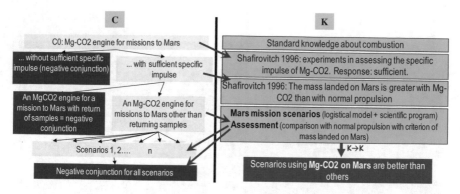

Fig. 4.15 "research" type and "development" type reasoning

> **Guide to interpreting the C-K diagrams** Light gray background: restrictive partitions and existing knowledge.
>
> Dark gray background, light characters: expansive partitions in C and the creation of knowledge in K.
>
> Arrows are operators C → K or K → C or even K → K. They illustrate diagrammatically the main stages of the reasoning

B. How to continue? The previous calculations constituted in K an additional knowledge used solely until now for the purposes of assessment. Within the logic of innovative design, this knowledge encouraged the "missions" to be structured differently. In fact, it appeared that these results, even the negative ones, were slightly better if Mg-CO₂ were used on Mars. That suggested a new mission partition: the initial concept was partitioned as "used only on Mars" (versus used elsewhere) (see Fig. 4.16). In this case, a new space had to be explored: *that of possible uses of Mg-CO₂ technology on Mars*. This partition created the acquisition of knowledge concerning mobility on Mars. The investigation revealed that mobility was not just the operational radius or speed but also susceptibility to unforeseen external conditions (storms, etc.) and the ability to build on scientific opportunities in particular. Hence a partition had to be drawn between *planned mobility* and *unplanned mobility*, and it was thus that the hopper concept emerged. Hence the set of successive expansions allowed *the identity of the object to be profoundly revised*, emphasizing the fact that the assessment criterion was no longer "the mass landed on Mars".

The consequence of this design effort was far from negligible, and there appeared to be real value in using Mg-CO₂; the project became financially viable as far as ESA was concerned.

C. For all that, "unplanned mobility" remained a concept hard to implement by a research laboratory specializing in combustion, or by the teams developing missions to Mars. The design strategy was therefore to add properties to the initial concept such that *learning in R or in D could be made possible*. Hence it was possible to work on a hopper capable of acting as a substitute for the rover earmarked for the next ESA Mars mission, Exomars 2009. It was known that this hopper should weigh less than 60 kg, complete its mission in less than 180 days,

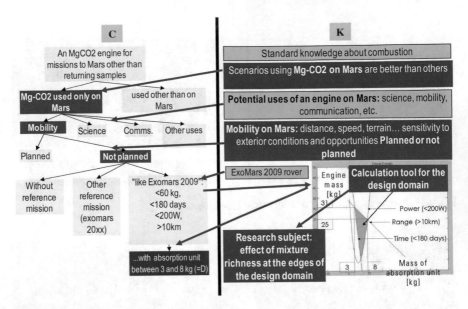

Fig. 4.16 Revision of the identity of the object. The hopper concept emerges. Reasoning continues until R&D starts

consume less than 200 W (power to be provided by solar panels) and cover at least 10 km. That did not mean that every hopper should meet these constraints; however, the assumption was that working on such a hopper would create valuable understanding for other situations.

Given the constraints of the rocket equations and an understanding of the technology of CO_2 absorption, these new objectives immediately put fairly precise dimensional restrictions on the absorption unit and the mass of the Mg-CO_2 motor, these constituting their "design domain". R and D could work on this design domain: D would develop a motor whose mass would correspond with the constraints of the "specifications sheet"; R would concentrate on the effects of modifying the combustion parameters (mixture richness, for example) at the boundaries of the domain.

The reader may check this example for V2OR assessment criteria. We give a few pointers:

Variety: the Mg-CO_2 system satisfied the variety criterion for the proposed avenues.

Originality: the hopper concept (vs. rover) or that of the unplanned mission (vs. scenario) were revisions of certain definitions.

Value: it is of interest to observe that the expansive partition of the missions gradually led to a profound transformation of the value criteria: no longer was the criterion that of the mass landed on Mars, but flexibility. An understanding of the mobility conditions on Mars were also sources of value.

Robustness: the work gradually identified a design domain for the motor and questions that R&D could tackle. Other criteria included data on the CO_2 absorption units, an understanding of the combustion of non-optimal mixtures, etc.

4.6 Workshop 4.1: Intensive Innovation and the Identity of Objects—Analysis Tools

After having studied "rule-based" forms of design (in Chaps. 2 and 3), and after having emphasized their performance and capacity for innovation, we might wonder whether the current reflections around innovation are not governed by a simple effect of fashion. After all, in confronting the current crisis would it not be sufficient to apply rigorously the principles and methods of rule-based design? The workshop will discuss this thesis revealing a new sort of innovation, namely the change in identity of objects.

4.6.1 Acceleration of Rule-Based Innovation

We will of course note an acceleration of rule-based design, reflected by shortened product lifetimes (manufacturers' vehicle ranges have gone from an average replacement roughly every 8 years in the 1990s (Weil 1999) to 6 years today). Development times (or "time to market") are falling sharply in many industries while the proportion of income generated by new products is increasing. Several markets draw inspiration from fashion, with collections having very short lifetimes (watches, mobile telephones, etc.).

4.6.2 Analyzing Objects' Disruption of Identity

This acceleration could doubtless be dealt with using the methods encountered in the previous chapters. However, there is another type of innovation: disruption in the identity of an object.

Definition Given a designed object that generally meets the conditions of a dominant design (see Chap. 3), we might say that there is disruption if the dominant design of the object is brought into question; the disruption might therefore involve, separately or simultaneously, the four languages of dominant design (functional, conceptual, embodiment, detailed) or the conditions implicit in dominant design (business model, customer value).

By definition, this innovation cannot occur in rule-based design (the latter requiring the existence of an established dominant design—see Chap. 3).

Exercise: show that the objects below (see Fig. 4.17) *represented disruptions of identity when they appeared.*

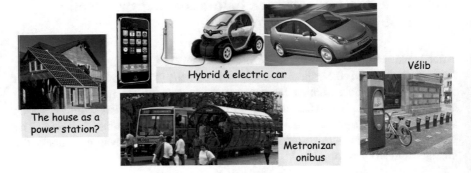

Fig. 4.17 Some examples of the new identity of objects

The analysis of a disruption of identity is not always obvious. The object can progress in a "masked" fashion, taking on the characteristics of a product in dominant design, while subverting certain characteristics:

- For a long time the Prius will be perceived as a vehicle "with a hybrid engine" even though it will profoundly alter the relationship with the environment and its type of driving (the Prius would be sold as being "fun to drive"). Similarly today, the electric car may be perceived as a vehicle "with an electric motor" or as an "electrification of mobility" leading to new types of vehicle. The Renault Twizy represents a more visible disruption of identity since it is a vehicle which (clearly!) is not a car, or at least not a car in the conventional sense.
- The first iPhone drew the same questions: under the dominant design of the mobile telephones and smartphones of the era, the iPhone was assessed as "less good" (see the Fig. 4.18 below for a humorous product comparison).

Fig. 4.18 The difficulty in analyzing disruption of identity

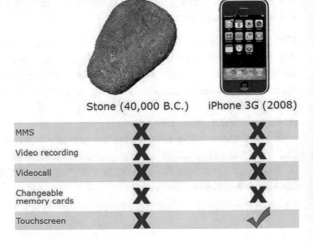

	Stone (40,000 B.C.)	iPhone 3G (2008)
MMS	X	X
Video recording	X	X
Videocall	X	X
Changeable memory cards	X	X
Touchscreen	X	✓

We may also observe that the commercial failure of a product does not necessarily mean a failure of the revisions of identity imposed on the product. The failure of Essensis (a yoghurt-based beauty cream launched by Danone in the 2000s) does not mean the failure of new identities for cosmetic/nutritional products in general.

4.6.3 Generalized and Repeated Disruptions

Disruption of identity is by no means a new phenomenon: at the end of the 19th century new objects appeared, such as the aircraft or automobile, which were also new identities. The difference, however, is in the generalization of the phenomenon and in its repetition. We suggest two exercises:

Exercise: take any arbitrary industrial sector and look for some potential change of identity.

Particular attention may be drawn to two types of sector:

1. would a capital-intensive sector be shielded from disruption (which might make investments obsolete)? Semiconductors provide an example to the contrary: a microchip factory costs in the order of several billion euros, and the inexorability of Moore's law requires that most of the tools and processes be replaced every 2–4 years. The power industry, which is also very capital-intensive, seems to be experiencing change in this direction.
2. Would the luxury and traditional sectors (food in particular) also be shielded from disruption? Several examples here also show powerful disruption of identity (molecular gastronomy, for example).

Exercise: look for repeated disruptions of identity in an industrial sector

Below is an example in accessories related to mobility. Suppose a mobile telephone(see mobile) designer contemplates at the year n what he should be developing for the generation of products. A collection of products or demonstrators for the general public might be presented every couple of years. We observe that every two years this designer would have to follow a cone of performance set by the (latest) dominant design, and simultaneously consider the next dominant design (Fig. 4.19).

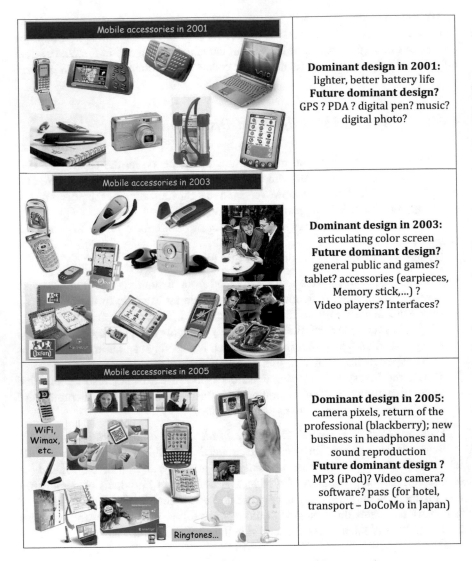

Fig. 4.19 Repeated disruptions in the identities of objects - mobile accessories

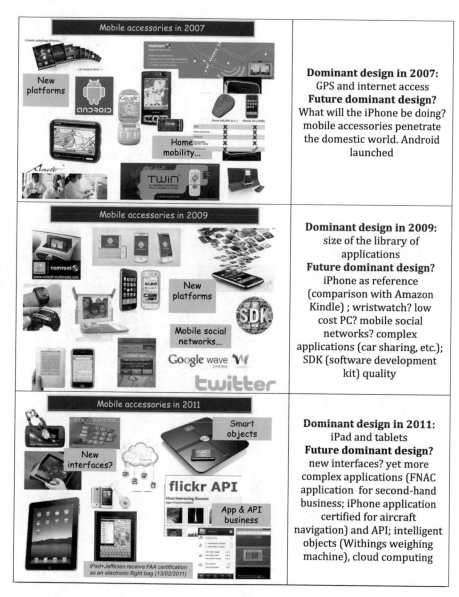

Fig. 4.19   (continued)

4.7 Workshop 4.2: Smart Shopping Cart and Other Exercises

The purpose of this workshop is to familiarize us with the C-K theory. We follow a two-step approach:

4.7.1 Use of the Theory on a Brief

This exercise consists in devising "smart shopping cart" propositions by using the C-K theory. The evaluation criteria are those presented in this Chapter: V2OR. It is important to do this exercise while having means for knowledge growth: for example, internet access which helps acquire new knowledge easily.

Other topics are possible: smart umbrella, uses of a vehicle when stationary, etc. The reader shall first verify that there is a concept involved. It is evident in the cases presented here. For example: the smart shopping cart is at the same time interpretable and neither true nor false. Note that a "connected objects" topic is not a concept; connected objects already exist. In this case we must formulate an associated concept: "connect unconnected objects", for example.

During the workshop we verify mainly the rigour of the reasoning carried out:

- Are the concepts really concepts? (very often the C space is initially used to reorganize knowledge)
- Are the written partitions in C really related to a proposition in K space?
- Do we obtain expansive partitions?
- Is there knowledge growth? In particular, was the internet used?

We give two examples of simplified C-K trees applied to the smart shopping cart concept in exercise 6 below.

4.7.2 Simple C-K Exercises

In order to practice the basic C-K operations, we can do the following exercises:

4.7.2.1 Exercise 1: departitioning

a. *Find a "good idea" corresponding to brief C0 = "smart shopping cart" (without using the C-K method here!)*
b. *Using the C-K method, "departitioning" this "good idea":*

- *Link the "good idea" to the C_0 concept (write the good idea in the form $C_0 \cdot P_1...P_n$)*
- *Identify associated knowledge bases (i.e. basis from which $P_1, P_2,...P_n$ are taken)*

c. *Generate variants of the good idea, derived from the proposition $C_0 P_1...P_n$*

For demonstration purposes, on the topic of "departitioning", one can refer to the Telia case (Le Masson et al. 2006) which presents a departitioning example on the "innovative services for 3G telephony" brief (the exercise was conducted in 2001, when 3G telephony had not been launched yet—apps had not yet been designed).

In the case of "shopping cart", an example of a "good idea" is: "a shopping cart which moves around the supermarket by itself, searches for the products we need for the week, goes through the cashier and arranges itself in the trunk of the car." We do successive departitions to expand on the following ideas: "in the trunk of the car" ➔ this could also be the house, the refrigerator, etc.; more generally: we are led to model the cycle of purchases with all transshipments and intermediary storage spaces; "it arranges itself" ➔ this could also be, for example: "helps arrange in a smart way, without effort, etc." and more generally we are led to model the "arrangement" logic (who arranges and with what purpose: preservation of objects, easily identifiable items, etc.). The reasoning also urges us to enrich the knowledge base and generate a large set of partitions.

4.7.2.2   Exercise 2: systematic expansive partition

a. *In K, identify a property P_a of the shopping cart*
b. *Generate an expansive partition on C_0 based on this property P_a*
c. *Based on this expansive partition do an expansion in K*
d. *Same exercise, taking a "smart" property P_b in K*

For example:

- for P_a we take: "a shopping cart is used to collect products at the supermarket"
- we generate this expansive partition: "a shopping cart which negates 'being used to collect products at the supermarket'."
- we continue on the expensive knowledge, with several possible paths:

  - A compromise path: it is not used to physically collect products but rather it collects them virtually (a barcode reader used in the supermarket aisles).
  - A stricter negation: it is not used to collect but for other things (e.g. it is used only to transport a set of preselected product, it is a cart for the modern drive-through supermarkets).
  - Even stricter: the value of the cart lies in the fact that it is not used to collect products. Thus we obtain a cart that helps us resist temptation! (to stay within budget, or when we are on a diet, when we have an allergy, etc.)

Some comments on exercise 2:

- it shows that an expansive partition can be done systematically
- we are able to expand quickly on the definition of smart cart
- the "higher level" the "definition" of P_a, the greater the expansion.

4.7.2.3   Exercise 3: systematic expansive partition on the languages of systematic design

a. *In K, reconstitute the languages of systematic design of the known smart cart*
b. *Repeat exercise 2 by taking for P_a any proposition of the K base formed in it his way*

For a functional P_a: see above.

For a conceptual P_a: the cart corresponds to an integration model which makes it possible to have a high capacity for each cart but a limited volume for an entire group of carts.

For an embodiment P_a: a cart without wheels.

This exercise shows a systematic expansion logic based on the languages of rule-based design.

4.7.2.4 Exercise 4: design based on a K base expansion

a. *In K, make a list of knowledge domains usually associated with "shopping cart"*
b. *Find a type of knowledge that is not included on the list below, K^**
c. *Based on this new base K^*, partition concept C_0.*
d. *Same exercise with a "smart" attribute*

This exercise demonstrates a K-driven reasoning.

4.7.2.5 Exercise 5: competition-innovation exploration

a. *Based on the C-K tree applied on concept C_0 (smart shopping cart), choose a concept C_1^* derived from C_0*
b. *Becoming knowledgeable: google C_1^*. This step reveals in very general terms a "competitor" that has had a similar idea*
c. *Use the knowledge acquired on the "competitor" and his proposition to continue to partition C_1^*.*

In (a), choose preferably a concept that sounds most original. In (b): in most cases a demonstrator similar to the proposed concept appears; sometimes we must do a thorough research in the depths of the web but it is very rare for a concept obtained after a few hours of work not to have been found by someone else. In (c) several paths are opened: either simply "*C_1^* is different from the competing solution*" or, more interestingly: carry out an analysis of the "competing" proposition and position the alternative concept based on this analysis: "*C_1^* is better than the competing solution*".

This exercise helps us identify two critical points of innovative design:

1. A seemingly good idea rarely appears in an isolated manner.
2. If the idea has already been explored by a "competitor", this does not eliminate the concept but rather offers additional knowledge to help us continue on a nearby path.

4.7.2.6 Exercise 6: analysis of given C-K trees

It is assumed that a team has already done the tree proposed below (See Fig. 4.20).

a. *Why is the proposition "the display is not provided by the supermarket" an expansive partition?*
b. *Same question for "plug in for shopper display".*

c. *Using the proposed K base, do a new expansive partition on the proposed tree* (Fig. 4.20).

This exercise helps us practice innovative design based on an existing tree. We can do the same exercise on the second C-K graph above (Fig. 4.21).

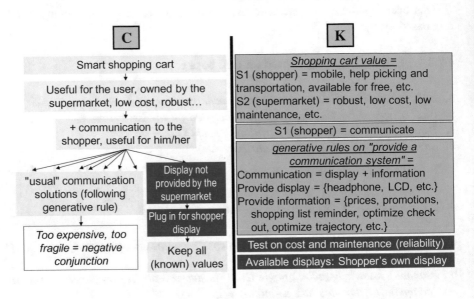

Fig. 4.20 Example No. 1 of a C-K graph on the "smart shopping cart" concept

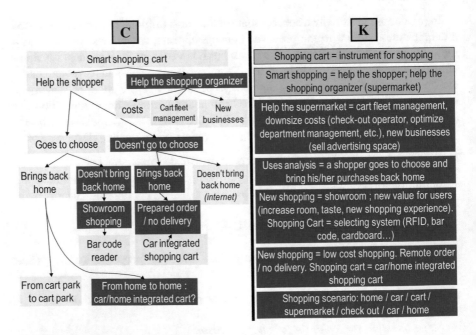

Fig. 4.21 Example No. 2 of a C-K graph on the "smart shopping cart" concept

4.8 Case Study 4.2: Ideo

This case study is based on a video filmed in 1997 by ABC News as well as on the available literature (particularly (Hargadon and Sutton 1997)), and on empirical research and collaborations with the Center for Design Research of Stanford, which is closely linked to Ideo.

The case takes place at IDEO, a company based in Palo Alto, California, which is an expert on innovative projects. The video shows how a team of fifteen people "brings the supermarket shopping cart into the 21st century" in only one week.

The reader may refer to the video. Then he/she shall answer the following questions:

1. *Is there a "process" for creative activity at Ideo and, if yes, describe it*
2. *Evaluate the process*

Below we suggest some elements for answering these questions, as well as some additional thoughts.

4.8.1 Process Description and Analysis

We use the design space/value management framework (this Chapter) to analyze each phase. Thus the process follows the stages shown below (see Fig. 4.22):

Note that the constraints of the television show affect the process (undoubtedly inciting a certain bias towards the "product" by making more difficult the development of more conceptual aspects). The business relation with the customer (the customer that pays for the Ideo service) is not taken into account in this case (again due to the use of video), although said relation would play a critical role in the process in a real case (as shown in the literature on the subject, which confirms the hypothesis that it is necessary to manage cohesion in innovative design).

Let us underline the main singularities of the process:

1. Contrary to the processes often evoked, brainstorming does not occur when the project is launched; it is preceded by a significant knowledge-acquisition phase (said knowledge focuses mainly on uses, making sure to include various types of users: customers, department managers, maintenance staff, cashiers, etc.).
2. There is a strong value management, which steers the process. It organizes the definition of design spaces, their duration, their resources, their designation (type of knowledge to acquire), it assures the rate of divergence and convergence, and it avoids the effects of fixation on an idea that would be too consensual. It is most visible and the most counterintuitive after the brainstorming phase (to a point that, after a first viewing of the video, viewers often subconsciously "forget" the scene filmed!): while we often read that after coming

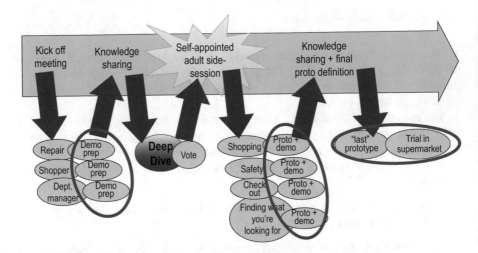

Fig. 4.22 Analysis of Ideo's "shopping cart" project in the design space/value management framework. We distinguish an initial phase of exploration (of uses) (in three separate groups), then a brainstorming brainstorming phase, a third phase of "functional" exploration (four teams in parallel), and finally a synthesis phase (which leads to further learning as the cart is shown to some users in a supermarket)

out of brainstorming we can proceed to a selection of the best ideas by following a collective selection process (such as voting), at Ideo such a process does take place but *certainly does not result in the selection of ideas explored afterwards*. After voting, a "side session" takes place, during which a very small team (consisting of the project manager, the company director and a few other people) studies the ideas put forward and the research conducted so far, and "authoritatively" designates the four exploration spaces that will follow—exploration spaces, which do not involve the ideas expressed. This phase is counterintuitive from the point of view of classical brainstorming; it is perfectly legitimate in a process of value management responsible for a balanced V2OR exploration.

3. A constant effort for knowledge sharing. Here too, at first viewing, these activities are often ignored. Nevertheless, in spite of the very tight deadlines, the team dedicates significant time to knowledge sharing, particularly after the initial phase of exploration of uses and then the phase of discussion on prototypes (see Fig. 4.23). On the other hand, this phase is very well organized: the teams prepare the reestablishment of acquired knowledge in the form of synthesis demonstrators, photographs, modeling (purchase cycles, etc.).

4. An important role played by prototypes. Ideo uses prototypes in nearly all phases of the process and always in a targeted fashion. We identify three different types of prototypes:

   a. After the initial exploration, the teams create synthesis prototypes which reflect their explorations in an effective way; these prototypes essentially designate the known, they are syntheses in K (and not in C).

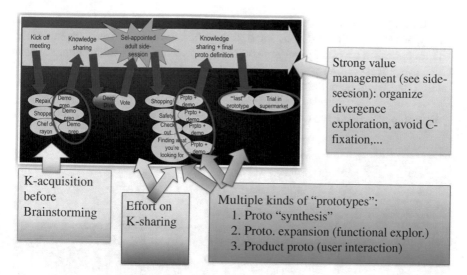

Fig. 4.23 Specificities of the IDEO process

b. After brainstorming, when contrasting paths are explored, each team creates an exploration prototype which extends as far as possible a functional logic (the "maximum" shopping cart for the "check out" function, another for the "find what we're looking for" function, and so on). These prototypes are concept prototypes in the sense that they express very explicitly that they are not completed but that they bring to life (particularly by real stageplay) carts that so far are widely unknown.

c. Final-product prototypes: at the end of the filmed process, the explored paths converge towards one shopping cart incorporating several alternatives. This last prototype, contrary to the previous ones, has many of the advantages of a smart cart from a user's point of view (it is "functional"). It makes the user react but without the user being bothered by approximations in the film. However, this cart is still a concept (it does not verify, for example, the criterion "true" from the manufacturer's point of view!).

Thus we bring to light a very sophisticated process, corresponding to the principles laid out in this Chapater.

4.8.2   Process Evaluation

It is possible to evaluate the process in V2OR (exercise).

Here we propose a more general evaluation that focuses on the way in which the process allows designers to resist different types of fixation effects (see Chap. 5). The grid combines a cognitive approach (which aims to find variety/originality criteria on one hand and value/robustness criteria on the other) with an approach involving fixation social factors, which will study the manner in which the process allows people to "break the rules" (overcome the effect of social fixation on C) and help them achieve sufficient legitimacy to create new rules (see Fig. 4.24).

By completing the table we obtain an evaluation of the IDEO process (Fig. 4.25).

	Cognitive factors of the fixation effect	Social factors of fixation effects: overcoming resistance to rule breaking
C-expansion (K→ C, C→ C)	→ **Cover the whole conceptual potential of the initial concept?** *Criteria*: fluency, variety, **originality**	→ **Involve and support people in a rule-breaking process?** *Criteria*: well-being, participants satisfaction (ie feel comfortable in C-space), number of broken rules
K-expansion (C → K, K→ K)	→ **Activate, acquire and produce relevant knowledge?** *Criteria:* knowledge variety, new "value" and "robustness" criteria, identification of K gaps	→ **Manage collective acceptance and legitimacy of rules (re) building?** *Criteria*: experts commitment, stakeholders commitment (top management, designers, users, partners, etc.) to further exploration (research program, etc.)

Fig. 4.24   General evaluation framework for an innovative design process

	Cognitive factors of the fixation effect	Social factors of fixation effects: overcoming resistance to rule breaking
C-expansion (K→ C, C→ C)	**+** Capacity to produce rule-breaking propositions	**+** Capacity to collectively break rules (teamwork, knowledge sharing effort, including customer's representatives, provocative proto.)
	- Limited variety, limited understanding of the structure of the unknown (attraction by the product)	**-** No involvement of new "rule-breakers" (or K-providers supporting rule-breaking) during the process
K-expansion (C →K, K→ K)	**+** K acquisition on (all) users of the existing products	**+** Involvement based on "convincing" prototype (or "story telling"), to be developed into one "product"
	- Limited capacity of scientific K creation (in case of science-based products)	- No design strategy (only one product); no involvement of new stakeholders (no "hooks")

Fig. 4.25   Evaluation of the IDEO process

References

Agogué M (2012) *Modéliser l'effet des biais cognitifs sur les dynamiques industrielles: innovation orpheline et architecte de l'inconnu.* MINES ParisTech, Paris.

Agogué M, Kazakçi A (2014) 10 years of C-K theory: a survey on the academic and industrial impacts of a design theory. In: Chakrabarti A, Blessing L (eds) *An Anthology of Theories and Models of Design. Philosophy, APproaches and Empirical Explorations.* Bangalore, pp 219–235. doi:10.1007/978-1-4471-6338-1.

Agogué M, Le Masson P, Robinson DKR (2012) Orphan Innovation, or when path-creation goes stale: missing entrepreneurs or missing innovation? *Technology Analysis & Strategic Management* 24 (6):603–616.

Amabile TM (1996) *Creativity in context.* Westview Press, Boulder, Colorado.

Amabile TM, Conti R, Coon H, Lazenby J, Herron M (1996) Assessing the Work Environment for Creativity. *Academy of Management Journal* 39 (5):1154–1184.

Arrighi P-A, Le Masson P, Weil B (2012) Breaking the Dilemma between robustnee and generativeness: an experimental assessment of a new software design suite. In: *International Product Development Management Conference*, Manchester, UK, 2012. p 20.

Arrighi P-A, Le Masson P, Weil B (2013) From New Product Development (NPD) to New Design Process (NDP)? How new Computer Aided Design (CAD) tools lead to embedded learning and exploration in effective processes. In: *International Product Development Management Conference*, Paris, France, 2013. p 21.

Arrighi P-A, Le Masson P, Weil B (2015) Managing radical innovation as an innovative design process: generative constraints and cumulative set of rules. *Creativity and Innovation Management* 24 (3):373–390.

Ben Mahmoud-Jouini S, Charue-Duboc F, Fourcade F (2006) Managing Creativity Process in Innovation Driven Competition. In: Verganti R, Buganza T (eds) *13th International Product Development Management Conference*, Milan, 2006. EIASM & Politecnico di Milano, pp 111–126.

Benguigui J-M (2012) Les 10 ans de la théorie C-K: Revue de littérature. Paper presented at the *AIMS*.

Boden MA (1990) *The creative mind. Myths and Mechanisms.* George Weidenfeld and Nicolson Ltd.

Boden MA (1999) Computer Models of Creativity. In: Sternberg RJ (ed) *Handbook of creativity.* Cambridge University Press, Cambridge, pp 351–372.

Braha D, Reich Y (2003) Topologial structures for modelling engineering design processes. *Research in Engineering Design* 14 (4):185–199.

Cabanes B (2013) *Les ruptures dans le spatial: imaginaires, objet et concepts. Cartographie et méthode d'investigation de double inconnue.* Université Paris-Dauphine, MINES ParisTech, ENS Cachan, Paris.

Chakrabarti A, Sarkar P, Leelavathamma, Nataraju BS (2005) A Functional Representation for Biomimetic and Artificial Inspiration of New Ideas. *Artifical Intelligence in Engineering Design, Analysis and Manufacturing* 19:113–132.

Cohen P (1966) *Set Theory and the Continuum Hypothesis.* Dover Publications, Mineola, New York.

Csikszentmihalyi M (1999) Implications of a Systems Perspective for the Study of Creativity. In: Sternberg RJ (ed) *Handbook of Creativity.* Cambridge Universtiy Press, Cambridge, pp 313–335.

Dehornoy P (2010) Théorie axiomatique des ensembles. In: *Encyclopeadia Universalis.* Encyclopaedi Britannica, Paris, p Corpus.

Dym CL, Agogino AM, Eris O, Frey D, Leifer LJ (2005) Engineering Design Thinking, Teaching, and Learning. *Journal of Engineering Education* January 2005:103–120.

Elmquist M, Le Masson P (2009) The value of a 'failed' R&D project: an emerging evaluation framework for building innovative capabilities. *R&D Management* 39 (2):136–152.

Elmquist M, Segrestin B (2007) Towards a new logic for Front End Management: from drug discovery to drug design in pharmaceutical R&D. *Journal of Creativity and Innovation Management* 16 (2):106–120.

Gardey de Soos P (ed) (2007) *Conception innovante à la RATP: la méthode KCP. Cinq cas pratiques de conception innovante collective.* Les rapports de la prospective, n 146. RATP, Paris.

Gillier T, Piat G, Roussel B, Truchot P (2010) Managing Innovation Fields in a Cross-Industry Exploratory Partnership with C–K Design Theory. *Journal of product innovation management* 27 (6):883–896.

Guilford JP (1950) Creativity. *American Psychologist* 3:444–454.

Guilford JP (1959) Traits of Creativity. In: Anderson HH (ed) *Creativity and its Cultivation.* Harper, New-York, pp 142–161.

Hargadon A, Sutton RI (1997) Technology Brokering and Innovation in a Product Design Firm. *Administrative Science Quarterly* 42 (4):716–749.

Hatchuel A (2005a) New rationalizations of innovative design. An introduction to C-K theory. Copenhagen.

Hatchuel A (2005b) Quelle analytique de la conception? Parure et pointe en design. In: *Le design en question(s)*, Centre Georges Pompidou, Musée national d'Art moderne, 16–18 novembre 2005 2005b. p 13.

Hatchuel A (2006) Quelle analytique de la conception? Parure et pointe en design. In: Flamand B (ed) *Le design. Essais sur des théories et des pratiques.* Editions du Regard, Paris, pp 147–160.

Hatchuel A (2008) Du raisonnement de conception. Essai sur le "forcing" en théorie des ensembles. In: Hatchuel A, Weil B (eds) *Les nouveaux régimes de conception.* Vuibert, Paris, pp 133–149.

Hatchuel A (2013) Deconstructing meaning: industrial design as Adornment and wit. In: *10th European Academy of Design*, Gothenburg, 2013.

Hatchuel A, Le Masson P, Reich Y, Weil B (2011a) A systematic approach of design theories using generativeness and robustness. In: *International Conference on Engineering Design, ICED'11*, Copenhagen, Technical University of Denmark, 2011a. p 12.

Hatchuel A, Le Masson P, Weil B (2004) C-K Theory in Practice: Lessons from Industrial Applications. In: Marjanovic D (ed) *8th International Design Conference*, Dubrovnik, 18th-21st May 2004, 2004. pp 245–257.

Hatchuel A, Le Masson P, Weil B (2005) The Development of Science-Based Products: Managing by Design Spaces. *Creativity and Innovation Management* 14 (4):345–354.

Hatchuel A, Le Masson P, Weil B (2006) The design of science based-products: an interpretation and modelling with C-K theory. In: Marjanovic D (ed) *9th International Design Conference*, Dubrovnik, 15th-18th May 2004, 2006. pp 33–44.

Hatchuel A, Le Masson P, Weil B (2011b) Teaching Innovative Design Reasoning: How C-K Theory Can Help to Overcome Fixation Effect. *Artificial Intelligence for Engineering Design, Analysis and Manufacturing* 25 (1):77–92.

Hatchuel A, Weil B (1999) Pour une théorie unifiée de la conception, Axiomatiques et processus collectifs. CGS Ecole des Mines/ GIS cognition-CNRS, Paris.

Hatchuel A, Weil B (2002) C-K Theory: Notions and Applications of a Unified Design Theory. In: *Herbert Simon International Conference on Design Sciences*, Lyon, 15–16 March 2002, 2002a.

Hatchuel A, Weil B (2003) A new approach to innovative design: an introduction to C-K theory. In: *ICED'03, August 2003*, Stockholm, Sweden, 2003. p 14.

Hatchuel A, Weil B (2007) Design as Forcing: deepening the foundations of C-K theory. In: *International Conference on Engineering Design*, Paris, 2007. p 12.

Hatchuel A, Weil B (2009) C-K design theory: an advanced formulation. *Research in Engineering Design* 19 (4):181–192.

Hatchuel A, Weil B, Le Masson P (2013) Towards an ontology of design: lessons from C-K Design theory and Forcing. *Research in Engineering Design* 24 (2):147–163.

Hendriks L, Kazakçi AO (2010) A formal account of the dual extension of knowledge and concept in C-K design theory. Paper presented at the *International design conference - Design 2010*, Dubrovnik, Croatia.

Hendriks L, Kazakçi AO (2011) Design as Imagining Future Knowledge, a Formal Account. In: Grossi D, Minica S, Rodenhäuser B, Smets S (eds) *Logic and Interactive Rationality*. pp 111–125.

Hooge S (2010) *Performance de la R&D en rupture et des stratégies d'innovation: Organisation, pilotage et modèle d'adhésion*. MINES ParisTech, Paris.

Jansson DG, Smith SM (1991) Design Fixation. *Design Studies* 12 (1):3–11.

Jech T (2002) *Set Theory*. Springer Monographs in Mathematics, 3rd millenium edition, revised and expanded edn. Springer, Berlin.

Kazakçi A, Hatchuel A, Le Masson P, Weil B (2010) Simulation of Design reasoning based on C-K theory: a model and an example application. Paper presented at the *International Design Conference - Design 2010*, Dubrovnik – Croatia.

Kazakçi A, Hatchuel A, Weil B (2008) A Model of C-K Design Theory based on Term Logic: A Formal C-K Background for a Class of Design Assistants. Paper presented at the *International Design Conference - Design 2008*, Dubrovnik, Croatia, May 19–22.

Kazakçi AO (2013) On the imaginative constructivist nature of design: a theoretical approach. *Research in Engineering Design* 24 (2):127–145.

Kazakçi AO, Tsoukias A (2005) Extending the C-K design theory: a theoretical background for personal design assistants. *Journal of Engineering Design* 16 (4):399–411.

Kroll E (2013) Design theory and conceptual design: contrasting functional decomposition and morphology with parameter analysis. *Research in Engineering Design* 24 (2):165–183.

Kroll E, Le Masson P, Weil B (2013) Modeling parameter analysis design moves with C-K theory. Paper presented at the *International Conference on Engineering Design, ICED'13*, Séoul, Korea.

Le Masson P, Dorst K, Subrahmanian E (2013a) Design Theory: history, state of the arts and advancements. *Research in Engineering Design* 24 (2):97–103.

Le Masson P, Gardey de Soos P (eds) (2007) *La RATP et les enjeux de la compétition par l'innovation - un séminaire d'initiation à la conception innovante*. Les rapports de la prospective, n 145. RATP, Paris.

Le Masson P, Hatchuel A, Weil B (2011) The Interplay Between Creativity issues and Design Theories: a new perspective for Design Management Studies? *Creativity and Innovation Management* 20 (4):217–237.

Le Masson P, Hatchuel A, Weil B (2013b) Teaching at Bauhaus: improving design capacities of creative people? From modular to generic creativity in desing-driven innovation. Paper presented at the *10th European Academy of Design*, Gothenburg.

Le Masson P, Hatchuel A, Weil B (2016) Design Theory at Bauhaus: teaching "splitting" knowledge. *Research in Engineering Design* 27 (April 2016):91–115.

Le Masson P, Weil B (2010) La conception innovante comme mode d'extension et de régénération de la conception réglée: les expériences oubliées aux origines des bureaux d'études. *Entreprises et histoire* 58 (1):51–73.

Le Masson P, Weil B (2013) Design theories as languages for the unknown: insights from the German roots of systematic design (1840–1960). *Research in Engineering Design* 24 (2):105–126.

Le Masson P, Weil B, Hatchuel A (2006) *Les processus d'innovation. Conception innovante et croissance des entreprises*. Stratégie et management. Hermès, Paris.

Le Masson P, Weil B, Hatchuel A (2010) *Strategic Management of Innovation and Design*. Cambridge University Press, Cambridge.

Mullen B, Johnson C, Salas E (1991) Productivity loss in brainstorming groups: a meta-analytic integration. *Basic and Applied Social Psychology* 12 (1):3–23.

Reich Y (1995) A Critical Review of General Design Theory. *Research in Engineering Design* 7:1–18.

Reich Y, Hatchuel A, Shai O, Subrahmanian E (2010) A Theoretical Analysis of Creativity Methods in Engineering Design: Casting ASIT within C-K Theory *Journal of Engineering Design*:1–22.

Salustri FA (2005) Representing C-K Theory with an Action Logic. In: *Proceedings ICED '05*, Melbourne, Australia, 2005.

Shafirovich E, Salomon M, Gökalp I (2003) Mars Hopper vs Mars Rover. In: *Fifth IAA International Conference on Low-Cost Planetary Missions, 24–26 Septembre 2003*, Noordwijk, the Netherlands, 2003. ESA SP-542, pp 97–102.

Shai O, Reich Y (2004a) Infused Design: I Theory. *Research in Engineering Design* 15 (2): 93–107.

Shai O, Reich Y (2004b) Infused Design: II Practice. *Research in Engineering Design* 15 (2): 108–121.

Shai O, Reich Y, Hatchuel A, Subrahmanian E (2009) Creativity Theories and Scientific Discovery: a Study of C-K Theory and Infused Design. In: *International Conference on Engineering Design, ICED'09*, 24–27 August 2009, Stanford CA, 2009.

Shai O, Reich Y, Hatchuel A, Subrahmanian E (2013) Creativity and scientific discovery with infused design and its analysis with C-K theory. *Research in Engineering Design* 24 (2):201–214.

Sharif Ullah AMM, Mamunur Rashid M, Tamaki Ji (2011) On some unique features of C-K theory of design. *CIRP Journal of Manufacturing Science and Technology* in press.

Subrahmanian E, Reich Y, Konda SL, Dutoit A, Cunningham D, Patrick R, Thomas M, Westerberg AW (1997) The n-dim approach to creating design support systems. In: *ASME-DETC*, Sacramento, California, 1997.

Suh NP (1990) *Principles of Design*. Oxford University Press, New York.

Suh NP (2001) *Axiomatic Design: advances and applications*. Oxford University Press, Oxford.

Takeda H, Veerkamp P, Tomiyama T, Yoshikawa H (1990) Modeling Design Processes. *AI Magazine* Winter 1990:37–48.

Taura T, Nagai Y (2013) A Systematized Theory of Creative Concept Generation in Design: First-order and high-order concept generation. *Research in Engineering Design* 24 (2):185–199.

Tomiyama T, Yoshikawa H (1986). Extended general design theory." Centre for mathematics and Computer Science, Amsterdam, the Netherlands, p. 32.

Torrance EP (1988) The Nature of Creativity as Manifest in its Testing. In: Sternberg RJ (ed) *The Nature of Creativity*. Cambridge University Press, Cambridge, England.

Ward TB, Smith SM, Finke RA (1999) Creative Cognition. In: Sternberg RJ (ed) *Handbook of Creativity*. Cambridge University Press, Cambridge, pp 189–212.

Weil B (1999) *Conception collective, coordination et savoirs, les rationalisations de la conception automobile*. Thèse de doctorat en Ingénierie et Gestion, Ecole Nationale Supérieure des Mines de Paris, Paris.

Weisberg RW (1992) *Creativity Beyond The Myth of Genius*. W. H. Freeman Company, New York.

Yoshikawa H (1981) General Design Theory and a CAD System. In: Sata T, Warman E (eds) *Man-Machine Communication in CAD/CAM, proceedings of the IFIP WG5.2-5.3 Working Conference 1980 (Tokyo)*. Amsterdam, North-Holland, pp 35–57.

Chapter 5
Designing the Innovative Design Regime—C-K Based Organizations

The representation of the very unusual innovative project described in Chap. 4 is not at all realistic: the isolated innovative project does not really exist; even in entrepreneurial situations, where the company project appears to coincide with the innovative project, history has shown that start-ups rarely succeed with their initial projects. It seems to be repetition and interdependencies that lead to performance (Drucker 1985a); see also the studies on the Chalmers School of Entrepreneurship (Carreel 2011b). We will therefore study the logic of collective action which, apart from the unusual innovative project, constitutes an innovative design regime. The reasoning is the same as for the rule-based design regime: we ask what is the *infrastructure* that ensures a stable design regime. However, a difficulty immediately arises: we have seen that, for rule-based design (as for all bureaucratic systems) it was the rule base that created the infrastructure and its performance. Given that the innovative design regime regularly seeks to revise its own rule base, on what might it be founded?

We shall follow the three dimensions of a design regime (performance, reasoning, organization), in that order. Note that, as in Chap. 3, we begin with the overall performance of the regime before investigating the processes of reasoning and organization that might reach this level of performance.

5.1 Performance in Innovative Design

Here we are interested in the performance of a design group (team, department, firm,...) after the project. As in Chap. 3, we are talking about a firm or sector (i.e. more precisely, the firm's ecosystem).

© Springer International Publishing AG 2017
P. Le Masson et al., *Design Theory*,
DOI 10.1007/978-3-319-50277-9_5

5.1.1   Outputs: Sustainable Revision of Object Identity

The outputs of rule-based design correspond to the gradual extensions of a dominant design (performance cone and coverage); in innovative design, these outputs correspond to an ability to revise the dominant design, i.e. to create new object identities.

How might this phenomenon be measured? We saw in Chap. 3 that a dominant design corresponded to a Lancasterian definition of goods, i.e. a point in a vector space of characteristics (characteristics fixed in nature, with only the level changing). Also, certain recent work has sought to measure changes in identity by analyzing the *new characteristics* associated with a type of product (Ün 2011) (El Qaoumi et al. 2016). This type of measure presupposes a few methodological prerequisites:

- We seek to measure characteristics in a Lancasterian[1] sense, i.e. characteristics with an "economic effect", determining the consumer's function of utility (an arbitrary change of goods is not a characteristic; a Lancasterian characteristic is related to how useful the goods are to the consumer). From the point of view of innovation, this amounts to saying that we wish to measure only those new features that have an influence on the commercial relationship—this is a generalization of the definition "an invention having found a market" (see introductory chapter). A measurement solution consists of *using the comparisons carried out by consumer prescribers* (e.g. Que Choisir in France) who will make an accurate determination of the characteristics involved in consumers' choice criteria. The comparative tables of products therefore provide the Lancasterian characteristics at some instant *t* (moreover, this is the method recommended by Lancaster himself Lancaster 1966).
- We seek to monitor the evolution of the characteristics of a type of product, and can therefore make use of the investigations carried out by Que Choisir on mobile telephones or clothes irons. However, these analyses will be very limited if there is no standard product supporting the prescribers' surveys. Consider two extreme cases: the complex products studied by the specialized prescribers embodying numerous characteristics (automobile), and the world of use in which the evolution of standard products does not accurately account for the renewal of characteristics (hence hiking accessories cannot be studied just on the basis of the characteristics of boots or rucksacks).

---

[1]Recall, Sect. 3.1.2: for Lancaster, characteristics determine consumer choices and there are inputs in the consumer preference function. For a consumer, an item of goods is a point in a fixed space of characteristics. As far as Lancaster is concerned, the nature of the characteristics is stable; on the other hand, a new item can attain a higher level for certain characteristics or can attain new combinations of levels of characteristics. For Lancaster, however, there is a finite list of characteristics.

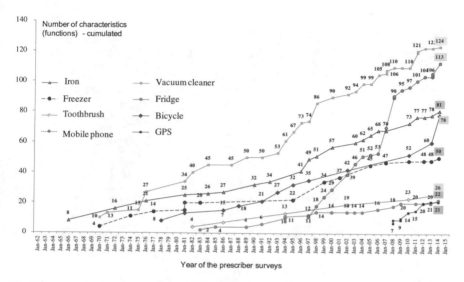

Fig. 5.1 An example of measuring outputs for an innovative design regime. *Source* El Qaoumi (2016)

- The accuracy of the measuring instrument is limited: (a) the tables proposed by the prescribers are necessarily limited in size! This can, however, correspond to a cognitive limit on the part of the "consumer", in which case that means that the instrument is able to *incorporate* this cognitive limit—it is not exactly the characteristics that are being measured, but rather the characteristics *that the consumer is able to take into account*; (b) only those characteristics that can be measured and compared by the prescribers are recorded; (c) they are measured with a delay relative to their appearance on the market. We note that the bias is asymmetrical: the instrument underestimates the emergence of certain characteristics, and a characteristic is recorded only if its economic effect has become sufficiently great to justify dedicating a line in the table to it, and if a measuring instrument can be made sufficiently reliable for the purpose. Hence *the emergence of new characteristics is underestimated by the instrument*.

Under these conditions, and in the case of a standard product investigated by a prescriber, the appearance of new characteristics can be measured over time. We therefore obtain the graphs (Fig. 5.1).

These data and associated studies reveal two interesting results:

1. There is no stability in the space of characteristics: this result brings Lancaster's fundamental assumption into question (recall that this assumption enabled Lancaster to maintain the theory of general equilibrium, at the cost of adding his theory of consumption based on characteristics). Rather, we observe the regular emergence of new characteristics (with economic effects) (El Qaoumi et al. 2016).

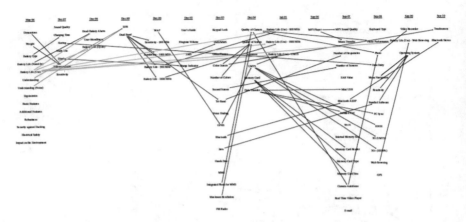

Fig. 5.2 The effects of propagation of new functions: new characteristics call up new characteristics

2. Moreover, a study of the nature of the characteristics involved reveals what the raw figures don't show: the new characteristics are not mutually independent, and do not obey the Poisson type of emergence often used in economic models of innovation. On the contrary, interdependence between new characteristics can be observed—like the "MP3 player" characteristic which, a few years later, would bring about the emergence of the "stereo" or "data storage" function (see Fig. 5.2, which represents the propagation of a new function over other new functions which appeared later). Moreover, these interdependencies can be interpreted as forms of learning on the part of the user (Brown et al. 2011).

5.1.2   Inputs: Skills and Imagination

Concerning inputs, just as for rule-based design we may still use the intensity of R&D or the number of designers as aggregate indicators of the design effort.

However, we can fill in the details of this measure: in rule-based design the assumption was made that the quantity of resources was a satisfactory indicator of design capabilities. Clearly, it would also have been necessary to include the level of skills. More generally in innovative design, however, the issue is not so much the quantity as the *nature of the skills*. Also, it might be useful to study the types of skill mastered by a company or collective. Recent work looking at company patents has led to a study of *the growth in the variety of skills* in certain businesses (Hesham Mohamed 2012) (see Fig. 5.3).

Furthermore, as we saw in Sect. 4.2, initial concepts can also be resources. The notion of the "good initial concept" as presented in the previous chapter might seem somewhat abstract. Even so, investigations of innovative design regimes have

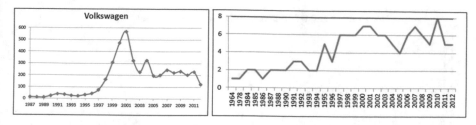

Fig. 5.3 An increase in the variety of skills in the major companies? *source* Hesham Mohamed (2012) *Left*, total number of patents filed by Volkswagen: there are big variations in the total number of patents; *right*, number of sections of international patent classification in which Volkswagen filed at least one patent over the year; surprisingly, there is no correspondence with the trend seen in the quantity of patents filed and we observe a trend in the sections where Volkswagen is granted patents (even when the total number of patents filed decreases)

demonstrated the critical role played by concepts "expressed" in the contemporary mechanisms of growth ("desirable", or "imaginary"). Marine Agogué has shown how fixation of the imagination on a small number of concepts might lead to a halt in the industrial dynamic (Agogué 2012b) (in particular see Agogué et al. 2012b). Conversely, the powerful industrial dynamic of the semiconductor sector (see the curves in Chap. 3) coincides with a generalized "unlocking of rules", i.e. an ability of the companies in this sector to constantly revisit concepts and express a desire for new technologies (Le Masson et al. 2012d). Moore's Law is the first of these "unlocking" laws: it implies that the technologies of the day are obsolete as a matter of course, such that the proposition "technologies for the new generation of semiconductors" is always a concept! Finally, we could see that opening up the space of concepts gave rise to the start of young industries. Such is the case, for example, of building using hemp (see Le Masson et al. 2012a).

Measuring these sector-wise imaginative ideas is not simple, and presupposes complex analyses of projects in progress (see the publications cited below). Some recent work (El Qaoumi et al. 2016) have suggested more aggregated methods of estimation using prescribers, quantifying for example the number of original "innovative concepts" per year by some benchmark prescriber for the sector (The *Auto-journal* for the automobile sector, for example). Over a long period such measures give an indication of the variety and renewal of a sector's imaginative ideas over time. In the graph below, constructed from the "trends" pages of the *Auto-journal*, we can pick out a phase of very slow renewal from 1959–1981 (roughly 1 every 7 years) then a faster renewal from the 1980s (roughly 1 every 2 years) and then a new acceleration from 2003 (roughly 1 per year). This renewal illustrates a "demand" for concepts—though not necessarily concluding that this demand was satisfied (Fig. 5.4).

This work on imaginative ideas also highlights new players that we shall study in detail in Sect. 5.3.

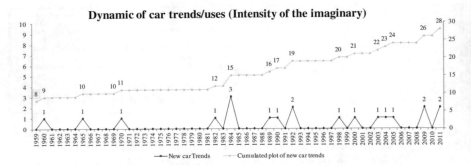

Fig. 5.4 Estimating the dynamic of imaginative ideas. Estimates based on the original "innovative concepts" listed by a prescriber in the world of automobiles (data per year)

5.1.3   Performance: The New Industrial Dynamic

5.1.3.1   Returns on Dynamic Design, Lineages

As we saw in Chap. 4, measures of returns project by project do not account for the effects of re-use; efficiency also involves managing the gradual acquisition of project skills at both company and sector level. In innovative design, the logic of dynamic returns is especially critical (see Workshop 1, Chap. 3 covering this idea): since any innovative design project will produce more knowledge than is strictly necessary to obtain a product (see the logic of the crazy concept in Chap. 4), static returns rarely perform well[2] (denominator too large); on the other hand, the ability to use knowledge "produced in excess" (see Chapel 1997, Le Masson et al. 2006): "re-use of excess knowledge") is an essential element in augmenting design skills as projects gradually progress.

As in rule-based design, these dynamic returns represent a coincidence of products and skill, leading to the rediscovery of forms of dominant design; however, these are much more susceptible to evolution and are in particular "endogenized" by the players, i.e. the players have the ability to alter the dominant designs and are not constrained by them. Several examples of such lineages can be found; in particular, see the cases of Tefal or Saint-Gobain in (Le Masson et al. 2006).

These returns are at the heart of "non-linear" growth curves (exponential growth). In the previous cases, Tefal's growth and profitability or the success of Saint-Gobain (particularly in terms of market share) are testament to the

---

[2]Recall that the static return is the ratio of project income (numerator) to project costs (denominator). In the case of an innovative design project, more knowledge is produced than is necessary for the project itself, hence costs are generally very high; therefore the denominator is large, and the return is poor.

effectiveness of developing successive product lines. To take a more famous example, Apple's success may also be interpreted as a talent for developing lineages of products that perform particularly well.

5.1.3.2 Possible Pathologies: Orphan Innovation

On the contrary, however, innovative design creates new pathological industrial dynamics—every bit as remarkable. We cite:

- The difficulties faced by contemporary backers of innovation (very poor profitability of risk capital over the last few years (Dantas Machado Rosa and Raade 2006: Kandel et al. 2011)).
- Very high rates of failure among innovations: in certain sectors, the number of new products launched (after the most demanding of validation tests) and then withdrawn from sale as a result of failure got up to 65%; out of three products developed *to market*, two failed (despite all the tests and intermediate selections which had already led to the rejection of more than 90% of the initial ideas)! This pathological behavior is broadly explained by the fact that in an innovative design situation the validation protocols can become obsolete—disruptive innovative products would in fact have required the development of special new tests (e.g. see the work carried out with Nestlé on this topic (Gapihan and Le Mestre 2008; de Metz 2010)). We are typically confronted by "K-reordering" costs that are frequently underestimated.
- Eternally emerging technologies: the fuel cell (inverse electrolysis of water) has been held up as the energy of the future for over a century! For these technologies it seems that the design efforts to be agreed upon have been underestimated and might be concentrated on certain dimensions, neglecting others that may still be critical (the logic of substitution vs. exploration of new values and new functionalities).
- Technological bubbles with disastrous effects: disruptive concepts can engender collective crazes, leading initially to an influx of resources, but these resources are concentrated over a small number of paths; they do not maintain the V2OR equilibrium and tend to restrict exploration. The logic of the business plan imposes a direction in accordance with the initial plan and hence a powerful limitation on the learning process. Finally, the poor profitability of these investments leads to their withdrawal. By the end of this cycle not just the ecosystem has failed to acquire new resources (both financial and intellectual!) but in addition it often becomes a lot more difficult to encourage new investment in it (for a model of the anticipatory mechanisms in the innovative design situation see e.g. Le Masson et al. 2012a) (see the Gartner curve below, illustrating the effects of these technological bubbles) (Fig. 5.5).
- Orphan innovations: safety on two wheels, independence of the elderly, energy and biomass, housing, prevention of malnutrition, etc. In many sectors it is not

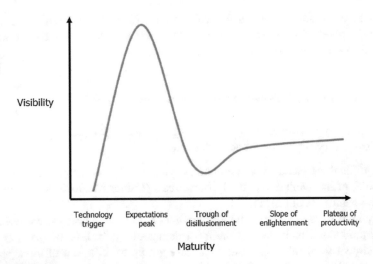

Fig. 5.5 The cycle of craze and abandonment associated with technological bubbles ("Hype Cycle" from Gartner consultants)

the "resistance to innovation" that is surprising but rather unsatisfied expectations; there is in these fields a desirable concept, a form of "social demand", and often R&D resources (aid, subsidies, clusters, and the support of partners); and yet very few innovations ever see the light of day, being often concentrated in a narrow field (a lot of work on the independence of the elderly focuses on sensors to detect falling) and fail to satisfy. Highlighting this phenomenon, Agogué et al. have dubbed it "orphan innovation" (Agogué et al. 2012b). Thus we have exploration but the dynamic returns have not been realized, learning has concentrated on a small number of pathways and variety has not been satisfied (see below Sect. 5.2.2.2 for an industrial example).

Note that in most of the situations above, it may be possible for there to be heavy investment in R&D with very poor returns in terms of innovative concepts.

5.1.3.3 The R&D Paradox Revisited

Updating these varied returns (increasing, decreasing, or even negative) allows us to revisit the R&D paradox.

Recall (in the introductory chapter) that the R&D paradox highlights the lack of connection between the intensity of R&D and the growth performance of a company. Having analyzed the logic of performance in the rule-based and innovative regimes respectively, the disparity can be explained thus:

- in rule-based design, the underlying logic of minimizing the learning process and the profitability of unique investments (conventional minimum ROI), leads to a linear curve; the logic of dominant design can lead to decreasing static returns (the most profitable static returns occur at the start of dominant design) (see the platform returns in Chap. 3 in the section on platforms). One explains the points on the lower right of the graph (high investment, low earnings).

- in innovative design, exaggerated learning mechanisms give rise to points on the upper left (low initial knowledge, high earnings). Conversely, the mechanisms seen above (pathological cases) lead to massive investment (orphan innovation, anticipation, etc.) with low returns (and hence more points on the lower right of the graph).

The R&D paradox is therefore caused less by rule-based design than by innovative design, which gives rise on the one hand to the emergence of champions, and on the other to pathological states of orphan innovation. The current paradox in R&D could be the sign of a rise in innovative design (and the difficulties of managing it).

These performance measures highlight a very rich phenomenology. They also pose the question as to what design strategies to adopt, and it is this that we shall look at next.

5.2   Reasoning and Tools

5.2.1   Issue: Collective De-fixation

What are the concepts and knowledge bases favorable to innovative design for an individual or company? In the case of rule-based design, we have seen that conceptual and generative models were the key resources; what are the corresponding resources for innovative design?

At the risk of over-simplification, we can say that the *resources* for design can be characterized by the cognitive difficulties that they help to overcome. In rule-based design, the history of the associated theories showed us that the difficulties (Le Masson et al. 2011) were twofold:

(1) a tendency to "short-circuit", i.e. to make premature use of available knowledge (wheels already built: see historical study in Chap. 2; or the immediate re-use of habitual rules);
(2) the knowledge itself (often with too many gaps, poorly suited to the designs to be made).

Whence we get the two major elements of a rule-based design: *generative models that avoid short circuits while guaranteeing reliable convergence; and conceptual models that ensure a good use of knowledge and the production of "well structured" knowledge.*

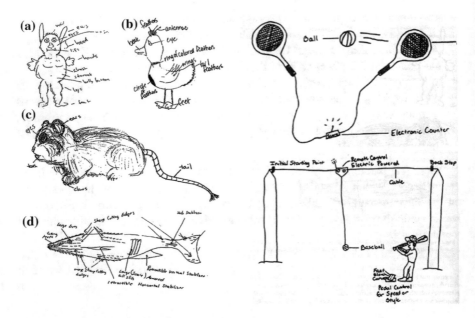

Fig. 5.6 The limits of individual creativity (Ward et al. 1999). *Left*, experiments due to (Ward 1994): the imaginary animals drawn during a creative session all have heads, eyes and limbs. According to Ward, individuals tend to follow "a path of least resistance" in creativity. *Right*, experiment due to (Smith et al. 1993) show examples of the effects of fixation: we show the example of a game above; games invented subsequently include, most of the time, an electronic counting device, rackets and some physical activity. Individuals tend to re-use knowledge that has been recently activated

In innovative design, the issue of design resources is that of reinforcing the ability to overcome the effects of fixation: the rule-based regime provides the means for a design "limited to the dominant design"; innovative design supports the *expansion and recomposition of the system of rules itself*. Hence it is the ability (when relevant) to distance oneself from the known rules that provides the quality of the innovative design "resources". Work on psychology (see the start of Chap. 4 in particular) shows that this "distancing" is not "natural" and that, even in a "creative" situation with as few constraints as possible, individuals and groups are victims of the effects of fixation.

As for the individual aspect, the figures show the results of some celebrated experiments on individual creativity (Fig. 5.6).

As for the group aspect, it is important to recall a phenomenon well known to researchers in the field of collective creativity: *brainstorming does not increase the number and originality of the ideas put forward by a group*. On the contrary, for several decades work on the psychology of creativity has highlighted what researchers call the "productivity gap": if we compare the creative performance of a

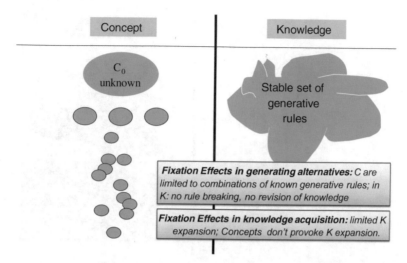

Fig. 5.7 Individual fixation in C-K theory. C-K theory invites a distinction between a fixation in C-space (individuals explore a small number of pathways and are fixated by the "structure" of the object, an attribute known as following the "path of least resistance") and a fixation in K-space (no new knowledge, use the recently activated knowledge, available immediately)

brainstorming group with that of a group of individuals of the same profile but creating ideas separately, the brainstorming group produces significantly fewer ideas than the sum of individual ideas. Among all the explanations, cognitive explanations have been put forward: in particular, the group tends to work only on ideas built on the basis of common knowledge, and not on the sum of available knowledge (Stewart and Stasser 1995; Paulus 2000). Moreover, we may note that brainstorming, invented by Osborn (1953) while he was working for the advertising agency BBDO, was not invented to increase the number of ideas being tossed around but to encourage them to circulate. The copywriters at BBDO did not need more ideas, but rather just needed their own ideas to circulate.

C-K theory summarizes the phenomena of individual and collective fixation (see Fig. 5.7) and finally identifies four types of fixation: in C- and K-space; individual or collective (see Table 5.1) (source: Hatchuel et al. 2011).

The objective of resources of innovative design will therefore be to assist in overcoming these four effects of fixation (see Fig. 5.8). Hence a *process* of innovative design can be assessed on the basis of its ability to overcome these four fixations. Our assessment criteria are thus enhanced by the process, criteria that we have used in the Ideo case study of Chap. 4, and that we shall use in the KCP case study later in this chapter. They are summarized in the figure below.

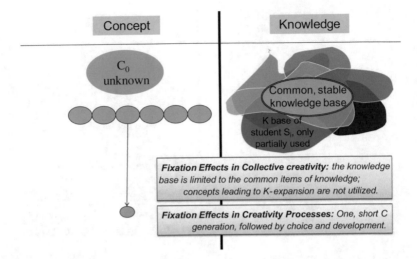

Fig. 5.8 Collective fixation in C-K theory. The group reinforces the fixation, in both C- and K-space. In C-space, reticence in suggesting expansive Cs, for which the relational and organizational costs appear somewhat high (social judgement). The collective tends to favor mechanisms of "choice" ending up with the premature selection of a single pathway. In K-space, reticence in making use of bits of knowledge that are not available to everybody (cost of sharing knowledge), reticence in rediscussing the rules established by certain experts, and reticence in creating new rules

Table 5.1 The four main effects of fixation that the resources of innovative design must help to overcome. Also given are the assessment criteria associated with an innovative design process attempting to overcome these four fixations. Observe that the first column contains the V2OR criteria

	Cognitive factors of the fixation effect	Social factors of fixation effects: overcoming the rule breaking resistance
C-expansion ($K \rightarrow C$, $C \rightarrow C$)	$\rightarrow$ **Cover the whole conceptual potential of the initial concept?** *Criteria*: fluency, variety, originality;	$\rightarrow$ **Involve and support people in a rule-breaking process?** *Criteria*: well-being, participants satisfaction (i.e. feel comfortable in C-space), number of broken rules
K-expansion ($C \rightarrow K$, $K \rightarrow K$)	$\rightarrow$ **Activate, acquire and produce relevant knowledge?** *Criteria*: knowledge variety, new "value" and "robustness" criteria, identification of K gaps, K-reordering	$\rightarrow$ **Manage collective acceptance and legitimacy of rules (re) building?** *Criteria*: experts commitment, stakeholders commitment (top management, designers, users, partners,…) to further exploration (research program…)

5.2.2   The Structure of C and K-Spaces Conducive to Innovative Design

What are the structures in C and K-space that are favorable to reasoning in innovative design?

5.2.2.1   Structure of K: A Connected Space (in the Sense of the Splitting Condition) that can be Re-ordered

Recall that in rule-based design, the K base was structured using conceptual models. In innovative design, we shall come across the logic of modeling but we will also discover some new virtues in the models.

Two critical properties are expected of the K base but do not appear at first sight to be readily compatible: helping with expansive partitions ("out of the box"), and helping to restore order to the knowledge (K-reordering) ("rebuild the (new) box"!).

Assisting with Expansive Partitions

On the one hand, the knowledge base should be conducive to expansion—it needs to ensure that it can "get out of the box". It must have "holes" (if we assume that items of knowledge do not have mutually established relationships, trying to link two unconnected islands of knowledge is equivalent to formulating a concept in C-K theory) and it must allow these to be identified. "Seeing the holes" means knowing what we don't know, being aware of residual contradictions, open questions. This is having "*the state of the non-art*".

The work on design and mathematics (Sect. 4.1.4.3) showed the importance of forcing. We then also showed the importance of the *splitting condition*, which ensured that the K base was structured in such a way that the generic filter (the new element) was indeed "new", separately from all that "was known" (see Jech 2002; Le Masson et al. 2013a). We shall look at this in greater detail (see Fig. 5.9).

The splitting condition ensures that the generic filter is no longer in the initial model. This condition is as follows: for any constraint p there are always two incompatible constraints q and q' which refine p. This condition is not satisfied in two characteristic cases:

(1) for a constraint p, there is a unique series of constraints q which refine p (determinism)

(2) for a constraint p and for q and q' refining p there exists an r which refines q and q' (q and q' are said to be compatible) (q and q' are modular, or the addition of q or q' is independent of the choice of r).

This idea of splitting condition provides some characteristics for the knowledge bases in engineering design. We are therefore led to distinguish two structures in the K base depending on the type of engineering—rule-based engineering,

- Definition: splitting condition: for every p, there are q, q' such that q<p and q'<p and q, q' incompatible
  - There is no constraint r<p that is "insensitive" to q vs q' choice
- Not SC: there is one p such that:
  - Either: there is always one unique q
    - if p, then q1, then q2,... = a deterministic chain of conditions
  - Or: if there is q, q', then they are compatible (ie: there is r such that r<q and r<q')
    - = modularization before r

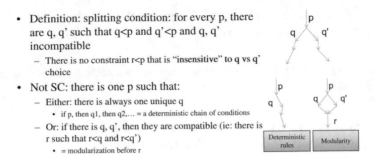

Fig. 5.9 The "splitting condition" and its interpretation: non-determinism and non-modularity

which disregards the splitting condition (modular deterministic base) and innovative engineering, for which the splitting condition is satisfied (non-modular and non-deterministic base).

Let us examine the meaning of a K structure[3] for which the splitting condition is satisfied, i.e. which is neither "modular" (no independence) nor deterministic:

- *Non-modularity*: any attribute P_j, derived from K to be added to the concept, draws in others that are specific to P_j, i.e. different from those that would have been obtained without P_j. The new attribute cannot be added as an "ornament" without propagation or consequences. The knowledge base must therefore be sufficiently "enmeshed", all items of knowledge being "linked" to all others. Several types of knowledge have this property:

  - Conceptual models are useful since they link often heterogeneous attributes to one another (Ohm's law U = R.I: voltage is related to resistance and current) and hence allow quick relationships between high level attributes to be established.
  - However, other types of proposition also have this property of non-independence. Hence a reasoning that makes use of knowledge about a "unique usage situation" or an "environment of use" or "an unusual life cycle": this is definitely not a conceptual model since the economic effect expected from the conceptual model does not necessarily exist—there will be too many attributes. Nevertheless, this type of knowledge introduces a logic of non-independence: in the singular situation, certain attributes will concatenate, allowing languages to be changed rapidly. For example, while working on novel image sensors, the designers at STMicroelectronics worked on the following situation: a camera mounted "on the helmet of an all-terrain cyclist on a mountain path" The situation in quotes allows the following partitioning attributes to be concatenated: robustness, types of

---

[3]Note that in the remainder of the text, we often use the expression "K base" to denote the set of knowledge in K-space (for a certain player, at a certain moment).

mechanical shock, types of inclement weather, types of image to be taken (all-terrain tour), types of image exchange (between the participants on the tour, during and after the tour, etc.), security (theft of helmet, etc.), health (weight, help in the event of an accident, etc.), etc.

- Metaphors, analogies, comparisons have the same mutual circulation property (heterogeneous, distant, unexpected) via logical relationships.

- *Non-deterministic*: nevertheless, however, the new attribute does not necessarily define the list of attributes to be added; on the other hand, at any instant, using Pj, several pathways are possible. This property means that *purely deterministic behavior can be avoided*: there is no relation of the form "if A then B", no absolute constraint. Any partition opens a new space of partitions; however, it does not automatically imply a unique consequent attribute.

  - This property is often utilized by conceptual models: the same attribute can be ensured in several ways (for any voltage U_0 there must be an associated current I_0 but the abstract nature of conceptual models leaves alternatives for obtaining a certain current I_0), or the model itself offers several DPs for the same FR (one can play with R or I to get U_0); or several possible models may exist (playing in particular on the domains of validity): for example, Ohm's law works as long as leakage currents can be ignored, but a partition can be constructed on the basis of "negligible leakage currents" or "non-negligible leakage currents"). Setting out the domains of validity of conceptual models is a normal strategy in engineering sciences for avoiding "deterministic" effects. (the world is not "always" according to law L; under *certain conditions*, and outside these conditions, it may obey a (sometimes poorly understood) law L').

  - *Taxonomies (even disordered), plant collection, and especially counterexamples or anomalies*, are also the means by which a knowledge base can satisfy the condition of non-determinism. They imply that "not every A is B" but may be B', B" or not-B, sometimes just "exceptionally" in some unexpected or inexplicable way.

Encouraging the Restoration of Order (K-Reordering)

The knowledge base must also contribute to the K-reordering. Let us first examine this K-reordering in the case of complex numbers or that of forcing. There are two phases of operation:

- An *extension* phase: in the case of forcing, we construct M[G], the new model which "starts off" from G creating all the "combinations" of G with the "elements" of M; in the case of complex numbers, starting from X (the unknown satisfying $X^2 + 1 = 0$) we construct all the polynomials of type $a_0 + a_1 X^1 + \cdots a_n X^n$ with real coefficients. This type of extension relies on the "operations" preserved during the extension: in forcing it is a question of inclusion, intersection, etc. and more generally, the set of ZF axioms are preserved and hence allow the creation of new elements; in the case of complex numbers, we can create polynomials in X since addition and multiplication are allowable in the field of polynomials with real coefficients.

- a phase of *restoration of order*: all "new" objects are not actually different; some are redundant. Hence $X^2 + 2$ seems to be new but since $X^2 + 1 = 0$ we have $X^2 + 2 = 1$ etc. which is not new! Similarly $X^4 + 2X^2 + 2 = (X^2 + 1)^2 + 1 = 1$ and so on. We therefore have *rules of distinction and recognition* and more generally, *relational rules between new objects* that have to be respected. These rules themselves encourage design in the sense that it is they that allow *the structure of the new set of designed objects to be reconstituted*. In the case of complex numbers, these rules are addition and multiplication in the field of complex numbers (hence we show that all operations on polynomials are equivalent to operations on equivalence classes of type $a + bX$ where a and b are real, addition being defined by $(a + bX) + (a' + b'X) = (a + a') + (b + b')X$ (which might be intuitive) and (less intuitively) multiplication being $(a + bX) \cdot (a' + b'X) = (aa' - bb') + (ab' + a'b)X$. In this second phase, we do not just check a great number of objects but *we also make sure of the overall cohesion of all these objects and their ease of use*.

In the case of non-mathematical objects, we find these properties for K:

- The extension properties come from the ability to combine the new element with the known: this presupposes, for example, a capacity for *rapid validation* (the legless camping chair is also validated for festivals, the garden, young people, old people, etc.); however, after validation it might be a case of a new form of design, rule-based this time: it is not the new chair that is validated in multiple environments, but the rules of an associated rule-based design are constructed such that variants compatible with of the legless chair can be efficiently designed. Having design languages available that are already well-structured can ensure a faster extension of the new object over a wide range of situations and uses. More generally, the issue of innovative design is not just the rapid generation of the new object but also the *generation of the associated languages of rule-based design*.

Twizy—an example of how the rules of design are regenerated in vehicle mobility

Twizy, Renault's new electric vehicle, is an unusual quadricycle; its design was real success. It also illustrates the design effort required to ensure compatibility with the ecosystem of mobility and the new vehicle. Let us give a few examples:

- choice of an electrical system compatible with the 220 V power supply sockets in the home.
- compatibility with the electric vehicle systems already available, e.g. Autolib.

- redesign of compatibility with the highway infrastructure: request for homologating the Twizy for driving on urban freeways.
- redesign, with internal Renault quality and after-sales service, marketing rules in certain countries.

In the Twizy case and Renault generally, the ability to extend or redefine quality standards is a major asset of innovative design. This ease of extension makes for the quality of the knowledge base in an innovative design regime.

Rules of composition, recognition, conjugation or validation thus appear as rules that are *preserved* (or effectively extended) by innovative design. A knowledge base favorable to innovative design will succeed, for example, in *very soon setting out its invariant part* (e.g. the conditions for validating the new definition; thus we might require that the "legless chair" be defined in the four rule-based design languages for chairs, etc.).

With regard to these preserved rules, it is understood that they may run counter to the previous *partition* criteria: by defining the invariant ontology we might be tempted to avoid propagations and strengthen determinisms (invariants as well). "Setting the definitions in stone" runs the risk of putting a brake on expansion.

The contradiction disappears if we note that these invariant models are also the *conditions for a quick and easy restoration of order* and in this way contribute to the capacity for collective innovative design. They work less as a constraint than as a general framework; a general framework that interprets the new as a singularity, not as a disruption, as an anomaly rather than an aberration. They constitute extended generic forms *which are no longer only the commonality of known things but also the generic of the unknown*.

A good example is Mendeleev's periodic table, which predicted the possible structure of as yet unknown atoms.

In conclusion, these structures in K space encourage both de-fixation (the ability to avoid unique definitions) and the capability for restoring order.

5.2.2.2 Structure of C Space: Provocative Examples and the Notion of Heredity

In rule-based design, we studied generative models as resources for giving structure to the unknown (and resisting forms of fixation: short-circuits in reasoning, the temptation to re-use known solutions too directly and inappropriately. Recall that in C-K theory, generative models are, a priori structures of concepts (see interpretation of the theory of systematic design using C-K theory, Sect. 4.1).

In innovative design, we shall see that the C space may also be structured such as to encourage de-fixation, with two objectives: encouraging a form of *exhaustivity* in the exploration of the unknown, and making disruptive situations manageable (see the two effects of fixation in C space identified in the context of C-K theory above).

De-fixating Using Provocative Examples—Experiments in Cognitive Psychology

Recent work in neuropsychology has established that individual de-fixation can be driven in particular by *provocative examples*, which push individuals in creative situations to put forward original solutions (Agogué 2012b; Agogué et al. 2011, 2014a). They show that the provocative example works *provided it is chosen outside the area of fixation*, and under these conditions the example allows individuals to break out of this area (though not get stuck in reproducing a new fixation with variants on the provocative example); conversely, an example chosen *in* the area of fixation will, on the contrary, reinforce the fixation, and individuals who can suggest ideas outside this area will suggest even fewer. Helping to overcome fixation in C space may therefore consist of *fabricating* "good provocative examples". Further, designers must be helped to fabricate their own provocative examples. The question is therefore one of giving the C space a particular structure that allows the designer to generate his own "provocative examples", even adjusting the desired level of disruption (Figs. 5.10 and 5.11).

Adjusting the Level of Expansion—The Idea of Heredity in Design

A very favorable C-tree structure rests on the idea of *heredity in design*. For a concept C_0, we say that an attribute has a strong heredity if its negation (with the expansive partition refusing to associate this attribute to the concept C_0) leads to high costs associated with restoring the knowledge base. These are the oldest codons of the "genetic heritage" of the object. Conversely, the most recent codons of weaker heritage will require less effort if they are redesigned.

The logic of heredity consists of structuring the concept tree by making use first of all of partitions with a strong heredity (attributes for which negation would lead to the highest reorganizational costs in K space). Proceeding thus on the entirety of the object's "genetic heritage" we find a tree structure with a spinal column that departs from the concept and concatenates the restrictive attributes from the most hereditary (the hardest to question, at the root) to the least hereditary (see Fig. 5.12). For each restrictive attribute along this axis, we find, perpendicular to the axis, an expansive partition whose expansive reach depends on its location on the tree (strong expansion expected for partitions close to the root, with weaker expansion as we move away) (Figs. 5.12 and 5.13).

We give below an example of a C tree structured by heredity around the concept of "an engine for a green aircraft for 2025" (source Brogard and Joanny 2010) (see Figs. 5.12 and 5.13).

Observe that this structure automatically generates expansive concepts which can be taken as provocative examples for multiplying the original ideas.

We stress the double role of heredity:

(1) heredity gives structure to imagination, more accurately a form of partial order in the unknown. It outlines a set of alternatives in the unknown, ordered according to their expected expansion in K space. Note that all these paths are expansive

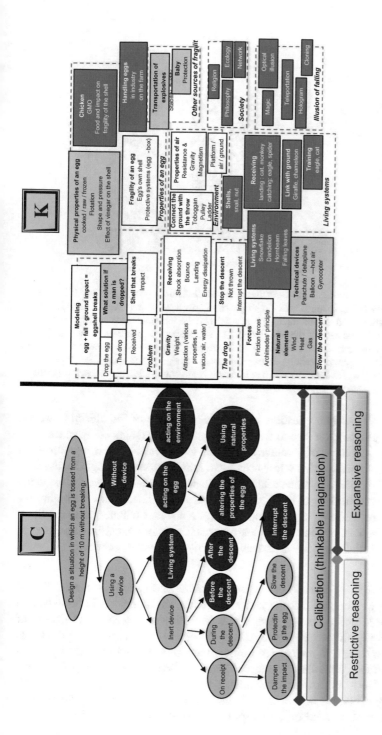

Fig. 5.10 Measuring fixation (Agogué et al. 2014a). Using C-K theory it is possible, in a creative exercise (here, "design a solution whereby a hen's egg will not break if dropped from a height of 10 m"), to construct a priori the set of all imaginable solutions and within this set to distinguish the solutions involving a restrictive reasoning (problem solving using common knowledge around the data of the exercise in creativity) and those involving an expansive reasoning. It is possible to analyze the results obtained empirically from one or more individuals using this reference. For example (not illustrated here) 81% of the ideas dreamed up by an individual are restrictive

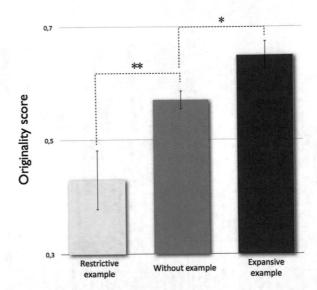

Fig. 5.11 Two different types of example: fixating and de-fixating (Agogué et al. 2014a). With control conditions satisfied, (control group with pre- and post-test), scores were obtained for three groups: the first group had no example, the second had an example chosen from among the restrictive solutions (in the case of the falling egg: "one participant suggested a parachute") while the third had an example chosen from among the expansive solutions (in the case of the falling egg: "one participant suggested training an eagle to catch the egg"). The scores for originality in the three groups are given above. There is a statistically significant difference between groups. *: $p = 0.05$ (Type I risk 5%); **: $p = 0.005$ (risk 0.5%)

since they all call into question some classic attribute of the object; however, questioning them gives rise to fairly considerable propagation in K space.

(2) Via duality, the logic of heredity also states what needs to be reconstructed at each level of partition: very low down, we know that quite a significant set of attributes is conserved (stakeholders, the business model, usage models, value chains, the architecture of the aircraft, technical principles, etc.); the higher we go, the more we have to redesign. In a more abstract manner, the hereditary structure itself allows the invariants on which opinions must be expressed to be stated (redefining the technical principles, aircraft architectures, value chains, etc.).

Note that *the logic of heredity in C space functions as a generative model*. Recall that in Chap. 3 we showed that the idea of a generative model could be extended beyond rule-based design to any design rule that allowed a certain criterion H of design reasoning quality to be improved. In the case of rule-based design, H was "minimize expansions". In the case of innovative design, H could be "the optimum level of disruption". With the generative model it becomes possible to *adapt the level of disruption* to the resources available (time, the ability to produce knowledge, the ability to involve new designers, etc.), to the company strategy, to external constraints (the need to fend off a disruptive proposal from a competitor, etc.).

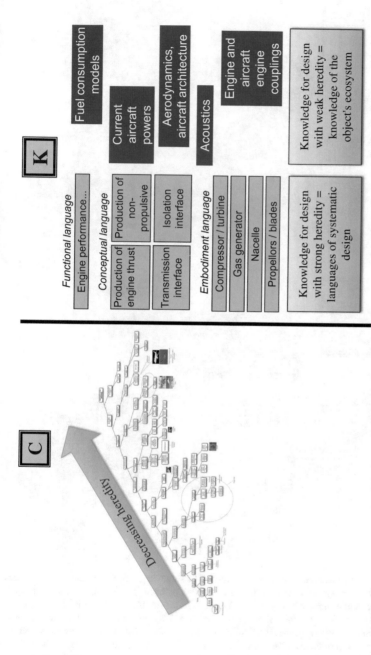

Fig. 5.12 C-K graph with heredity structure (general graphic). The complete C-K graph is given here. The K base on the *left* shows the knowledge bases for the rule-based design of an aircraft engine with three levels: functional, conceptual, embodiment; to the *right* of the K base is the knowledge associated with the highest level partitions (typically, models of aircraft architecture, fuel consumption models for an aircraft, fuel consumption model per airline operational model, dependence on how the aircraft is flown, etc.). C graph: different variants structured according to heredity. *Bottom left*, the current engine slightly less thirsty; the higher we go, the more we have to revise the design parameters

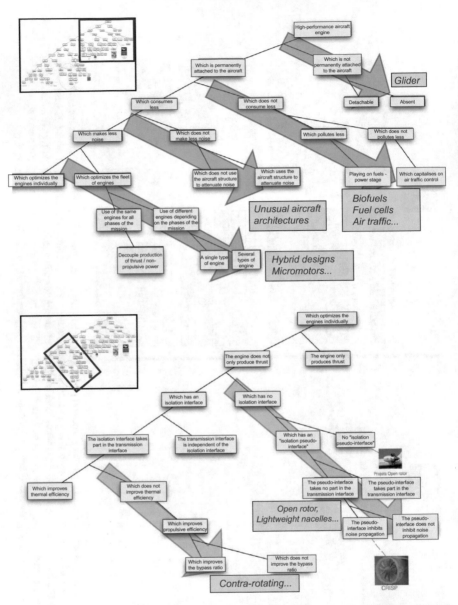

Fig. 5.13 C-K graph with heredity structure (details). The three diagrams above show some details of the C tree. The first (*top*) gives a tree with the least heredity; paths explored are glider (no engine…), new airline operational models, biofuels, original aircraft architectures, hybrid engines or micro-engines. The bottom graphic illustrates "rule-based" solutions, all engine parameters optimized (addition of stages, gear-driven turbofan, triple spool and counter-rotating turbines, etc.); finally, the center graphic gives the paths that preserve the architecture of the aircraft and of the aeronautical and airport ecosystem, but which radically alters the engine (e.g. uncowled engine of the "open rotor" type)

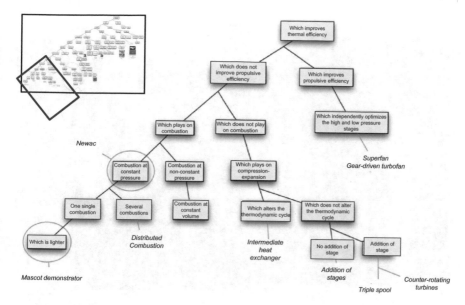

Fig. 5.13 (continued)

5.2.2.3   An Example: Courses at the Bauhaus

The schools of art and design are places where, for some considerable time, the question as to the conditions conducive to innovative design reasoning has been posed, in particular the question of appropriate "structures" of knowledge (and concepts). These schools trained their students so that they would become creative individuals. Obviously, the question of the "bases" to be provided for the creative endeavor lay at the heart of this type of training. Although the schools were there, and although creation is not an innate "selectable" talent, the courses necessarily fell back onto "supporting" knowledge and reasoning.

The most famous of these schools, the template for many, was the iconic example of the Bauhaus. A study of the introductory courses given by Itten (1975), Klee (1966) and Kandinsky (1975) reveals a teaching method geared towards a very ambitious innovative design (teaching students to conceive "new styles for their time" and not to "follow the rules of a particular style") and very high standards in teaching content (hence Itten's statement that "the imagination and creative ability must first of all be freed and strengthened", "the objective laws of shape and color (taught at the Bauhaus) help to extend a person's creative gifts", "liberating the study of color harmony from its association with shape", etc.). Here we shall give only a few of the essential elements of the content of these courses (for a study in greater depth, see Le Masson et al. 2013a).

Itten's course can, for example, be illustrated by the case of learning about textures (see Fig. 5.14), but other topics (color, form, line, etc.) would have the same logic.

It is striking to observe that Itten allowed his students to constitute a knowledge space with the remarkable property that it satisfied the splitting condition. This means that Itten suggested that his students construct a "forceable" knowledge base,

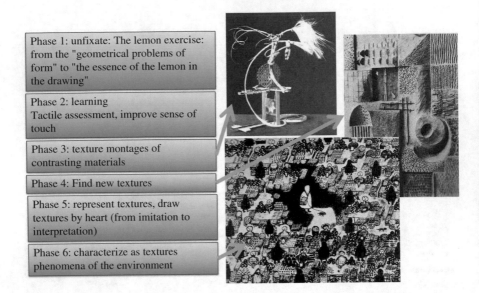

Phase 1: unfixate: The lemon exercise: from the "geometrical problems of form" to "the essence of the lemon in the drawing"

Phase 2: learning
Tactile assessment, improve sense of touch

Phase 3: texture montages of contrasting materials

Phase 4: Find new textures

Phase 5: represent textures, draw textures by heart (from imitation to interpretation)

Phase 6: characterize as textures phenomena of the environment

Fig. 5.14 Learning about textures at the Bauhaus: structuring a K base that satisfies the splitting condition. Exercise 1: de-fixation, avoid likening the object to its geometric form; exercise 2: learning, improving observation and perception; exercise 3: using texture as a means of design, exploring the generative capacity of combinations of textures (smooth roughness, gaseous fibrousness, shiny matt, etc.); exercise 4: research, with exercise: in the figure, all the textures are obtained from the same wood; exercise 5: going from imitation to interpretation, conjugating texture with its own personality; exercise 6: using texture as the main mode of expression, avoiding texture as ornament, exploring new formal relations between texture and composition

i.e. which would lead the creative individual to an original and new creation, beyond the normal combinations of the elements of the initial set. We see how Itten's exercises satisfy the two properties of the splitting condition:

- Non-determinism: faced with a concept, a student can no longer use the deterministic law. Because of the variety of contrasts, there is no longer a direct link between a color with a material then with a texture. Itten fought against the "laws of harmony" and clichés that attempted to define relationships (e.g. fibrous wood = warm, smooth metal = cold, etc.).
- Non-independence: for all that, all attributes and all combinations are not equivalent. This is not a matter of relativism. As far as Itten was concerned: "subjective taste cannot be enough for all problems of color". Relativism suppresses previously valued differences. Although texture may only be "secondary" or "modular", even though all work on wood might be similar, work using fibrous wood can no longer be distinguished from work on smooth wood. Hence Itten taught that one texture cannot be added independently of other textures and attributes. If a scene or montage could be made with texture, then the scene or texture are no longer "insensitive" to the choice of texture. For Itten any attribute (color, texture, material, etc.) influences the entirety of the work, and is propagated to all other aspects.

Take especial note that it is not the variety or the originality of knowledge that plays an intrinsic part (the course deals with fairly standard notions such as texture, color, etc.), but that it is the original combination of elements that gives rise to a creative movement.

Non-determinism and non-independence are thus at the very heart of Itten's teaching. We highlight the fact that these two properties are somewhat different from a logic of "machine elements" (modularity) or from the "laws of engineering science" (determinism) as could be taught in Itten's time in the German technical schools. We may also remark that Itten did not teach using a stable knowledge base but instead taught his students to build their own knowledge base meeting the two critical properties of non-determinism and non-independence.

5.2.3 *Strategies*

Recall that in rule-based design the data in a rule base very largely determine the definition of the CQT targets and their selection depending on the expectation criteria of the net present value (see Chap. 2) and the coverage of a functional space (Chap. 3), incorporating the uncertainty both by guiding projects (reduction of technical risk in a known market) and by market segmentation (reduction of commercial risk in a known technical space). Allocation of resources relies on the logic of static returns. In innovative design, strategies must take account not of uncertainty but of the *unknown*, and must also include the logic of dynamic returns. It is to this that we now turn our attention.

5.2.3.1 Issue: Risk Management in the Unknown (Not Risk Management in the Uncertain)

It is important to distinguish the unknown from the uncertain. In rule-based design, performance is measured in terms of the deviation from the initial target, with given resources. Since this offset is generally modeled as an uncertainty, project performance is equivalent to a reduction in uncertainty on the basis of known elements (see Chap. 2), so that it converges towards the expected value of the NPV. Moreover, uncertainties are decoupled, with market pull on the one hand and techno push on the other—and the reduction in uncertainty is built on known elements (no technical exploration without market assumptions and no commercial exploration if the technical know-how is not available). Finally, thanks to decision models in the uncertain, we know that there is value in actions that reduce uncertainty (see Sect. 2.2).

This logic is no longer applicable to innovative design, since this is no longer about the management of uncertainty:

1. On the one hand, "uncertainties" build up: the market and technical know-how are explored *simultaneously*. When they are known, it is these elements that structure the "techno push" or "market pull" exploration. Given the reduction in

uncertainty between the two, "techno-market" decoupling avoids having to explore the different forms of interdependence between technology and market (the discovery of a new client-value having the potential to explore new phenomena, themselves opening up the exploration of new client-values, etc.).

2. On the other hand, this is not a question so much of uncertainty as of the unknown: uncertainty equates to the probability of occurrence of an event, but an event for which its *nature* is known—if there is an 80% chance that it will rain tomorrow, we cannot be certain that it will rain, but we do know what rain is like. The unknown involves the nature of the event itself: if we wonder about life on Mars or exoplanets, it is *the very nature, the form of this life*, that is unknown. In innovative design, the logic of design enables the functional languages (hence potential markets) to be revised along with the design parameters (hence technologies). Thus we have a *double unknown*, market and technology.

The expected performance is therefore *not a reduction in uncertainty but an ability to structure the double "market-technology" unknown* (under the constraint of resources).

We shall study various strategies by extending the framework of decision theory in the unknown (as we saw in Sect. 2.2) to two innovative design situations.

We know that the general strategic model for exploring the unknown will consist of initiating design spaces in which learning is possible, and consolidating what is learnt in terms of value management to better define subsequent exploration and development (see Sect. 2.2).

Among these design spaces, there are multiple forms of possible couplings, from the sequential logical processes taking advantage of successive learning processes (dynamic returns) to parallel forms of exploration playing on the complementarities and interdependencies among the paths being explored.

In this general model we can distinguish two highly contrasting forms of risk management depending on the state of the knowledge base. The first is the most intuitive: this involves managing the double unknown on the basis of very limited knowledge (or more generally, we wish to greatly extend the K base in unknown and not easily determinable directions). The second might seem paradoxical: these are double unknown situations in which the known is, however, very important— C-K theory immediately lifts any ambiguity: there may be a willingness to proceed to expansive partitions while the knowledge base is very rich. We shall study these two situations before moving on to deal with more complex cases.

5.2.3.2   Strategies for Structuring the Unknown

The first form involves assuming that not only is there a double unknown but that the unknown part is very limited. Although there may be a lead, it is generally insufficient to initiate a project. Hence *alternatives must be generated and the best strategy found*.

This situation is very different from the test as modeled in classical decision theory, in which the nature and possible states for the trial are all known; here we have only partial knowledge of the state of nature and possible results.

In the case of uncertainty reduction, the best test was the one that provided the most information on the states of nature—formally it should ideally have been determined by the states of nature, i.e. totally correlated with the states of nature (for specialists: *in* the hyperplane of random variables, functions of the states of nature; or as near as possible). However, in this particular case we can assume that the "lead", i.e. the "idea", is a form of *"carrying out the first test"* which provides some information on the states of nature (the forecasts said it was going to rain, but we have no idea of the associated probability nor of any possible states of the weather). This first idea contains some implicit assumptions (a particular market m_i, a specific technology t_j, etc.) which can be likened to the realization of random variables (markets m_1, m_2... m_i... m_n; technologies t_1, t_2,... t_j... t_p). A further exploration in this case consists of *conceiving the space of possible trials* (conceiving the possible markets, possible technologies, etc.). Consequently, we seek to generate the set of possible trials by assuming that we have available the realization of one of these possible trials. In other words, it is as if there were a projection (the concept *given* a market or technology, etc.) and that we were seeking to conceive the origin of this projection. Hence we have to carry out the *inverse of a projection*, i.e. an "orthogonal" *expansion* of the available realization; we are no longer trying to find out as much as possible about residual uncertainties (e.g. those associated with the feasibility of the initial "idea") but rather to generate a space of alternative ideas; to that end we need to look for those that are "orthogonal" in direction to the known idea.

This reasoning justifies the *"crazy concept"* logic: this path is the one most "orthogonal" to the known paths. Other paths are also interesting, but the learning that comes with them will be in some sense "redundant" in relation to what is already known. (see Fig. 5.15 the difference between the case of the trial to reduce uncertainty and the exploration required to provide structure for the unknown).

Some examples:

- The WITAS case by SAAB Aerospace (see Le masson et al. 2010b, Cambridge university press) is a typical case of exploring the Crazy Concept: thinking about future military drones during the 2000s, "a traffic surveillance helicopter" was studied (crazy at that times since it seemed unlikely that anything can be learned about the flight problems of a robot aircraft in a military situation);
- Urgo laboratories wanted to work on smart bandages, so they explored a crazy concept: "wound diagnosis—before knowing that a bandage was required" (crazy since it seemed unrelated to learning about the protection of wounds) (Carreel 2011a);
- in order to work on future generations of sensors for digital cameras using CMOS technologies (which had been mastered by the company's factories), STMicroelectronics initiated a study of "a camera sensor which does not use CMOS technology".

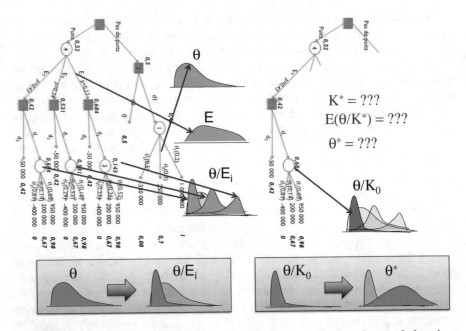

Fig. 5.15 Decision theory in the uncertain versus decision theory in the unknown. *Left* we have simply rotated the decision tree through one quarter turn, this time with three states of nature. θ represents the states of nature with a schematic distribution; E represents the possible states of the trial with a probability distribution; for each of these trial results there are new probabilities for the states of nature θ/E. On the *right* is the situation in the unknown: we know a trial K_0 and the probabilities of the states of nature associated with this trial θ/K_0. However, we do not know the other possible trials K*, the other possible distributions θ/K* and the probability distribution over θ*. In the case of uncertainty we are looking for an E that is most highly correlated with the known θ (this allows the greatest reduction in uncertainty); in the case of the unknown, we seek a K* that is the most "orthogonal" to the known θ/K_0 since the latter is the least correlated with the known; in other words, the entire exploration will be dedicated to the unknown and not relearning what is already known. The two diagrams illustrate the effect of exploration on the distribution of the alternatives: in the event of a reduction in uncertainty we pass from a "broad" to a sharper distribution (reduction of standard deviation); in the case of exploration of the unknown we pass from a sharp to a broader distribution (enlarging the space of alternatives)

These crazy concept strategies would go on to give rise to extremely powerful effects of structuring the unknown (we shall investigate the case of non-CMOS sensors in Sect. 5.3).

5.2.3.3 Generic Technology Strategy

A second family of strategies attempts to take account of the fact that there might be a double unknown situation while there may be a great deal of knowledge from the start about a set of leads. The problem arises from the fact that this broad set is very "flat", i.e. no lead opens up a sufficiently high market expectation to justify an initiation.

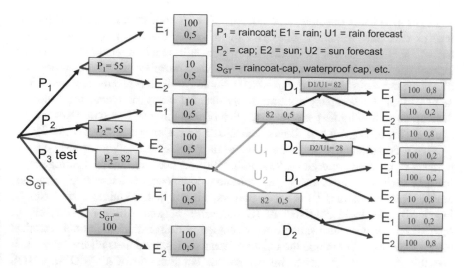

Fig. 5.16 Generating a "generic technology" alternative in a decision tree

We are apparently in the Savage decision theory context, but no distinction is possible between states (and no learning process to inform the choice: provided there are many states, there is no trial strategy that sufficiently reduces the uncertainty). As in the previous case, it is then a matter of conceiving a new alternative while taking advantage of the known alternatives and, in particular, combining different scenarios among all of them. We shall see that this involves conceiving a generic technology which is compatible with the greatest possible number of states of nature.

Let us take another look at the example given in Chap. 2.2, Sect. 2: let us assume that alternative P_1 is to sell raincoats, and that alternative P_2 is to sell caps. If the probabilities are 50–50 for rain-sunshine, the expected utility is 55. We can always start a study to understand weather forecasts; we have already seen that this study comes out at 82–55. However, we may also consider generating the alternative P_4: designing a combined raincoat/cap. This alternative comes out at 100 (see Fig. 5.16).

The alternative thus generated *creates interdependencies* between alternatives intended to be independent; they create these interdependencies depending on the value associated with the combination of value thus obtained about the states of nature.

The example given here is simple. In practice there are often many alternatives and the markets associated with their combinations are not obvious. However, this type of strategy is the basis of the design of generic technologies, and not just one market can be addressed, but several; even if each of these markets is of low probability, if the technology allows a sufficient number of them to be addressed, then the probability that no market will emerge becomes very small. Let us assume a generic technology capable of addressing 10 markets, but 10 markets of very low probability (15%). The probability that at least one of these markets will emerge is $1 - (1 - 0.15)^{10}$ i.e. greater than 80%. *Although none of these markets may be well "known", the technology has an almost guaranteed market.*

Let us give some examples: the domain of semiconductors is one in which this type of reasoning regularly gives rise to industrial success. Historically, *planar technologies* are generic technologies which still make it possible for silicon technologies to be compatible with a vast number of applications. At STMicroelectronics, *MEMS* is a generic technology which is responsible today for a large part of the company's earnings. A detailed study of recent technological developments highlighted several examples of reasoning of this type (Kokshagina et al. 2012a, b; Leguay and Rousseleau 2012; Barthelemy and Guémy 2012).

This was particularly so for high data-rate wireless technologies: the potential markets were many, and all of low probability (at least six highly heterogeneous markets such as vehicle radar, fast wireless download on interactive terminals, optical communication, hard drive players, airport wifi terminals, etc., can be imagined). The logic consisted of *not choosing* between these markets, but of seeking how to use the available technologies to design that additional technical element that would combine the existing technologies to address all the markets. It was this that led the teams to investigate the combination of a CMOS-biCMOS double processor with high level so-called "back-end" (analog) technologies. The design of an architecture and process to make these disparate building blocks compatible made it possible to access all improbable markets—the technology was finally marketed in the first instance in the automobile domain before then being used in other domains as well.

The studies at STMicroelectronics also show the dangers in this industrial context of decision type reasoning in the uncertain: encouraging the most promising market leads to too much development in this unique market; the technology obtained turns out to be quite specific; if the market has not materialized (an event of relatively high probability in situations of disruptive innovation), the possibility of re-use is very limited. Cases of this type show the limits of a traditional *try and learn* strategy.

5.2.3.4 Hybrid Strategies—Lineages

These two basic strategies can be combined, and therefore give rise to complex strategies. This combinational logic can be illustrated by a simple case (here we are extending part of the Saint-Gobain case study Le Masson et al. 2006); we stress the reasoning behind risk management, the published part stressing the organizational aspects (Le Masson et al. 2010b).

The innovative field focuses on athermic windshields. The company identified three technological concepts (see Fig. 5.17) at different stages of maturity.

Traditional risk management reasoning (rule-based design) leads to a selection of the alternative with the highest expected utility. However, this logic is only valid if the alternatives are independent and the subsequent design does not alter the space of alternatives. The C-K diagram (see Fig. 5.18) shows a large number of expansive pathways with potential capacity for learning (white on dark background in K base) (furthermore, this C-K diagram shows a good V2OR structure). Hence it is preferable to apply an innovative design strategy.

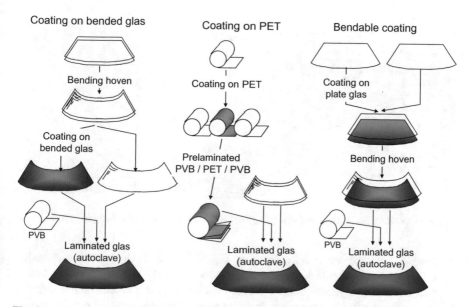

Fig. 5.17 Three technologies for manufacturing windshields with athermic coatings

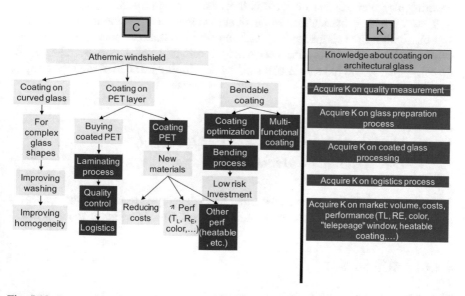

Fig. 5.18 Innovative design strategy: managing interdependencies within a portfolio of alternatives

The logic of the crazy concept and that of generic technology led, on the one hand, to the preservation of original avenues (holding onto the less well known paths) and on the other hand, to attempts to combine the various approaches with one another. It was for this reason that the company actually initiated three projects

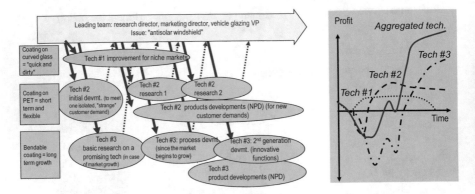

Fig. 5.19 Value management and design space for the management of a complex product line

in parallel (for the same budget) (see Fig. 5.19): a slightly mad low-budget approach using a conformable coating; a somewhat "regulated" low budget approach (flexible layer on curved glass) which was able to create some valuable knowhow for the three approaches (e.g. learning about the product quality criteria and their validation); and a middle way completing certain learning exercises and covering a greater number of markets than the safer approach.

Thus we have a particularly sophisticated value management and design space scheme, with a financing trajectory that allowed the three technologies to be conceived while at the same time being more profitable and less exposed than the trajectory that would have been obtained by developing just one of the technologies (see Fig. 5.19).

5.2.4   Tools and Processes—KCP, C-K-Invent, C-K References Method

Several tools, methods or processes have been developed to support the reasoning and strategies we have presented do far. We shall discuss a few of these below.

5.2.4.1   KCP Method

The KCP method enables a large company to engage in collective reasoning for innovative design. The difficulties of collective processes for "disruptive" innovation are well known: the exploratory seminar tends to produce a consensual, but fixed, vision (with little disruption); The "commando" is capable of serious disruption but involves a limited number of designers who often struggle to convince and reach beyond the circle of the first participants. For all that, it is not feasible to engage in long training programs so that all the participants can master the theories of innovative design reasoning. The KCP process "linearizes" a C-K reasoning and enables many designers to escape the effects of fixation.

Such a process is set up to build a field of innovation. The phases involved are as follows (we shall present the methodology and its assessment in greater detail in a triple workshop in this chapter):

1. The aim of the K phase (knowledge phase) is to share knowledge and prepare for the emergence of novel concepts. It involves no creative effort and discourages any precipitate search for solutions. So that experts and non-experts can work on novel concepts together they have to share any knowledge (technical, economic, commercial, legal, scientific, usage, competition, etc.) that might indicate obsolescence in the current solutions and which suggest new potential for development. If well chosen, this contribution of knowledge erodes any received wisdom and builds a common cognitive breeding ground which is sufficiently rich that new concepts can then be easily formed. This phase is always multi-disciplinary and can involve outside partners (users, customers, suppliers or others). Besides a "state of the art", this phase is also a *state of the non-art*. It attempts to highlight "anomalies", provocative examples, paradoxes or counterintuitive cases. Finally, it endeavors to cover the fields of competence necessary to deal with a later field of innovation.

2. The C Phase (concept phase) organizes a creative procedure "steered" in accordance with precise rules. The "collective" is separated into groups that need to work on the basis of "projector concepts". This involves surprising and highly contrasting suggestions that illuminate the innovations to be developed, just as projectors at night extend the visible space. These suggestions are formed in accordance with rigorous theoretical principles, guaranteeing creative work and the greatest possible innovative power. On the basis of a projector concept, each group explores and puts forward original alternatives. The group must also identify any gaps in its knowledge and work out new avenues of research. In this phase, the groups make a mutual presentation of their results, thus strengthening the creative power of the collective.

   These projectors are constructed using a C-K diagram (produced by the steering team) taking account of the effects of fixation identified in the collective. Their aim is to enable the collective to revise the identity of the object. It is recommended that the so-called "adornment" and "wit" projectors should be balanced out: these adornment and wit ideas were introduced by Hatchuel (2006a) to describe the types of revision of the identity of objects[4]; adornment revises identity via superposition that enhances but does not call any past identity into question (a vehicle with a panoramic windshield enhances the vehicle but does not compromise its identity); the "wit" revises the identity of the object by

---

[4]This distinction was inspired by the famous treatise *Agudeza y arte de ingenio,* written by the spanish author Baltasar Gracian in the 17th century (Gracian 1648).

re-examining certain properties that seem necessary for its definition (a mobility scooter shakes up the identity of an automobile, but certainly opens up numerous service concepts around the stationary vehicle!).

The presentation of projectors when the C phase workshops are initiated draws on the knowledge accumulated during the K phase. These can be illustrated. It is then important to take account of the fixating or de-fixating effects of the illustrations chosen (see Sect. 5.2.1 in this chapter and Agogué et al. 2014a)

3. The aim of the P phase (proposition-prototype phase) is to aggregate, recombine and develop the original suggestions from the C phase to organize a coordinated design process. This involves invoking an innovative design strategy which does not just boil down to product or service ideas. The design strategy consists of setting up a program to deploy, in a coordinated manner and staged over time, a set of actions from the rapid implementation of solutions to formulating research programs via the definition of new prototypes or via a search for new partners. This phase also attempts to register the exploration and structuring of the innovative field within the framework of the company's strategy, its constraints and its possibilities. The result is a shared and reasoned strategic timetable of innovation offering original alternatives, preparing for disruption and clarifying the contributions of each designer.

The originality of this method lies in the fact that a controlled and rigorous reasoning, making use of all forms of expertise, is not a hindrance to the creativity of the participants but rather enhances and builds it. Added to this cognitive effect is an important social dimension: the innovative paths that result from the work of the group are part of a shared view. Any new innovative suggestion will allow this shared representation to make visible to all the value and position of these paths relative to others. In addition, collective work encourages the innovative effort. The innovative power of KCP workshops has been confirmed by experiments carried out on disruptive concepts and technologies in several companies (future subways at RATP, new types of cockpit at Thales, future ecological vehicles at Volvo, sustainable turbines at TurboMeca, and railway maintenance at SNCF, not to mention others at Sagem, Snecma, AREVA, etc.). Several dozen of these processes have been successfully undertaken; the KCP method is now the standard company method at RATP, Thales and SNCF.

The KCP method has given rise to several reports and research articles (Defour et al. 2010; Gardey de Soos 2007; Elmquist and Segrestin 2009; Hatchuel et al. 2009); (Arnoux 2013).

Let us highlight a few characteristic features:

1. The KCP method is not the same as brainstorming, and is in several ways quite the opposite: not "forget what you know" but the obligation of a phase of exchange of knowledge; not "free exploration" but strongly guided exploration; not selection of the "best idea" but a design strategy that addresses the whole concept tree to take advantage of the interdependencies.

2. The KCP method can be interpreted in C-K space (see Fig. 5.20).

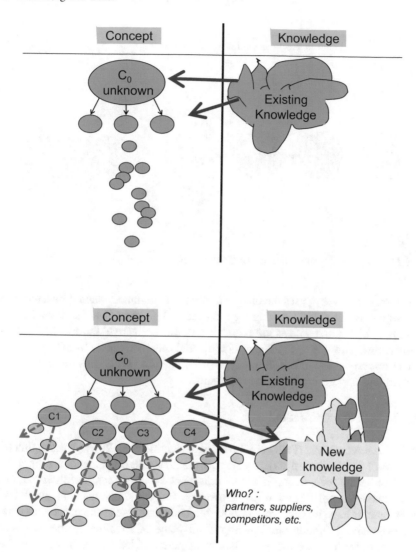

Fig. 5.20 *Top*, C-K diagram with fixation (see this Sect. 5.2.1); *bottom*, representation of de-fixations as a result of the KCP process

a. Technically, it may be noted that *the structure of the knowledge base after the K phase will tend to satisfy the splitting condition*: the multiplication of the knowledge spaces (multiplication of exceptions and alternatives) and allowing the emergence of unexpected interdependencies.
b. The "projectors" allow a better exhaustivity in C space and may be seen as equivalent to the logic of a provocative example (see this Sect. 5.2).
c. It can be shown (see KCP workshop) that the method enables the four effects of fixation identified in this chapter to be overcome (see Sect. 5.2.1). Each phase contributes.

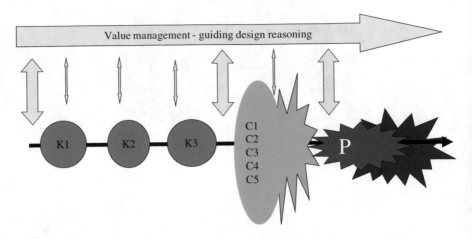

Fig. 5.21 Simplified representation of the KCP process

3. The method presupposes detailed and intensive guidance, steered by a team with design reasoning skills, having a good practical grasp of C-K theory and capable of identifying the effects of fixation for the collective trained in KCP and interacting strongly with the participants and other designers from the company. The representation of a KCP process (see Fig. 5.21) highlights the design space logic and underlying value management.

5.2.4.2   C-K Invent Method

The C-K invent method (Felk et al. 2011; Kokshagina et al. 2014; Kokshagina et al. 2016) is an application of C-K theory for creating patents. Patents are often considered to be the indirect result of technological development, and therefore come "after" the technical design work. However, the contemporary industrial issues around patents have become very important: a constant rise in the number of patents filed, expensive litigation and the important place of royalties in the cost of designed products (e.g. see the graph below of patents filed by sector between 1990 and 2007 in the WPO database (Fig. 5.22)). Also, several companies try to file patents without immediately having created the technology. This is the purpose of the C-K Invent method.

Modeling a Patent and Design Theory

The method relies primarily on a study of the criteria of patentability, and the two main ones in particular, viz. novelty and inventiveness. Under the terms of the European Patent convention of 1973 (latest revision in 2007) we have the following definitions:

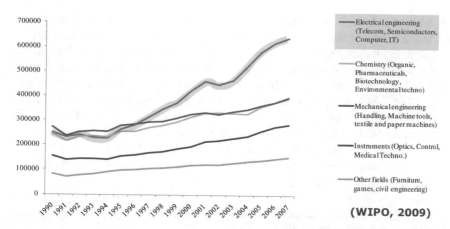

Fig. 5.22 Number of WPO patents filed per year and by industrial sector (*source* WPO 2009)

- *Novelty*: Art 54 (1) An invention shall be considered to be new if it does not form part of the *state of the art*. (2) The state of the art shall be held to comprise everything made available to the public by means of a written or oral description, by use, or in any other way, before the date of filing of the European patent application (see also *35 U.S.C. § 102(a)*)
- *Inventiveness*: Art 56 An invention shall be considered as involving an inventive step if, having regard to the state of the art, it is *not obvious to a person skilled in the art*. (see also *35 U.S.C. § 103*)

Let us examine the first criterion: this is in fact a criterion constructed with reference to a *knowledge model*. The state of the art as defined by article 54 above corresponds to the knowledge base K_0 of all public knowledge: prior patents, publications, knowledge incorporated in physical structures (e.g. machines, existing objects).

In contrast, inventiveness relies not on a model of knowledge but on a *model of reasoning*. The assessment guide from the European patent office states: "The "person skilled in the art" should be presumed to be a skilled practitioner in the relevant field, who is possessed of average knowledge and *ability* ... have had access to everything in the "state of the art", ... and have had at his disposal the normal *means and capacity for routine work and experimentation*. ... The skilled person is involved in constant development in his technical field and ... may be expected to look for suggestions in neighboring and general technical fields ...or even in remote technical fields" (guidelines for Examination in the EPO §11.3). The criterion is thus obtained by reference to the set of propositions that a man skilled in the art might construct by "non-innovative" reasoning using the knowledge base K_0. This is the set of combinations constructed on the basis K_0. This set is denoted $PSA(K_0)$, where PSA = person skilled in the art.

Theories of design have habituated us to this distinction between knowledge and reasoning. In C-K theory the novelty criterion corresponds to an expansion in knowledge and the inventiveness criterion corresponds to an "innovative" reasoning that can be equated to a reasoning that includes at least one expansive partition.

Let us now examine the two other criteria of patentability, which take us one further step towards stating the structure of the knowledge base associated with a patent. These two criteria are those of industrial application, and of "sufficient description". Industrial application leads to propositions that essentially include predicates referring either to (feasible) actions A, or to (useful) effects E. The sufficient description leads to causal propositions: actions give rise to certain effects via the intermediary of rules or laws (such that $A \rightarrow E$). Hence any patent is a proposition containing three types of attribute: actions A_i, effects E_j and rules R_{ij} linking actions A_i to effects E_j.

This analysis allows us to state the problem of creating patents. This involves conceiving propositions constructed using predicates A_i, E_j or R_{ij} and demonstrating the following properties:

- industrial application: the A_i must be feasible ("increasing the mass of the photon" would not be a feasible action), and the E_j must be useful (subject to some additional rules: no patents on living things, problems patenting software, etc.)
- sufficient description: the complete patented proposition $A_1...A_n \rightarrow E_j...E_p$ must be known in the state of the art K_0 augmented by some new knowledge mentioned in the patent δK_0: $(A \rightarrow E) \subset (K_0 \cup \delta K)$
- novelty: the complete patented proposition $A_1...A_n \rightarrow E_j...E_p$ must not have already appeared in the state of the art $(A \rightarrow E) \not\subset K_0$
- inventiveness: the complete patented proposition $A_1...A_n \rightarrow E_j...E_p$ must not appear in the set of propositions that can be generated by the man skilled in the art PSA (K_0): $(A \rightarrow E) \not\subset PSA (K_0)$.

Stages of the C-K Invent Method

The C-K invent method primarily consists of structuring the state of the art K_0, attempting to describe objects in a language of action-effect. With this knowledge base, it becomes possible to consider PSA (K_0) on the one hand and on the other, to develop almost systematically a set of expansive partitions enabling an exit from PSA (K_0) (see Fig. 5.23).

Several forms of expansive partitions are possible: either add A_0 or E_0 which are not in the base of initial knowledge (aligning the transistors in a microprocessor (new E) using a magnetic field (known A)); or "link" the A_i and E_j which are not linked in K_0 (recover the dissipated thermal energy (E) using the Seebeck effect for thermoelectric conversion (A)—A and E are known but have not been combined).

We can then systematically obtain series of A_i or E_j, making use particularly of "non sectorial" repositories of knowledge outside the sector's state of the art. Hence a Zwicky matrix can establish relationships between several actions and several effects.

Note that the expansive partition does not completely define the patent: it remains to state a proposition that satisfies the sufficient description and the industrial application—this is the condition required to obtain a conjunction in the

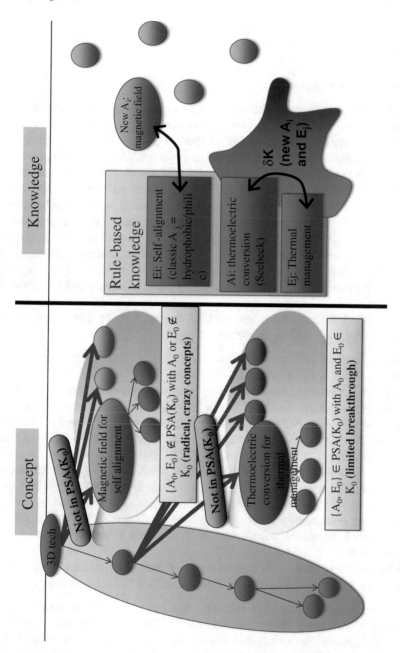

Fig. 5.23 Schematic representation of patent creation in C-K Invent. In C, *left* Schematic representation of PSA(K_0), set of concepts that can be formulated on the basis of K_0 with no expansive partition. In C, *right* two types of expansive partitions: either from A_0 or E_0 which were not initially in K_0 ("magnetic field" in the example) or from an as yet unknown relationship between the known A_i and E_j in base K_0

logic of patent creation. Also, the phase of generating expansive propositions is followed by a development phase based on the promising concept.

Illustration of the Method

The method was implemented by Yacine Felk during her doctoral thesis at STMicroelectronics in 2009. A comparison could be made. In an innovative field (of 3D technologies[5]) the teams first worked in a framework of developing innovative technologies, and then considered their patentability. This first phase did not result in patents: numerous ideas were aired (the combinations were remarkable: it involved assembling and connecting two elementary electronic components (the die) and a vast number of assembly and connection combinations can be imagined) but none of the combinations devised seemed sufficiently original to be patented. Patentability is difficult to assess (complexity of assessment and a large number of alternatives to examine) and the process does not necessarily lead to forms of expansion in K (novelty) nor to expansive partitions in C (inventiveness). On the contrary, the combinatorial mechanism leads instead to an examination of the space PSA (K_0).

Secondly, the teams made use of C-K theory without developing any technologies; they were looking to create patents.

1. They started by organizing the knowledge base so that known forms of action, effects and A-E relationships would appear.
2. In this K_0 base they then immediately generated expansive partitions, either by making use of new A_0 ($A_0 \notin K_0$) or new E_0 ($E_0 \notin K_0$), or by putting forward relationships A_i-E_j between the A_i and E_j, known but unrelated until now.
3. Using these expansive partitions, they constructed formulations satisfying the four criteria (especially the criterion of sufficient description, via an expansion in K). A lot of work went into this phase, often assuming an enhancement of the available knowledge, efforts in creating knowledge (including experiments, learning about unexpected technologies, etc.). This phase ended up with very elaborate formulations in which it was sometimes hard to recognize the initial expansive partition. Yet it was this partition that ensured the inventiveness.
4. Libraries of references on the relationships between Actions A_i and Effects E_j may be used to check a form of exhaustivity in the initial concepts.

This second method resulted in unexpected success: 25 patent suggestions were presented to the internal patents committee at STMicroelectronics and out of these 25, the company made 15 patent applications—an astonishingly high proportion of success. When questioned, the members of the committee emphasized the high quality of the

---

[5]The usual technologies are 2D: the microprocessor is built on a flat substrate using successive layers. 3D technologies would be able to make use of forms of "assemblages" of semi-finished components.

patents proposed: the patentability criteria were largely satisfied and further, the patents filed seemed to them to be of a "high level" in the sense that they covered a large number of technological approaches. In addition, subsequent work showed that these patents could lead to original technologies—again in contrast with the technological design strategy that had failed to end up with any disruptive technologies.

A Few Observations on the C-K Invent Method

1. There are few methods for creating patents. The TRIZ method (see Chap. 3), often mentioned in this context, is not a method for creating patents, but a method for analyzing existing patents. Also present in TRIZ is Action-Effect modeling. Rather, methods for creating patents can be found through genetic algorithms (Koza et al. 2004) which run through complex combinations. Nevertheless, the publications available tend to describe a way of "re-creating" existing patents rather than creating new ones.
2. The method highlights an essential property of concepts, barely discussed until now, namely that they are "non-commutative". This means that, contrary to intuition, in C space we cannot switch the order of attributes added to the concept. The first added attribute *gives rise to* the possibility of some later attribute. The attributes lead to forms of irreversibility since the consequence of their addition is the production of specific knowledge. In the case of patents, non-commutativity is illustrated by the fact that the partition that starts via technology also tends to hinder expansive partitions, while the partition that starts through the criterion of patent inventiveness leads to an expansive partition and to the production of original knowledge, finally culminating in disruptive technologies.
3. Other applications of the method have enabled it to adapt to specific contexts. The "3D" example above concerned an emerging innovative field, with few technologies and little in the way of skills on the subject within the company. The method was then tested in domains where this time there were already strong patents and issues of protection for technologies in the process of being created. The work showed that by adding certain strategic A_i or E_j to the concept, it is possible to precisely exclude certain technological avenues from the competition.
4. Modeling patents using theories of design also allows effective forms of analysis for existing patents in a technological field. The classical methods of analyzing patents tend to be through semantic searches, but they take little account of patentability criteria, in particular attributes that might ensure inventiveness. Conversely, one patent can be compared against others depending on the specific attributes that constitute their inventive character. Patents can ordered thus: at the "top of the tree" are the primary patents that depend very little on prior art, and at the "bottom of the tree" are those patents that draw on known parameters in the primary patents, then adding a few supplementary attributes. Certain companies, notably the French Yole Developpement, have developed services built on these principles for analyzing patent databases.

5.2.4.3   C-K Reference Method

The C-K Reference method is an application of C-K theory to mapping alternatives on an innovative field.[6]

Prospecting tools, or more recently, the development of tools for studying "path creation" as opposed to "path dependency", have led to the development of tools for representing future scenarios identified by the players in an innovative field at some given instant. The most recent work (Robinson and Propp 2008) has generalized the idea of scenarios to the identification of "potential paths" by combining known data. Hence they obtained the "endogenous futures". Intuitively, it is understood that these methods provide a mapping of the known alternatives and their combinations. However, by construction, it is not possible to take into account those paths that would not be identified by the participants or which would not be a combination of the identified paths. Maybe there aren't any, i.e. perhaps if the participants are sufficiently numerous all the paths are known? But perhaps there are fixation phenomena that drive the participants to work on only a few paths?—in which case, how are the unknown and non-combinatorial paths identified?

Design theories first of all enable an analysis of classical methods: they consist of generating "potential paths" based on a knowledge model (everything that is known about possible technologies, possible uses, possible markets, etc.) and a combinatorial reasoning model (combination of these various possibilities). In other words, these methods only generate restrictive partitions based on knowledge K_0; they generate no expansive partitions (surprising potential paths, etc.) nor do they produce any expansion in knowledge. The purpose of the C-K reference is precisely to map not only the restrictive avenues but also these expansive avenues, these "potential paths" that might this time be "paths into the unknown" (for a detailed presentation, see Agogué et al. 2012b).

The method consists of applying C-K reasoning. The initial concept is that of the field of innovation (safety on two wheels, independence of the elderly, etc.); initial knowledge is that of experts in the field. Reasoning is carried out by someone trained in C-K theory.

The stages involved are as follows:

1. Construction of a C-K diagram on the basis of a survey of existing knowledge in the field. This survey can be quite demanding (numerous interviews, scientific state of the art, technical aspects, patents, etc.). The first state of the art leads to the construction of trees to impel the experimenter to fill in the gaps in his knowledge, sometimes far from the initial domain and often *at the cost of a major remodeling exercise*. A few examples: the reference for the independence of the elderly depends in particular on remodeling "the elderly person" as a "vulnerable person", thus taking better account of questions of context and frailty in relation to this context (poorly lit area, homes with many changes of level, etc.); the reference for energy derived from biomass relies on an energy

---

[6]Several pioneering works can be cited on reference bodies and their organization in a hereditary framework e.g. Felk et al. 2009; Cogez et al. 2011; Ben Abbes 2007.

model that distinguishes three major components of a power system: its thermodynamic performance (conversion efficiency, etc.), its logistic and commercial performance (transport, storage, availability, exchange and sharing, etc.) and its associated social value (the notion of "family, home and hearth" with the fireplace at the center of a key social entity, energy appears as one of the foundations of collective systems, etc.) (for a detailed discussion of the model, see Brun and Polo)

2. Constructing the C-K diagram so as to make the best known paths appear on the left in C space in a hereditary manner (see in this Sect. 5.2.2.2), and the most original paths on the right (disruptions in the hereditary logic).
3. Putting the reference to the test through confrontation with experts in the field or by positioning known research projects on the subject (see Fig. 5.24)—this is a check to ensure that no known path (or any combination of known paths) has been ignored.

The illustration in Fig. 5.24 is taken from (Agogué 2012b), covering the case of independence for the elderly, and involves work carried out at the request of Cluster I-Care in the Rhône-Alpes region. The reference invokes European research projects on the subject and thus highlights various "paths into the unknown" neglected by the designers in the domain.

A few observations on the method:

1. C-K references were used to diagnose orphan innovative situations. Unexplored paths may be there for two reasons: either they have been identified but deliberately put to one side for reasons of risk and value, in which case this is a deliberate innovative strategy and not orphan innovation; or they have not been identified by any participant in the ecosystem. It is not value or risk criteria that have led to their being ignored. This is then a matter of orphan innovation, caused by a cognitive blockage (an effect of collective fixation). It is not always obvious how to decide between these two alternatives; however, in the case of independence for the elderly, it is surprising to observe that all the paths explored involve hereditary paths, i.e. design reasoning of the restrictive type, and that conversely, all the "paths into the unknown" are expansive paths. Such a coincidence is already surprising. Furthermore, more in-depth studies (from some of the players involved in such projects) show that the paths that have not been followed have generally not been considered. In particular, these players confirm that these paths have not been rejected as a consequence of a negative assessment. These two indicators therefore lead one to believe that this is a situation of orphan innovation.
2. C-K references have been used for experiments in creativity. We have seen that creativity consists of measuring the variety of responses given to an "open" question (see Sect. 4.2). However, measures of the variety and originality themselves are not obvious. In Torrance tests, the solution consists of progressively formulating a reference on the basis of a sufficiently large sample.

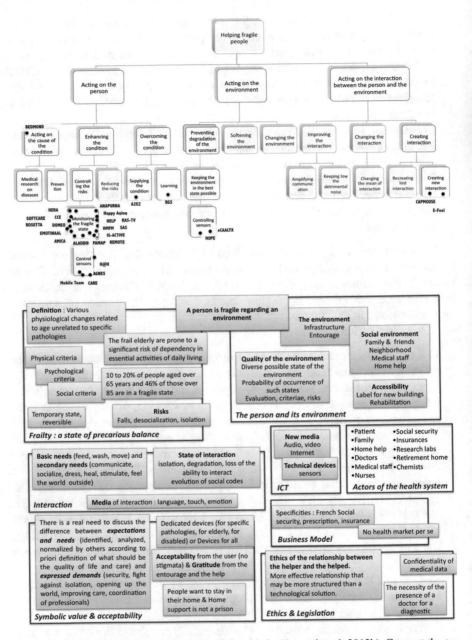

Fig. 5.24 C-K reference for independence of the elderly (Agogué et al. 2012b). C tree at the *top*, K base below. Note the effort involved in remodeling the notion of a frail person

This solution cannot be envisaged for small samples, for which certain "very rare" solutions will, statistically, virtually never be suggested. The underlying logic of the baseline allows a reference for small samples (Agogué et al. 2014a).

3. The mapping method has frequently been used for case studies in management sciences. Mapping enables data collection to be controlled: if the data do not allow reconstruction of some possible C-K reasoning, this means that the collection is almost certainly incomplete (see several examples cited in Agogué and Kazakçi 2014). See one or two cases: (Elmquist and Segrestin 2007; Elmquist and Le Masson 2009; Gillier et al. 2010; Le Masson et al. 2012d).

4. The method is also very useful in industrial cases. One might imagine that, in situations where knowledge has to be rapidly renewed, mapping quickly becomes obsolete. Experience has shown that, paradoxically, even in situations involving an intensive renewal of knowledge (science-based products such as semiconductors or avionics), the mapping remained remarkably stable, thereby meriting the term "reference". This is due to the fact that the knowledge produced is predominantly on already known conceptual paths and thus barely alters the concept tree. In certain allegedly "turbulent" domains, references remained valid for several years.

5. The reference demands a certain rigor in its implementation, and presupposes a user experienced in methods of design. Hence the method is often implemented by researchers employed by companies. We see the appearance of new players ("researchers") in organizations whose role is to rigorously implement the method of benchmarks. Such was the case with the advanced research at STMicroelectronics (for which we present certain special features in Sect. 5.3 and in case study 5.3 of Chap. 5 on conceptive research and conceptive absorptive capacity).

Other methods of innovative design derived from C-K theory are in the process of development today, often with industrial partners. This is especially true for the C-K Expert method (a method for involving experts in innovative design) (Amsterdamer and Molin 2011), methods for designing generic technology (Barthelemy and Guémy 2012; Kokshagina 2014), bio-inspired methods (Freitas Salgueiredo 2013; Freitas Salgueiredo and Hatchuel 2016).

5.3 Organization

Performance in innovative design can be likened to a sustainable revision of the identity of objects (in this Sect. 5.1); collective de-fixation is an issue of *reasoning* in innovative design, it is furthered by certain concept and knowledge structures, and may be supported by specific strategies and methods (in this Sect. 5.2). We shall now set out the *forms of organization that enable companies to rid themselves of fixation.*

In Chap. 3 we studied the organizations behind rule-based design (R&D, marketing, etc.) that can be equated to forms of *generative bureaucracy*, i.e. forms based on a precise system of rules but allowing a regular revision of these rules.

As with all bureaucracy, rule-base design is based on a rational model, that of rule-based design, in which the evolutionary domains of knowledge and invariant knowledge can be set out. These invariants are e.g. the nature of functional specifications, the nature of professional skills, etc., what we have called generative models. The objectives of innovative design are precisely to revise the identity of objects, i.e. functional revision and that of professional knowledge, existing generative models and hence revision of the invariants of rule-based design. For this reason, innovative design cannot rely solely on rule-based organizations. It presupposes other organizations, nonetheless compatible with rule-based design organizations. We shall study these now.

5.3.1   From R&D Organization to RID

First we shall study innovative design organizations within the company.

5.3.1.1   The Logic of RID

It is not possible for engineering departments and research laboratories to represent the spaces in which the identity of objects is regenerated. This is a direct consequence of what we have studied in chaps. 2 and 3. Hence the company requires a new function, the "innovative design" function (or "I" function) with its own logic of performance and its own methods—those seen in this chapter.

This innovative function is responsible for formulating research questions and development requirements specifications; in return, it is fed by the knowledge produced by research and by the questions thrown up during development when confronted by skills gaps (see Fig. 5.25).

See (Le Masson et al. 2006) for a detailed discussion of the origins of RID.

5.3.1.2   Managerial Principles of the I Function

The managerial principles of the I function can be set out by comparing with the principles of managing a research laboratory (R) and an engineering department (D) see Table 5.2.

The mission of research is to provide answers to questions; by implementing a rigorous methodology, the mission of development (or engineering department) is to suggest products (more accurately, rules enabling the manufacture of a family of products) meeting a particular set of specifications and making the greatest possible use of available skills. The mission of the "I function" is to build innovative fields, i.e. to prepare potential requirements specifications and the skills necessary for concepts considered strategic by the company (and for which the company does not yet have the rule-based resources essential for R&D).

Fig. 5.25 Principle of an
RID organization

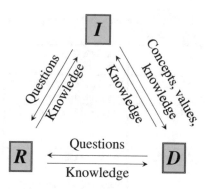

The points of entry for R, I and D are therefore very different: for research the point of entry is a "research question" (e.g. "what are the CO_2 emissions for engine xxx in a driving cycle xyz"); for development, the point of entry is a product requirements specification (e.g. a list of functions for a new vehicle seat), and for the I function, it is a "concept" associated with an innovative field (e.g. "vehicle services when it is stationary").

Outputs are respectively: scientifically validated knowledge, the definition and rules of product manufacture, and new rules for rule-based design.

Methods are also very contrasting: scientific method for research, (depending on the disciplines), methods of rule-based design for development (see Chaps. 1 and 2: project management, conceptual models, generative models, etc.), innovative design methods for the I function (design strategies, KCP, C-K Invent, etc.).

The relationship with time is very different: for research, temporality is fixed by the load/capacity ratio, where the load might be the number of experiments to be carried out (i.e. sample size) and the capacity might be the capability of the instruments producing the knowledge. For development, temporality is that which results from development planning, i.e. from the layout created on the basis of elementary tasks (see Chap. 2) and constraints on resources (availability of experts). In the case of innovative design, the temporal horizon is contingent upon exploration: the design efforts themselves will cause deadlines and temporal horizons to emerge. Thus one endeavors to find concepts that can be made in the short term ("quick and smart") but also concepts that might require more ambitious exploration.

Resources for research are laboratories, their apparatus and experts, both internal and external. Resources for development (discussed in Chap. 3) are the expertise and means of production of knowledge (see e.g. CAD tools). Resources for innovative design are exploration teams (internal or external), with the underlying logic of division of labor into design spaces (see Chap. 4).

The economic value of research depends on the value of the question asked—research can provide well crafted answers to pointless questions, though from the research point of view, this is good quality work. The value of development is in calculating the initial profitability of the project (or the portfolio of projects). While the project team can achieve a CQT objective, the project can nonetheless be a

Table 5.2 Comparison of management principles between Research, Innovative design (I) and Development

	Research (R)	Innovative design (I)	Development (D) (or rule-based design)
Mission	Standalone or proposed scientific question	Field of innovation	Functional requirements specifications
Aims	Validated knowledge	Regeneration of the rules of rule-based design	Realization of the project
Methods	Rules for the scientific production of knowledge (statistics, design of experiment, etc.)	Design strategies (lineages, platforms, etc.) ending up with regeneration of the rules	Development strategies (accumulation of conceptual models, robust generative models) and project management
Resources	Laboratories, teams by disciplines, universities, documentation	Coordinated exploratory groups	Project leader and skilled professionals
Horizon	Depend on the process of investigation (forward load planning depends on knowledge production resources)	Contingent and strategic: the process defines its own horizons depending on learning processes and strategy	Standard (development rules) and contingent (adaptation depending on commercial imperatives and techno-economic uncertainty) milestones
Economic value	Value of the question	Constructed during the process (discovery of sources of value, discovery of design resources)	Project value (see calculation of NPV)

commercial failure: project and development quality come from the realization of the objective, not commercial success. The value of innovative design lies in the rule-based design projects that such an approach makes possible (not only in the concepts produced but in the skills that result) (Table 5.2).

5.3.1.3 Illustration

Several already published examples of innovative design can be found: in particular, see the examples of Tefal and Saint Gobain in (Le Masson et al. 2006). See also the example of Turbomeca (Arnoux 2013).

Later on we shall encounter a detailed cases study of the "I" function at Thales Avionics as reported by the department manager, Denis Bonnet (Case study 5.2, Chap. 5) (presented at the seminar *Théorie et Méthode de la Conception Innovante* [Design Theory and Methods for Innovation], 8 January 2013).

5.3.1.4   I-Function and Rule-Based Design: RID in the Major Companies, RID in Start-Ups

It is not the purpose of the I function to supplant rule-based design (R&D in the case of the major companies) but instead to regenerate or generate it (in the case of start-ups). A chapter on the relationship between innovative design and rule-based design can be found in (Le Masson et al. 2006).

On the basis of an exploration of the innovative field, the I function can suggest some fairly strong disruptions to rule-based design, from the marginal modification of a design rule up to a complete revision of the system of rules. The marginal revision preserves heredity while the intention of the radical revision is, on the contrary, to distance itself as far as possible from the hereditary path. C-K type benchmark tools can enable the innovative function to "tune" the disruption: the mapping produced by these tools leads to the identification "on the left" of strongly hereditary paths, and hence little questioning of identity, while "on the right" are the paths that have been disrupted with respect to heredity, and concomitant strong alterations of identity. The path "on the left" tends to preserve and make denser the associated rule bases; the path "on the right" tends to create new "islands" (see illustration below) (Fig. 5.26).

Application: in the case of the open rotor (see illustration of the notion of heredity, in this chapter), projects feasible through development can be identified "on the left" while "on the right" we have projects that radically alter not just the engine but also the aircraft and its entire ecosystem.

Innovative start-ups present an extreme case of innovative design without rule-based design. Research works on the growth of start-ups have shown that their

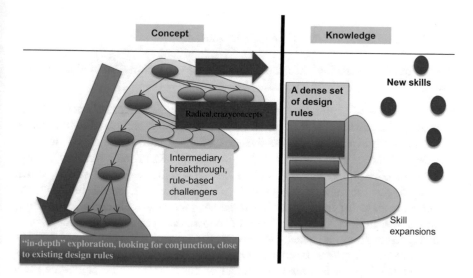

Fig. 5.26   Adjusting disruption with respect to rule-based design

growth is based especially on their "second" product, i.e. from the instant a capacity for innovation has been established (e.g. see Davila et al. 2010). The second product is indicative that the potential for growth is well in place. The issue of the sustainable growth of start-ups is the ability to develop a rule-based design in support of their first efforts in innovative design (see also the Avanti case in (Le Masson et al. 2010b)—Avanti is an innovative start-up that built its growth on "smart tools for DIY"). Hence to understand the growth of a start-up, one has to answer the question: what are the sources of rule-based design? Although not giving a complete answer here, we can indicate a few complementary elements:

1. The set of rules for rule-based design has to be created "from scratch"—in big companies, the set of rules results from continuous learning and adaptation in the engineering department—this is obviously not the case for a start-up.
2. Even in the absence of any internal reference, the basic principles of rule-based engineering seen in Chaps. 1 and 2 (especially) remain valid: the need to have rules on three language levels (functional, conceptual, embodiment), decoupling efforts, modularization and setting up platforms, implementation of conceptual models and generative models that can be applied to ranges of products.
3. Although the references for rule-based design may be absent internally, they are often very much in evidence externally! The start-up inhabits an industrial ecosystem that very frequently imposes its references straight away (functional references imposed by standards of use and consumption; technical references imposed by industrial standards or supplier networks, etc.). For a "start-up", adjusting the design is also knowing how to play with this reference ecosystem (there may be many, and certain ecosystems will sometimes be more flexible in modifying some of their rules, etc.).

5.3.1.5　I Function, Cohesion and Coordination

The innovation function addresses the double concerns of cohesion and coordination:

- *Coordination* of exploration passes via the division of labor between different fields of innovation and then into a field of innovation and coordination within the innovative projects. These divisions of labor correspond to the logic of Design Space—Value Management (see Sect. 4.3). The I function is also coordinated with the company's other skill sets: with *rule-based design*, which incorporates the new rules (products, skills, etc.), or which indicates the emerging fields of innovation that it cannot take care of itself; with *strategy*, which also indicates possible fields of innovation, sets the contingency levels and types of risk to be taken, etc. Finally, the I function is coordinated *with the ecosystem of suppliers, customers or final users*, through demonstrators, common research projects. user-involvement processes, etc. These points are illustrated by the examples of Thales or Saint Gobain.

- *Cohesion*, as shown in Chap. 4, is more critical in innovative design than in rule-based design: rule-based design relies on an established strategy and an already established *common purpose*; innovative design contributes to redefining the strategy and re-establishing the *common purpose*. The I function contributes to the cohesion of the collective in an innovative design situation by constantly "socializing" the concepts considered: concepts are shared with strategy and marketing either to obtain a mandate to explore certain topics from the company management or to feed thoughts on strategy; concepts are shared with engineering and research in such away as to provide the core business and expert leaders with visibility of possible future changes in key skills (thereby avoiding sudden obsolescence in certain skill-sets or disciplines); this visibility also occurs as a result of the involvement of expert leaders in the process of exploration. Finally, cohesion comes about with the whole ecosystem, either to include new players or to so that any designer/participants present can stimulate or monitor the dynamic processes of powerful innovation.

More generally, the innovation function multiplies the relationships between players outside the firm and outside the firm's industrial sector, players with whom the firm had no link until that instant. It is of interest to note that the dynamic of this new collaborative system (cohesion as well as coordination) is certainly relevant to the relationships formed subsequently (see the many studies on the effect of previous networks on subsequent networks—e.g. see (Ozman 2013) for a summary) but especially so to the dynamic of how object identity is revised.

5.3.2 R_c and D_c: The New Players in the Processes of Innovative Design

Innovation functions have been widely distributed among companies over the last few years, taking various forms and progressively enhancing their methods. In particular, forms of "specialization" have appeared and division of labor in innovative design, constituting a conceptual engineering that mirrors the rule-based engineering embodied by R&D. From this springs *conceptive research and conceptive development*:

- while research is a controlled production of *knowledge*, conceptive research is the controlled "production" of concepts, including the identification and expansion of concepts and the expansion of the associated knowledge (double expansion);
- development consists of proposing a conjunction that is *the best suited to a requirements specification provided by a customer* in monitoring timescales, costs and risks while minimizing the knowledge produced; conceptive development proposes a *conjunction for the greatest number of possible environments* (no specifications provided and no single customer identified) while making the best possible use of the available skills and with the same level of control over costs, timescales and *risks*.

Fig. 5.27 Principle of an
RID organization with
conceptive research and
conceptive development

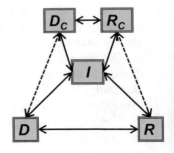

The innovation function organizes the direction taken by conceptive research and conceptive development as well as the relationship between research and conceptive research, and the relationship between rule-based development and conceptive development (see Fig. 5.27). We shall now study these two functions, Rc and Dc.

5.3.2.1   Conceptive Research: The Controlled Production of Concepts

Principles of Conceptive Research

Conceptive research is the controlled production of concepts. Non-conceptive research uses its methods (see "modeling" and "optimization" in the introductory chapter) for the production of knowledge, for modeling and optimizing existing objects; conceptive research claims similarly rigorous methods but *for conceptual exploration* of the unknown. For any given concept, conceptive research tries to structure the concept tree as systematically and as rigorously as possible, endeavoring to be "fixated" as little as possible by standard representations of the company's strategic decisions. To that end, the stabilized rules of design are revisited (conceptual models and established generative models) and during the reasoning process, these models are renewed (new conceptual models, new generative models).

> **Some examples of conceptive research**
> The mapping of "green aircraft engines for 2025" may typically be taken as a conceptive research activity, since it built up the set of concepts besides those concepts specifically worked on by the company (i.e. the open rotor).
>   As Felk (2011a, b) showed at STMicroelectronics, advanced research comprises an organized conceptive research part: for a concept such as "the next generation of imagers" (i.e. of sensors for cameras embedded in mobile telephones), the explicit purpose of conceptive research is to investigate concepts on which no-one else in the company is working: development of such a concept concentrates on new generations based on standard silicon technologies (CMOS) and research models the path of photons and signal

loss through successive layers deposited onto the silicon substrate. Conceptive research thus consists of working on the concept of "the next generation of sensors, not using CMOS". The result of this research is not a product or technology, but a tree of concepts (and associated knowledge).

Historically, some famous research labs were dedicated to conceptive research—see (Le Masson and Weil 2016).

Conceptive research creates the most complete picture on the basis of an initial concept, independently of the company's strategies, and spells out any possible revisions to conceptual and generative models. Note that we speak of "research" when the knowledge production process is "controlled"—this "control" is actually the main quality criterion of research. What is the logic of "control in the case of "conceptive research"? The control of conceptive research is based on two main criteria: the quality of the design reasoning (partitions, etc.), and the quality of the new models in K space.

Methods and Performance of Conceptive Research

Just as with research, conceptive research consists in providing rigorous answers to the questions it faces, except that the purpose of those questions is to produce concepts (and expansions of the associated knowledge) and are not "problems" that come under the disciplines of research. Yacine Felk's thesis, written in partnership with ST-Microelectronics, led to some critical elements in the methods of conceptive research (Felk 2011a, b).

Where research was able to make use of traditional methods of knowledge production (modeling, optimization, etc.), conceptive research has to demonstrate rigor in the scientific production of concepts. Hence one of its preferred methods is the C-K reference, from which rigorous mappings can be constructed. The issues of the adoption of conceptual exploration also led conceptive research to make use of the tools of C-K invent.

The objectives correspond to the traditional objectives of industrial research, namely usable knowledge, (partially) adoptable (patents), and produced at minimum cost:

1. *Usable knowledge (transferability)*: industrial research ensures the production of knowledge and its transfer to development (see chain linked model of Kline and Rosenberg Kline and Rosenberg 1986); conceptive research ensures the "transfer" of concepts to development or to some innovation function. This takes the form of a transfer of concept trees that structure fields of innovation. The value of the transfer is not that of the innovation made possible by some additional knowledge (as in traditional research) but that of risk management associated with an innovative concept: identification of alternatives, identification of generic technical concepts (see Sect. 5.2 on risk management in the unknown) and development of rich and complex design strategies.

2. *Appropriable knowledge*: in particular, industrial research gets it value from the patents it generates. This is true of conceptive research, where the issue is one of appropriating the value of the conceptual explorations that have been undertaken. Inverting the situation, however, research tends to patent technologies it has already conceived or improved (see the role of patent experts in the first central laboratories of AT&T or GE in the 1920s Reich 1985), but conceptive research has neither the time nor the resources for a detailed development of technologies before they get patented. It is often a matter of initiating the design reasoning for the patent and then exploring the design of associated techniques. Thus we have a fundamental inversion (creation of technology then patent, vs. creation of patent then technology). This inversion can be poorly understood, with broad explorations making it possible to file patents in highly varied domains but with very little relevance. Conceptive research therefore requires a special kind of management for filing patents that are relevant to a particular innovative field. For a (generally new) given field of innovation, it is the role of C-K invent to file essential key patents, though without dispersing efforts on secondary patents (see the description of C-K invent above).

3. *Effectiveness of the production of knowledge*: it is not the responsibility of industrial research to produce all the scientific knowledge that the company may need; since the 1990s it has become clear that the role of research is that of the company's "absorptive capacity" (Cohen and Levinthal 1989, 1990, 1994), to find, among the laboratories outside the company, the right experts for the resolution of the problem. The value of research lies in its ability to maintain the relevant networks with the experts in the major disciplines. The task is also just as important for conceptive research, but with a slight change of direction: it can be shown (Le Masson et al.2012c, d) that the normal capacity for absorption is built on reference disciplines or professions, indeed that innovative design can no longer be limited to a list of predefined skills, with its role being precisely that of identifying the skills of tomorrow. This requires some other form of absorptive capacity. The first capacity for absorption is epistemic, and is a priori built on a typology of disciplines and skills (episteme). The second is conceptive, and is built on a conceptual reasoning that then hinges on some original items of knowledge. The idea of conceptive absorptive capacity is illustrated in greater detail in case study 5.3 "Non-CMOS imagers at STMicroelectronics— Case study of capacity for conceptive absorption".

These three types of measure are illustrated in Table 5.3.

Coordination and Cohesion

Conceptive research, like the I function, involves a double organizational logic:

- Coordination, with other skill-sets and professions, development and research in particular, but also coordination with outside ecosystems (utilization of knowledge, stimulation of the production of knowledge).
- Cohesion: while it constantly explores the boundaries of disruption, conceptive research in fact works towards cohesion; the logic is one of risk management,

Table 5.3 Industrial research versus conceptive research

	Industrial research	Conceptive research
Usable output	Usable knowledge Measure: rate of transfer: quantity of knowledge produced effectively present on new products developed by the company	Usable concepts Measure: positioning of design projects, better risk management
Appropriable output	Appropriable knowledge Measure: patents	Appropriable concepts Measure: portfolio of relevant patents covering a design space
Minimizing the costs of production of knowledge	Networks accessing experts in the company's professions and disciplines (epistemic Absorptive Capacity) Measure: number of publications (guaranteeing the status of the in-house expert in the scientific community)	Ability to create and change moving dynamic networks, apart from the skills present in the company (Conceptive AC) Measure: measure of CAC: ability to break the rules with the strongest identity, ability to reconstruct knowledge yardsticks, ability to guide the production of knowledge in the ecosystem.

conceiving (identifying) and "socializing" the alternatives both within and without. To that end it assists in constructing an ecosystem around the company appropriate to the fields of innovation—it also helps the ecosystem explore collectively the critical fields of innovation, those that will build the industry of tomorrow. It steers external laboratories towards questions that will produce knowledge potentially useful for innovative concepts, i.e. for future innovative projects undertaken by the company that initiates them (or any other company that might be interested in the same innovative field).

5.3.2.2 Conceptive Development, Tackling the Unknowns with the Reliability of Development?

Principles of Conceptive Development

Conceptive development is a particularly remarkable activity since its purpose is to obtain the same reliability (cost, timescales, risk) as development even though there may be no clear market demand. It explores the unknown but must remain predictable and, as far as possible, re-use what knowledge is available.

In C-K terms, this is a reasoning that endeavors to make use of available or easily accessible knowledge, but which nonetheless seeks some disruption in C space; hence the disruption that requires the smallest deviation from that which already exists, but one that creates most value. The issue is that of finding the least expansive partition that brings about the greatest expansion. This is a δK, ΔC type of reasoning.

The creation of applications for a known technology is an archetypal case of conceptive development. Example: "Conceive of applications for existing fuel cells". This example is sufficient to show that these highly topical situations come up in practice against recurring difficulties, and are exposed to repeated frustrations since they are assumed to have been conducted using rigorous and demanding methods and principles of design that are generally not available. Applications of known technologies do not represent the only case of a generic concept: in the general case, applications *and* technologies can be partially unknown, yet their conception can be based on a logic of conceptive development.

The Notion of Generic Concept

The difficulty for conceptive development focuses mainly on the development of this highly "productive" expansive partition. To address this question we introduce an additional idea, that of the *generic concept*. On the basis of a design exercise, the generic concept allows not just one solution to be obtained, but a very large number of solutions (i.e. of original applications). The conjunction associated with a generic concept is therefore a generic technology such as we have studied previously (Sect. 5.2). We shall now investigate this notion, along with the methods and organizations it engenders (for more detailed information, see Kokshagina et al. 2013).

Definition: a property of the generic concept is that it does not focus on one particular application but on a whole set of (partially unknown) applications. Instead of having only the form $X.P(X)$, where P is a new attribute of X, it has, more accurately, the form $X.P(X).Q(X)$ where Q denotes an additional expected attribute of X involving the relationship between X and other objects (notably classes of applications or classes of use). When a conjunction is obtained with such a concept, that means that an X has been found that satisfies P and all situations corresponding to Q.

$Q(X)$ is an attribute that ensures that X can be "compounded" with an entire set of existing items of knowledge Y_i in K space and, if $X.P(X)Q(X)$ is true (conjunction), forming in K space new objects $X.Yi$ with $X.P(X).Q(X)$ having some interesting properties. These new objects result from the combination of the unknown X with the known Y_i. In the event of a conjunction, $Q(X)$ ensures that K will include not just a "solution" X but an entire set of combinations incorporating X.

> **A formal example (on mathematical objects)**
> If knowledge is reduced to the rational numbers[7] **Q**, then "X such that $X^2 - 2 = 0$" is a concept (actually there is no solution to this equation in **Q**, even though this equation can be written uniquely in **Q** with known notions: the numbers 2 and 0, squaring, subtraction, equality, etc.).

---

[7]Rational numbers are of the form p/q where p and q are integers.

Non-generic conception: by using algorithms (e.g. e sequence of intervals converging to the solution: the solution lies between 1 and 2, then between 1 and 3/2 then between 5/4 and 3/2, and so on), it is possible progressively to find a quantity x which is not in **Q** and which satisfies the equation (we call this $\sqrt{2}$). Thus we have conceived a single new "number". Note that this new "number" actually has poor properties since we do not know if it can be "compounded" with the numbers in **Q**: since we only know how to add or multiply rational numbers, we cannot add or multiply the new number $\sqrt{2}$.

Generic conception: the techniques of algebraic extension lead to the conception not of X.P(X) but X.P(X).P'(X) where P' = "a field containing **Q** and the solution to the equation". The procedure is very different for this second concept: these are extensions to the field and it is no longer a study of convergent sequences; the result is not reduced to a single new solution but to the set of all numbers of the form $a + b\sqrt{2}$ where $a \in$ **Q** and $b \in$ **Q**, it being possible to add and multiply these numbers (technically, this is a field). We have thus conceived a number *and* the associated combinational operations—hence we have conceived an infinity of numbers.

Theoretical perspective: Q(X) opens up an important enhancement to C-K theory. In C-K theory a concept is a proposition of the form "there exists an X.P(X)" such that X.P(X) $\notin$ K(X). P(X) is an attribute that "removes" X from K(X). The introduction of Q(X) opens up other possibilities for working on certain properties of the concept while remaining in a very general framework (universality of K models): Q(X) is an attribute that is no longer concerned with the relationship with the existing K(X) but with the *future K(X)*. This is already implicitly true with P(X) since conceiving X.P(X) involves constructing a K'(X) (future) which contains an X such that X.P(X) is true. With Q(X) we require that the future K'(X) contains not solely such an X but also numerous relationships between this X and properties of the current K(X), relationships determined by Q(X). Hence Q(X) imposes a certain structure on the future K'(X). *Hence Q(X) controls the operation of K-reordering.*

The mathematical example above illustrates this point: in the first case (non-generic conception) we require only that the future K'(X) contains the previous field and a solution to the equation—hence K'(X) is no longer a field but is the union of a field and a singleton; in the second case we require K'(X) to conserve the structure of a field. We understand that the effort of conception is not the same! Iteratively, a second exercise in conception will definitely not take the same direction in both cases: in the first case, elements will be added one after another with no relationship between them; in the second case, we shall proceed to successive extensions to the field.

The Specific Nature of Conceptive Development: Transforming Concepts into Generic Concepts

As already indicated, in development the starting points are of the form $X.P(X)$ where P corresponds (simplifying) to the requirements specifications associated with a target market. In conceptive development, we have to make the assumption that there is no longer a target market sufficient to completely guide development. On the other hand, there may always exist concepts $X.P(X)$ (a new generation of technologies for image sensors, smart fuel cells, etc.). The first task of conceptive development is to "inject" into such concepts attributes associated with the applications of the concept, meanwhile preserving the assumption that there is no target market. In other words, to $X.P(X)$ must be added $Q(X)$, a series of attributes associated with potential applications. *This transformation of a concept into a generic concept is the operation that specifically defines conceptive development.*

Hence there will be two very different moments of conception in the design of X for a unique application A_{i0}: on the one hand, the design of a $X.P(X).Q(X)$ where Q (X) is related to several A_i including A_{i0}; then on the other hand, the design $X.P(X)$. Ai given that $X.P(X).Q(X)$ is now true; In other words, the last properties separating Q from A_i remain to be conceived—this operation appears as a sort of adaptive adjustment of the technology to some partially unknown external context.

Illustration: This operation is particularly visible when conceptive development consists of conceiving applications for a (nearly) known technology. We also understand why such an operation is generally very difficult: the concept is then of the form $X.P(X).Q(X)$ where P represents the already designed technology, i.e. an extremely long and complex set of properties often incorporating attributes that implicitly refer back to specific applications.

Special cases: Certain forms of $Q(X)$ are known: technological development works thus when its objective is to improve some given functional performance considered as common to numerous potential applications. For example, one might be working on a new generation of MEMS devices ($X.P(X)$) with the aim of significantly reducing their power consumption ($Q(X)$). This last attribute makes the new generation compatible with all the applications that might claim a better energy efficiency (including applications as yet unknown). Or again, less technical: "a system for the emergency evacuation of a conference hall" ($X.P(X)$) "meeting the safety standard 'evacuation in under two minutes'" ($Q(X)$: we address all applications in which the safety standard is defined this way.

This second example also illustrates the fact that $Q(X)$ can have a reasonably good generic performance. If the standard were no longer 'evacuation in under two minutes' but rather 'double swing doors two meters wide', the generic character is reduced since the second standard covers the first, but not vice versa (it is possible to envisage a system for evacuation in under two minutes which does not necessarily use swing doors). We observe that the first standard is expressed in functional terms, while the second is expressed in terms of embodiment.

Hence we might note that the languages of rule-based design can be useful in *developing generic concepts* and even for *ranking their generic character*.

*Assessment Criteria and Strategies for the Development of Generic
Concepts*

More generally, the development of Q(X) must enable the following:

(1) End up with a greater number of applications.
(2) Control costs, i.e. do not demand new knowledge, at least, not too much.

"Ending up with a greater number of applications" corresponds to several ways
of constructing Q:

(a) Make use of several known applications (we shall speak of a *concept common
to* these applications).
(b) Make use of a critical attribute for known (or partially) unknown applications.
This attribute is, a priori, of some interest for a class of applications more
general than the sole applications known initially. In this case Q(X) denotes the
set of applications sharing the critical attribute Q(X).

Example: There is a famous case: "an engine producing mechanical power and
*compatible with all machines using mechanical energy provided by a rotating
shaft*": it was Boulton who suggested this concept to Watt; at that time there was
a need to design a new generation of steam engines quite apart from their use in
mines. With such a concept, the generic character extended not just to the
spinning machines known at the time but also to applications as yet unknown:
engines for boats, trains, etc. Boulton's concept stimulated Watt to develop new
technical features for his steam engine, initially designed to pump water from
mines. The new machine was thus "double acting" and, using a novel drive
mechanism, achieved rotational motion (for more detailed explanations see
Kokshagina et al. 2013)

(c) Construct Q to get away from probabilistic reasoning. As we have seen, reasoning for risk management in rule-based design consisted of choosing between
several technical alternatives depending on the market scenarios or potential
applications with which they may be associated (see Chaps. 1 and 2): the
scenarios are mutually exclusive and have a certain probability of occurrence;
the decider chooses the technique that will maximize its usefulness over all
these probabilities. However, in a generic design situation, the issue is not one
of choosing, but of conceiving a technique; decision reasoning becomes a
conceptive reasoning, and it can be shown (Le Masson et al. 2013b) that
reasoning on probabilistic scenarios leads to the conception of a technique that
will be good in all possible scenarios. The attribute Q therefore becomes "being
compatible with applications that cannot occur simultaneously".
There are several possible forms for Q. For example, Q may involve multiplying the potentially interesting applications (and hence add up the chances);
however, Q may also involve removing contradictions and respond to applications that might otherwise remain conflicting. The archetypal example is the
"raincoat-cap" we saw earlier (Sect. 5.2): it rains or it is fine, but not both at the
same time. This concept can be written as "a personal system for protection
against the weather which is *useful both when rain falls and the sun shines*".

Point (2) (minimizing the production of new knowledge) shows that the strategies mentioned in 1) also presuppose an ability to draw on what already exists. Intuition can be misleading on this point: adding the constraint Q does not necessarily lead to an increase in costs and additional exploration (with respect to the concept X.P(X)), but it may instead steer development in original directions; by ensuring enhanced forms of compatibility, this guidance reduces the risks (criterion 1) but may also enable technologies to be re-used (criterion 2).

Point (2) also suggests other strategies (which are not incompatible with those of point 1). For example, it might be of interest to start off with technological building blocks that each open the way to particular functions and to reason about the conception of combinations of these building blocks that might not yet be known. The combination of two blocks would then open up the path to all applications combining the functions of the two blocks (the union of the functions of the two blocks but also all the new applications generated by the combinations). This union would make use of the available knowledge, i.e. knowledge associated with the building blocks themselves. The design effort would not focus "only" on the combination. Hence we have a strategy that might be of low cost and might, however, open up large spaces of applications (see the detailed cases in Kokshagina et al. 2012b).

Illustration of this latest reasoning: one of the industrial successes of STMicroelectronics is a new radio frequency signal processing component (e.g. WiFi) (this case was mentioned in the "generic concept" strategy in Sect. 5.2.3.2). There were three building blocks at the start: one block for the processing of complex digital signals (a "computer" block) (F1), one transceiver block for high frequency signals (F2), and one block for reducing power consumption (F3). Hence three blocks each with a function. There are potential applications for which two, or even three, of these functions could be of interest (e.g. 3 functions = possible wifi hub for large public spaces such as airports; 2 functions = radar for vehicles; fast wireless data download (films) on local terminals; new active interfaces). And the possibilities for "combining" these blocks while preserving their functionalities are unknown. The generic concept is precisely that of developing a technology X that addresses all combinations of F1, F2 and F3 drawing on existing technologies. *Ex post*, this new technology appears as a "platform" for which the three previous technologies are the "modules". However, recall that design can be "upside down": first come the technologies, then a design effort to "assemble" them in "modular" fashion.

Associated cost equation: an operation such as "platform design" is justified economically under certain conditions. Let there be an algebra A(F1, F2, F3). For example A is the set of parts of {F1, F2, F3}, namely {F1}, {F2}, {F3}, {F1, F2}, {F1, F3}, {F2, F3} and {F1, F2, F3} (We could consider a more sophisticated algebra, i.e. the set of all linear combinations of the type $(\alpha F1, \beta F2, \gamma F3)$ where α, β, $\gamma \in [0, 1]$). The development cost of T giving access to A, T_A, must be less than the cost of developing specific technologies, not yet known, associated with each of the parts. In our example, let:

$$C(T_A) \leq C(T_{F1,F2}) + C(T_{F1,F3}) + C(T_{F2,F3}) + C(T_{F1,F2,F3})$$

We note that $C(T_A) \neq C(T_{F1,F2,F3})$ since knowing how to combine the three functions is no guarantee that we can "withdraw" one function to make a combination of two out of three.

When Q(X) Must Itself Be Designed

The different strategies seen above fall under two very distinct categories:

- The cases wherein Q may be deduced from the K base (identifying properties common to several known applications, combining known functions, etc.). Hence strictly speaking there is no design of Q(X). It is a matter of selection, and we can then refer to the criteria we encountered above.
- The cases wherein Q is itself the result of a process of expansion.

This second case is more difficult, and we shall not deal with it in completely general terms. However, it does give rise to methods of generating generic concepts (Barthelemy and Guémy 2012) that we are able to describe below. In particular, for each method we describe the performance attained by the Q(X) thus generated (number of applications and re-uses of knowledge).

Method 1: generic concept generated by superposition of applications.
We start off with a partially known technology T_0 conceived for an application A_1 and we know another domain A_2 that might become a domain of application for T. The method involves working on the concept "T for applications A_1 and A_2 and all their combinations".

Example: STMicroelectronics has developed a haptic (touch) technology for an application that simulates textures on a computer touchpad or touch-screen (A_1) and is dreaming up a new application for the sensation of "relief" on the buttons of a mobile telephone touch-screen. The concept is *not* "a haptic technology for feeling buttons" but rather "a haptic technology for creating the sensation of textures on a touchpad *and* the buttons of a touch-sensitive keypad on a mobile telephone".

Performance attained by Q(X): during this exploration Q(X) is designed to be common to A_1 and A_2 (and hence to any composition of A_1 and A_2) or even common to other applications A_i as yet unknown. Moreover Q(X) allows the knowledge associated with T_0, A_1 and A_2 to be re-used.

Method 2: generic concept generated by a "projector" application.
Starting from a partially known technology T we seek to develop it for a potential application A* in a very unusual and highly specific environment.

Example: STMicroelectronics has developed a technology for making 3D images based on a dual exposure technique. We want to apply it to a mountain bike (ATB) helmet. The issue is obviously not the market for ATB helmet cameras (although it did not exist at the time, this case was investigated before the success of the GoPro). On the other hand, exploration of the concept revealed several previously unknown properties that generate some relevant Q(X) (e.g. "with correction for positional microvariations relative to the two viewfinder systems").

Performance attained by Q(X): Q(X) thus constructed is not reduced to A* but extends to several domains uncovered while working on A*. Hence in the case of 3D images, the exploration leads to work on all applications in a rough environment —the ATB being one, but not necessarily the only one nor the most pertinent. Q(X) allows re-use since we started from the initial technology T_0.

Method 3: generic concept generated by attraction of designers (logic of design sharing).

Starting from a partially known technology T we are looking to involve *designers* from new environments for possible applications A_i. Working with designers will also reveal Q parameters common to these domains of application *and* will take account of the design abilities specific to the domains concerned.

Example: STMicroelectronics has developed a power management technology for standalone wireless sensors. Applications could cover the control of complex industrial systems, sports, agriculture and the environment, health, etc. Working with designers in these different fields reveals not so much the functional characteristics common to each of these environments (e.g. a battery life of at least x hours, etc.), but rather the levers of action needed by the designers in each of these environments (e.g. the ability to set an alarm for remaining battery life). The generic character thus extends to applications associated with these designers, and uses knowledge of the technology *and the abilities of the designers from the different domains* (not just their knowledge, but also that they have the ability to design at least cost).

Performance attained by Q(X): this work leads to a Q(X) that incorporates applications envisaged by all the designers who have been drawn to it; by construction, this Q(X) is associated with the skills of these designers.

Contingency of Conceptive Development

Conceptive development strategies are demanding and complex to implement, and are not applied in all techno-economic environments. For example, if there is a clear, reliable and economically sensible demand (in other words, if this context is one of rule-based development), then a conceptive development strategy is certainly sub-optimal. More generally, such a strategy cannot be justified if the value chains and architectures are stable. In what techno-economic environments is conceptive development relevant? And more generally, what are the criteria that discriminate between different techno-economic environments?

Recent work has highlighted certain decisive characteristics of the techno-economic environment for the success of a generic concept type of strategy (Hassen 2012). Two determining criteria have been identified:

(1) A generic concept strategy can be justified if there is no target market with a sufficiently high probability of success, but if many markets can be envisaged at the same time, all with low probability.

(2) The strategy works best if the available technical building blocks are strongly "aggregative", i.e. the already combined technologies make subsequent

combination with other technologies *easier* (see cost equation above). Typically, combining a computer with a gyroscope makes it easier to add another gyroscope onto a silicon chip; conversely, in a vehicle's driver space, being able to combine a seat-belt pre-tensioner with the airbags does not make it any simpler to add a safety system for children. In this second case, we speak of segregative technology.

Note that these characteristics do not have to be "naturalized", indeed:

(1) For criterion 1, the markets envisaged for a technology are not "naturally" numerous. If a technology appears to be aimed at a market that is too risky to justify normal development, the design involves extending the list of markets envisaged for this technology in order to then go on to development.
(2) For criterion 2, the technologies are not naturally "aggregative". Conceiving aggregative technologies can become a strategic objective. The issue is, for example, in the creation of original "techniques of aggregation" (design of assemblies, systems for ensuring compatibility, etc.).

Organization and Performance of Conceptive Development

Conceptive development presupposes significant work on Q-type attributes. Hence the initial phases (identifying the concept) are the most sensitive.

The kinds of expertise required are, on the one hand, a firm grasp of the techniques available (hence professional expertise, especially multi-disciplinary, knowledge of combinations that the expert knows cannot be achieved), and on the other hand, a knowledge of the various applications concerned (this is more a sales or marketing competence—though a knowledge of "functions" rather than a knowledge of the sales volumes and sales probabilities of each particular application). The key player in conceptive development is therefore a highly experienced generalist, skilled in techniques and applications, and always on the lookout for latent concepts.

The selection criteria for such development projects can no longer be built on the NPV (which is itself built on a business plan based on a single application). *In "aggregative" contexts and markets for which business opportunities are weak, the "single-application" business plan is not a good criterion.* The issue is one of finding the concept X.P(X).Q(X) that can address *many applications* with a minimum of technical exploration.

5.3.3 Colleges and Architects of the Unknown: New Designers Outside the Firm

As we saw in Sect. 3.3, rule-based design relies on an industrial organization in the sector that stabilizes the system of rules (norms and standards), improves the designers' skills (teaching) and enhances it (research). Besides the firm that designs and the consumer that uses the end-product, there are many "third parties" or intermediaries that contribute to the (rule-based) design process. In the case of innovative design, the role of the ecosystem and its "intermediaries" is also very important. Several studies have shown that this role is in fact fulfilled by the specific players (individual players or institutional players), that were referred to as colleges and architects of the unknown (Le Masson et al. 2012d; Agogué et al. 2013c). We we shall analyze these players below.

5.3.3.1 A Logic of Action Derived from Models of Design Regimes: Exchange of Concepts

To understand the logic of action of these players, we highlight a few elements modeling an economy of design. Recall one of the first advances in the 1970s: at that time research works undertaken by so-called evolutionary economists showed that the assumption of perfect rationality among the economic players, necessary for demonstrating general equilibrium, actually corresponded to an infinite knowledge resource: the optimum between different choices can be found provided all the resources necessary for the calculation are available (a perfect knowledge of the alternatives and their value). However, this assumption is often unrealistic (Herbert Simon's "satisficing" solution and notion of limited rationality). Evolutionist models therefore modeled forms of limited access to knowledge (cost) and (for example) reported on the phenomena of path dependency (re-use of the same routines, less expensive than exploration).

These models incorporate costs due to access and forms of knowledge exchange. Design theories, which work on conceptual expansion, underline an implicit assumption in these same evolutionist models: innovative "ideas", initiatives from entrepreneurs, evolution even of the environment (new societal trends, etc.) would be "given", generated by "entrepreneurs" capable of covering all imaginable disruptive services and products. This time we are dealing with the assumption of an *infinite resource of concepts*, that recent work in psychology (the idea of fixation) or formal approaches in design theory have opened up for discussion. In an innovative design economy, it is important also to take account of the limited ability to generate and exchange concepts. It is precisely the (more or less) good "management" of this limited capability that can also explain growth differentials. Thus we have a model of knowledge and limited concepts, with dynamic growth of these two spaces over the course of time (see diagram below).

We can pick out a few cases quite simply. To this end, note that, in an ecosystem at some given instant, "available" knowledge can be associated with concepts of so-called "**realizable**" goods; these are goods that have not necessarily been produced and exchanged, but for which a possible combination of available items of knowledge might be seen to emerge; by symmetry, "available" concepts refer back to "**imaginable**" (or desirable goods) that have not yet given rise to realizable goods (these are the "ideas" of innovative entrepreneurs, for example) (see Fig. 5.28).

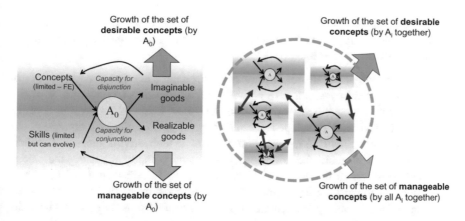

Fig. 5.28 Modeling contemporary industrial dynamics: spaces of limited access concepts and knowledge; dynamic of all realizable goods and desirable concepts. *Left* single-player model; *right* multi-player diagram

If realizable concepts and desirable concepts are overlaid and evolve slowly, we get a rule-based design regime; if the evolve rapidly together, we get an innovative design sector; if realizable and desirable diverge, we get forms of denial of innovative design or suffocation of rule-based design (the sector proposes "realizable" innovations that no longer correspond to what users find "desirable", for example.

A few models incorporating limited conceptive abilities and exchange of concepts have already been proposed (Agogué 2012b; Le Masson et al. 2010a). These are based on the introduction of a value potential or on the introduction of partially unknown goods. They report on the variety of the industrial dynamic: not only do they model the standard forms of dominant design, but they also provide insight into the dynamic of intensive innovation (like semiconductors, capable of very regularly renewing their techniques and values) or, on the contrary, industrial dynamics held up because of orphan innovation.

In these models (not detailed here) one variable is especially critical: *the exchange of concepts*. The received wisdom on industrial strategy frequently leads one to believe that concepts would be a matter for the most confidential of strategic plans; instead, the models show the importance of allowing concepts to circulate, thus having the effect of stimulating the imaginations of each player in order to "de-fixate" themselves and regenerate skills within the ecosystem.

5.3.3.2   Some Examples

The importance of the exchange of concepts is confirmed by empirical studies in two types of very contrasting situations:

- In the case of great industrial success stories (Intel, etc.), we observe a remarkable management of the ecosystem. Intel organizes two platforms to manage its ecosystem thus:

  - A downstream platform, responsible for managing the ecosystem of its microprocessor users. These activities would develop technologies to facilitate interfacing the microprocessors with new fields of application. During the 1990s Intel developed the USB port, which allowed computers to communicate with numerous components in their environment, and which also brought about the development of many components using the calculating power of the computer (this platform is described in detail in Gawer and Cusumano 2002). In the downstream ecosystem, this platform corresponds to an ability to circulate original concepts that assist potential users in de-fixating themselves.
  - A platform upstream responsible for managing the ecosystem for the designers of microprocessor fabrication processes. This management (described in Le Masson et al. 2012d) takes on a highly collective form since all the microprocessor designers, suppliers and research laboratories come together under the ITRS (International Technology Roadmap for Semiconductors), which meets three times a year to exchange information on the technical concepts required to maintain the momentum of Moore's law. The work involves a huge number of designers and spawns a regular publication, freely available online (this organization is described in Schaller 2004 and modeled in Le Masson et al. 2012d). Thus we have a collegiate management for exploring the unknown. Again, we note that this platform contributes to the circulation of original concepts and allows each participant (process suppliers in particular, but researchers as well) to de-fixate themselves.

- Conversely, in the case of orphan innovation (see Sect. 5.1.3.2, for this idea), we find many designers, determined entrepreneurs, researchers and a powerful societal demand—rather it is those design managers in a particular ecosystem that are greatly missed (see Agogué et al. (2012b) for a more detailed description). Furthermore, innovation seems to appear with the emergence of original players (centers, clusters, associations, etc.) who attempt to organize the exploratory efforts of players in the ecosystem. These are not entrepreneurs themselves, but "architects of innovation" who facilitate designers' innovative efforts. (Agogué 2012b; Lefebvre 2013).

Hence we see the appearance of *ecosystem designers* who may take the form of distinct third-parties (architects of the unknown) or more collegiate forms (colleges of the unknown) (Le Masson et al. 2012d; Agogué et al. 2013c).

Several studies have described some of these players. The reader may examine the following cases, for example:

- The common laboratory SAFER in Sweden, which puts researchers at Chalmers university in touch with vehicle designers for working on new safety concepts (Agogué et al. 2013a),
- An agricultural cooperative that enables collaboration between ecologists, researchers and farmers to come up with new ways of farming cereals taking greater account of biodiversity (Berthet et al. 2012),
- Clusters seeking original initiatives for the independence of the elderly (Agogué et al. 2012b) or safety on two wheels (Agogué 2012b)
- Associations giving themselves the mission to push innovation to ensure the future of ecosystems (Ariel example in Agogué 2012b),
- The "building with hemp" association which organized a new industrial sector (Le Masson et al. 2012a),
- More powerful forms such as ITRS (Le Masson et al. 2012d) or the French *Institute de la vision* (Institute for sight),
- Older forms such as the Lunar Society which brought together the main innovators of the first industrial revolution at the end of the 18th century (Agogué 2012a; Schofield 1957, 1963), or the Franklin Institute which laid the groundwork for the development of an industrial environment in Philadelphia (1824–1865) (Sinclair 1974).

5.3.3.3 Logic of Action and Primary Roles

The logic of action of these players can be summed up in one sentence: get concepts circulating. This distinguishes them from the ordinary players, who essentially contribute to the circulation of knowledge. At a time of open innovation, many intermediaries make a point of finding bearers of *solutions*, i.e. of knowledge, for any problem posed. The intermediary ecosystem designers get involved when there is no well-defined "problem", when the issue is still that of exploring the unknown and the intermediary brokers of knowledge can no longer intervene.

We now see what that means in terms of action. First, recall the characteristics of some of the already widely studied ecosystem players, qualifiable as "intermediaries of the known". A recent summary (Agogué et al. 2013a, 2016) distinguishes three main types of player that have been well described in the literature: brokers of open innovation, technology transfer managers, or ecosystem intermediaries. These classical intermediaries play four major roles (see Table 5.4): to connect, to involve, to avoid conflicts and to stimulate innovation. However, they can play these roles only *in cases where the unknown is limited*. In the "unknown" situation, the models no longer apply: the relevant stakeholders cannot be identified *ex ante* (connecting in the unknown), mobilization cannot occur a priori around a legitimate vision (unknown vision), there is no pre-existing common interest to enable possible conflicts to be overcome (unknown common interests), and there are no questions or well defined problems at the outset (unknown problem) (see summary Table 5.4).

In other words, these players can act provided the part of the unknown to be explored is limited, and that "fixations" do not need to be overcome. Prescribers,

Table 5.4 Roles and limits of classical innovation intermediaries (from Agogué et al. 2013a, 2016)

Main Functions	Intermediary as a broker for problem solving	Intermediary as a broker for technology transfer	Intermediary as an ecosystem bridger	Can intermediaries be active in the unknown?
Connect	Connect seeking companies with problem solvers (e.g., Nambisan and Sawhney 2007)	Establish connections between academic or industry science and external players in the market (Turpin et al. 1996)	Create and maintain a network for ongoing multilateral exchange (van Lente et al. 2003)	Can they connect parties when relevant stakeholders are not identified?
Involve/commit/mobilize	Enlist scientists by defining common rules supported by internal "champions" (Sieg et al. 2010)	Perform marketing activities in order to attract potential investors (Thursby et al. 2001)	Mobilize resources: Human capital, financial capital, and complementary assets (Bergek et al. 2008)	Can they mobilize joint innovation while being in conflict and competition?
Avoid/solve conflicts	Define the right problem; avoid the conflict between overdrawn expectations ("Holy Grail") and limited solution capacities (Sieg et al. 2010)	Balance heterogeneous (conflicting) interests of stakeholders, in particular financial and non-financial objectives (Shohet and Prevezer 1996)	Create legitimacy for a new technological trajectory, create a common agenda for actors with different (opposing) interests (Hekkert et al. 2007)	Can they overcome conflict without pre-existing common interest?
Stimulate innovation	Articulate and combine knowledge (Bessant and Rush 1995), re-engineer knowledge (Klerx and Leuwis 2008)	Actively engage in the exploration of new technology uses and the transfer of knowledge (Becker and Gassmann 2006)	Support learning processes, foster feedback, stimulate experiments and mutual adaptations (van Lente et al. 2003)	Can they stimulate innovation without pre-defined problem or research questions?

technology brokers or problem brokers can work as long as the products, questions and solutions (but also the interests and skills of each) are identified and stable. If the subjects of collaboration, interests, partners and unifying visions are unknown, and are the focus of a major design effort involving a collective de-fixation, then the standard models fail to work well.

We can assume that the architect or college of the unknown ought to fulfill the same roles—to connect, to involve, to resolve conflicts and support exploration—but that they need to fulfill these roles in situations where fixations of all kinds must be overcome; it is from the exchange of concepts (or from the generation then exchange of concepts) that these de-fixations are expected.

We can summarize the roles and methods of the architects and colleges in Table 5.5.

5.3.3.4 Action Models, Depending on Types of Innovative Dynamic

In describing the manner in which the roles are fulfilled by these players, it is preferable to examine some contingent situations. Four cases are discussed: ecosystems subject to strong and repeated mutation, ecosystems in transition and in conflict, ecosystems in an orphan innovation situation, and emerging ecosystems. Note that each of these cases is studied in detail in the publications mentioned throughout the text.

Ecosystems Subject to Strong and Repeated Mutation:
Managing "Unlocking Rules" by Colleges of the Unknown

The archetype is ITRS, briefly described previously. We shall not revisit questions of performance of the sector (already mentioned in Sect. 5.1). The mission for the members of ITRS is clear (see the memorandum of understanding signed when ITRS was created in 2001, and unaltered since): "Sponsoring participants shall cooperate to *identify generic technology needs* for the global semiconductor industry without regard to particular products of individual companies and encourage on an industry-wide basis *potential solutions to future technology challenges*". In practice this means identifying the spaces for which there is as yet no response, "holes" in the technology, and where a solution might be necessary all the same. Not solutions, but problems are identified. This positioning is contrary to the most standard models of competition and optimization: the players tend, a priori, to re-use the technologies available, improving them as they go, thereby causing a path dependence; the rules of the sector are then "enclosing" (lock-in effect): performance and known technologies are the most shared, with new knowledge and disruptive technologies having little influence. The logic of ITRS, starting from unsatisfied "needs", tends to share the gaps in the esablished practices and tends to focus the collective effort on concepts that are still open. Hence we are dealing with an organization, a college, that makes use of "unlocking" rules.

Table 5.5 Roles and methods of architects and colleges (from Agogué et al. 2013a, 2016)

Primary functions	Pathology in a situation of the unknown	Action of colleges and/or architects
Connect	1. Find experts? 2. Mute experts?	1. Build "flexible" communication infrastructures (network of ad hoc experts) (see Siemens case in Agogué et al. 2013a) 2. Find a way of involving experts in questions of innovative design (see C-K expert method)
Involve	1. No participants 2. No "contributors" → risk of speculative technology bubble	1. Ensure the legitimacy of the collective work space la (see SAFER case in Agogué et al. 2013a) 2. Involve designers, recall the state of "concept" (undecidable; avoid the quid pro quo: this is not yet knowledge!)
Manage conflict	The unknown does not suppress the interests and power relations; risks of quid pro quo	1. Conceive of common interests (see CECT-CNRS case in Agogué et al. 2013a) 2. Re-open conceptual paths to allow complementary or independent interests to emerge (see "building with hemp" case in Le Masson et al. 2012a)
Support exploration	Risk of collective fixation	1. C-K reference works to encourage de-fixation 2. Direct work on provocative examples

The first of these "unlocking" rules is Moore's law itself, its main characteristic being that it shows that today's technologies will not be sufficient for tomorrow, that they will be obsolete, and that new technologies will therefore be required for the industry. We might consider Moore's law to be "enclosing"; however, from the point of view of technological innovation, it is also a generative law which regularly drives process designers to new feats.

We find such "unlocking" rules in the organizational practices of ITRS: working groups come together under the banner "we're not picking winners or losers". ITRS neither selects nor decides, but is content to record the concepts put forward by the participants. If possible, ITRS will pronounce on the more likely alternatives and on the alternatives that are very different from the technologies currently in place.

Each of the working groups meets to manage an ecology of innovative pathways (concepts) (Le Masson et al. 2012d). For each topic there are many pathways; new pathways open up over time, and can be ranked in order: certain paths are direct alternatives to the existing processes, give or take a few parameters (e.g. in the case of etching, current technology uses a laser, but improving performance means reducing the wavelength); others presuppose more radical technological changes of platform (in the case of etching, the thinking today is towards UV lasers, which require instruments that are very different optically); the final pathways are highly disruptive (etching is no longer done with an optical beam but rather by physical

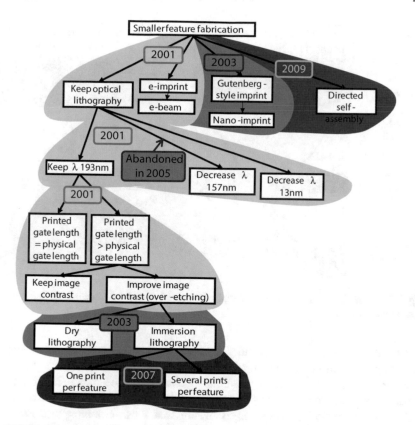

Fig. 5.29 Ecology of innovative pathways for the "etching" technical working group (TWG) at the ITRS

impression (like a Gutenberg press); the functional building blocks are no longer etched but auto-position themselves. (see illustration below). The working groups in fact create references of unknown technologies that will avoid the effects of fixation (Fig. 5.29).

Collegiate forms such as ITRS have to revise industrial transition models. Even though these transitions are often seen as chaotic phases where the former "regime" of rules succumbs to the pressure of disruptive innovators and the demands of society, we have a form of transition steered by the regime itself, with ITRS being in fact the engine of disruptive exploration and a space from which even social demands can be regenerated (see diagram below, taken from Le Masson et al. 2012d) (Fig. 5.30).

Ecosystems in Transition and Conflict: The Design of Common Interests

This section focuses on the work of Elsa Berthet and Blanche Segrestin. The interested reader is urged to look at their publications (Berthet et al. 2012).

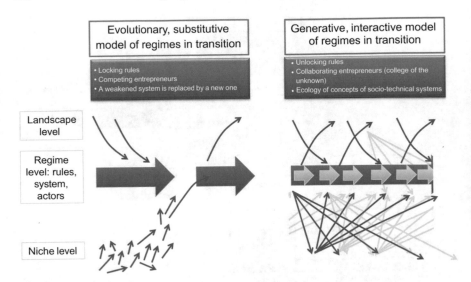

Fig. 5.30 Diagram of a regime in "disruptive" transition and of a regime in intensive and continuous transition

In a general sense, contemporary innovative concepts can become *conflicted*, including between pioneering stakeholders. For example, when nature parks come into conflict with the practices of animal husbandry in the mountains, the proponents of these different forms of environmental protection can clash fiercely. Such cases are inevitable and frequent; assuming that these common interests might emerge spontaneously and collaboratively is somewhat utopian. On the other hand, in this kind of situation, conflict can also be a sign of a gap in conception, a fixation. In this case, the effort of innovative thinking can lead to the design of common interests. Innovation has no need to wait for the conflict to be resolved, but instead can contribute to its resolution. This resolution is facilitated by third parties who can assist in defining the common interests. This example is central to the CEBC-CNRS research described in Berthet et al. (2012).

In this CEBC-CNRS case, opposition between ecologists with a partisan approach to biodiversity and the farmers became very lively. The ecologists showed that the little bustard (a species of bird to be protected) could be protected by cultivating alfalfa (which increased the food resources for the young bustards); the farmers wanted to use the grasslands for producing forage. This was therefore a conflict situation. The research center intervened to designate "grassland" as a "design" space to be explored, i.e. "a grassland corresponding to the interests of both ecologists and farmers" had in fact to be considered as a concept: such a suggestion did not exist, but could be conceived! With a local cooperative, the research center initiated a collective process involving a large number of participants (local authorities, agricultural specialists, naturalists, researchers, water syndicates, ordinary citizens, farmers and the cooperative). The exercise identified alfalfa as an ecological infrastructure that could, under certain conditions, become a space for the production of forage, help keep useful pollinators for the farmers or encourage orthoptera that the young bustards could feed on. In all these cases, these concepts allowed new functions to emerge. Furthermore, the exercise identified some original parameters: to ensure that numbers of pollinators and orthoptera were maintained, it appeared that the dates for when and where to mow had to be carefully coordinated. New tools and new forms of coordination were therefore required.

When this work was finished, the research center and cooperative were able to set up an original alfalfa concern and strengthen a program of original research to work on other developments.

The role of the third party organization can be summed up thus:

- It gave visibility to the interactions between players, and led them to come up with an acceptable line of approach for the agro-ecosystem. It encouraged and made possible the processes of exploration and learning.
- It gradually developed tools for monitoring, checking and assessing the processes. These tools were able to identify bottlenecks and fixations.
- it took on a form of governance, at least temporary. The third party saw who needed to be involved and had to put in place operating rules appropriate to the exploratory nature of the work.

Note that this third party organization is acting on the design of a new type of "common good". The "common good" is an essential idea in contemporary economic theory, but the designer of this common good has until now been very poorly identified and described. Elsa Berthet's thesis makes a major contribution to this idea.

Ecosystems in an Orphan Innovation Situation: References for Circulating Concepts

This section focuses on the work of Marine Agogué. The interested reader is encouraged to look at her work (Agogué 2013a, b).

Apart from the diagnosis of orphan innovation (already discussed in Sect. 5.1), the issue could be that of supporting the de-fixation of an ecosystem. Work in the cognitive sciences has been able to show the *importance of "provocative" examples* in helping individuals to de-fixate themselves (See Sect. 5.2.2.2). Let us pause for a moment on the experimental scenario set up to carry out a scientific experiment. This scenario is interesting from an organizational point of view since the experimenter (who set up the experiment after having designed it) finally appears as a player per se, and a player capable of stimulating the creativity of the individual involved in the experiment. For all that, the experimenter does not find the solution himself. It is precisely in this role that some ecosystems managers find themselves today—centers of competitivity, clusters, associations, etc.—responsible for assisting designers to become "innovative designers". Their logic is to formulate "controversial examples" capable of stimulating innovative design by de-fixating the players involved.

This role is fulfilled by the *architect of the unknown*, who works in the following manner: for an innovative field, he starts by developing a reference. From this basis, in a second stage he can:

1. Identify the relevant skills, fixating or de-fixating, and identify those players who have them.
2. Identify the fixating or de-fixating pathways. Some of these paths may be fixating (or de-fixating) for any ecosystem but some can be fixating for some players and de-fixating for others. This is why it is important to have not just the reference but also the fixations of each player.
3. Establish protocols for circulating de-fixating examples to each of the players that they might find useful (and the associated knowledge).

In practice, the results obtained from a few examples are convincing. However, they nonetheless show that, for organizations, demonstrating a controversial example is not always enough to trigger innovative design. De-fixation also comes via a *sharing of knowledge*, and might presuppose some *aid in undertaking reasoning in innovative design*.

Aside from controversial examples, the architects appear likely to strengthen exploration when they help project participants to enhance the concepts they are working on, and position them better from an academic or industrial point of view. Hence one of the action modes favored by the I-Care cluster in Rhône-Alpes, responsible for health technologies, was the setting up of a project panel to take the time to analyze the positioning of the project against the reference and discuss possible changes of direction with the participants.

Emerging Ecosystems: Managing Generative Expectations

On certain subjects and in certain ecosystems, industrial growth begins and more precise promises start to emerge, raising the expectations of the various stakeholders. In these phases of emergence, there is always a risk that collective innovative design will be the victim of fixation. One of the major risks is the speculative technology bubble, illustrated by the so-called Gartner curve shown in Sect. 5.1.

Managing these expectations is one of the issues addressed by the architects and colleges of the unknown. In the sense that they could be the occasion for attracting new resources (financial in particular) these bubbles seem to represent an opportunity. However, they also represent a real threat in the sense that it is not necessarily just financial resources that guarantee an innovative design performance but above all, expectations reinforce fixations: the promise that attracted the financial backers becomes a commitment that has to be met, and it ends up constraining exploration. When, inevitably, it appears that the initial promise has to be altered, all the committed resources can disappear rapidly, wiping out the design effort currently at work. Managing expectations therefore means keeping speculative bubbles under control. Where do the colleges go from here?

The success of building with hemp was analyzed as a successful exercise in keeping things under control (Le Masson et al. 2012c) (see also case study 5.3 detailed in this chapter). In this particular case of innovative design, the ecosystem included farmers, transformers (cooperatives treating the hemp to extract co-products, especially long and short fibers), cement manufacturers developing new products incorporating natural fibers, architects interested in using a forgotten "historical" building product, and the scientists responsible for assessing new materials. Still under the banner of the "building with hemp" association, these various players started to work together. This association was born out of the initiative of one of the cooperatives, which had seen the regular decline of hemp and the unfortunate tests carried out by some inventors seeking to incorporate natural fibers into concrete. Given this orphan innovation, the cooperative started by exploring possible ways of building with hemp, gathering around it several pioneering workers from many domains. Together, they created the "building with hemp" association. This association thus plays the role of a college of the unknown, gradually developing an enlarged set of concepts for building with hemp. These ideas started to interest certain industrial players: first applications aroused increasing expectations, and the fad grew. Along with the bubble, many financial backers put themselves forward, ready to invest in hemp. Moreover, certain alternative hemp producers started to mistrust a venture that was not always consistent with the principles of sustainable building from its beginnings. Conflict awaited the speculative bubble. It is interesting to note that the expectations actually had a serious disadvantage: since their construction was often superficial and trivial, only the less knowledgeable sponsors took part in the venture; conversely, the competent designers tended to distance themselves: they knew that the ideas put forward were only "concepts", requiring a lot of additional effort. The speculative bubble caused what economists call an adverse selection.

The role of the college is to deflate the bubble: during an interview at a major regional daily newspaper, at the height of the craze the president of the association reminded the financial backers that promises are not concepts, and that major design efforts had yet to be agreed upon—and that investment would show no return for many years. In so doing, he brought about a positive reaction: the less knowledgeable backers tended to shy away; the competent designers remained, convinced that the circle of designers that had been brought together were the best to confront the issues of designing and building with hemp.

Another initiative of the college recalls the management of conflict discussed above: when the first projects started, they could only satisfy certain operators; inevitably, these first projects had winners and losers. The logic that prevailed from then on was not that of some illusory compromise or ambitious solidarity. On the one hand, the college protected the projects in the process of development, but on the other, suggested original avenues for building with hemp, counterbalancing the first projects being developed. It was the multiplication of concepts that became the means by which conflict was managed. The success of the college was due less to one common interest than to the ability to arouse interest in multiple concepts.

Hence the college was characterized by a double effort:

(1) On the one hand, driving an improvement in conceptual proposals arousing the interest of the industrial and financial sectors. Expectations had to be gradually created and initial promises had to be successfully transformed into concrete projects.

(2) On the other hand, just as important but perhaps much less intuitive, an effort in continuously renewing expectations. This renewal avoided speculative bubbles and conflicts of interest.

Two types of expectation can be distinguished. The most normal expectations (decision theory) qualified as anticipatory: these expectations function as promises to acquire original concepts, cheaper and earlier, and are called "anticipative expectations". They attract players who are not designers. The second type is embodied in generative expectations, which are expectations that provoke the development of anticipative expectations; they do not stop the design but, on the contrary, allow other anticipative expectations to emerge.

The reader interested in the details of innovative design regarding building with hemp may refer to (Le Masson et al. 2012a) and case study 5.3 of this chapter.

5.4 Conclusion: a new governance for innovation

With rule-based design we studied a generative bureaucracy, far from the misleading and simplistic images of the entrepreneur or inventor. With innovative design, generative bureaucracy does not disappear; on the contrary, it finds more and more complex forms, it regenerates and expands.

Its performance is not read (solely) in "new products" or "good ideas" but the ability to design new words, to collectively regenerate the definition of objects in accordance with organizationally and socially sustainable processes. We observe not only very high levels on this performance scale (intensive innovative design with high learning rents) but also low levels (orphan innovation). Thus the paradox of R&D is explained.

Reasoning and methods go well beyond brainstorming and the selection of "good ideas": the issue is that of collective de-fixation. Formally, this refers to highly particular structures in C and K space. For K, we have seen in the case of rule-based design, the importance of conceptual models, these good "résumés" of the known, which define the residual unknown; these models are also important in innovative design, but in addition we also ask of them not to "enclose" their user, i.e. to provide him with the keys with which to "get out" of the model—like a box which, instead of enclosing, might offer the means to come out "from within". For that, these K structures must be *non-modular and non-deterministic* (or in more technical language, they must satisfy the splitting condition). For C, we have seen the importance of generative models in rule-based design, indicating what knowledge to use at what moment, such as to preserve the spaces of controlled expansion. The logic becomes generalized in innovative design: the idea of *object heredity* enables all the expansions that might be associated with an object to be systematically structured, from the most restricted to those most able to propagate. These ideas allow specific strategies of innovative design to be developed—explorations of quirky concepts and work on generic concepts. Several methods allow a collective de-fixation: the *KCP method* for enlarged collectives highly sensitive to multiple fixations; the *C-K invent* method for the creation of patents; the *C-K reference* method for mapping an innovative field.

The innovative design organizations are responsible for revising the foundations of the rule-based design organizations: the rules bases (functional, conceptual, etc.) that provide the structure for the activities of engineering departments, marketing and R&D. This is therefore a critical activity that presupposes a considerable degree of control and guidance, not a lack of control and guidance. Control and guidance do not focus on convergence onto a target but rather on the quality of explorations in terms of fixation. RID type organizations, conceptive research or conceptive development can therefore see their missions, roles, responsibilities, division of labor and objectivized performance in innovative design, controlled and strengthened.

However, innovative design organizations do not stop at the company gates. With rule-based design we showed that the sectorial organization played a critical role (schools and universities for training experts; research laboratories to renew conceptual models and measure their functional performance, etc.). Other forms appear with innovative design, taking responsibility for the collective exploration of innovative fields, including the most extreme situations (potential conflicts, poorly identified expertise, strong fixations, strong but biased anticipations, etc.). *Colleges and architects of the unknown* appear as original organizations. They occupy the interstices and spaces where markets or enterprises do not yet exist. These can be considered not as secondary additions to the enterprise or market, but as an *essential precursor, inventing new forms for them.*

5.4.1 The Main Ideas of this Chapter

- Sustainable revision of object identity
- Orphan innovation
- Effects of fixation (individual and collective)
- Non-determinism and non-modularity of a knowledge base (splitting condition)
- Restoring order in a knowledge base
- Hereditary structure of a concept tree
- Risk management in a double unknown situation: by structuring the unknown or by generic concept
- KCP method
- C-K invent method
- C-K reference method
- Conceptive research; capacity for conceptive absorption
- Conceptive development; generic concept
- College and architect of the unknown
- Unlocking rules, conception of common interests, circulation of concepts, generative expectations vs. anticipative expectations

5.4.2 Additional Reading

This chapter can be extended in several directions, on:

- questions of performance:
  - see questions of definition: radical, disruptive, breakthrough, etc. (Veryzer 1998; O'Connor 2008; Christensen 1997)
  - measurement: Coombs et al. (1996, Kleinknecht and Bain (1993), see the Oslo manual (OECD 2005; Talke et al. 2009; El Qaoumi 2012)
- cognitive approaches: Agogué et al. (2014a, b), Jansson and Smith (1991), Finke (1990), Ward et al. (1999)
- the splitting condition: Jech (2002); see Dehornoy's course on forcing Lenfle et al. (2016).
- Bauhaus courses: Klee (1922, 2005), Itten (1961, 1975), Kandinsky (1975); on their analysis: Le Masson et al. (2013a)
- generic technologies: Bresnahan and Trajtenberg (1995), Shinn (2004); their creation (Kokshagina et al. 2013; Le Masson et al. 2013b; Kokshagina 2014)
- notions of heredity: Brogard and Joanny (2010), Felk (2011a, b)
- methods derived from C-K theory:
  - see Agogué et al. (2012a) for a summary
  - KCP: general publications (Elmquist and Segrestin 2009; Hatchuel et al. 2009; Arnoux 2013); industrial cases: RATP (Gardey de Soos 2007); Turbomeca (Arnoux 2013); Thales (Defour et al. 2010)
  - C-K invent (Felk et al. 2011; Koh 2013)
  - C-K references (Agogué 2012b, 2013a, b; Agogué et al. 2012b)

- RID organization:
  - On other organizational proposals: ambidextrous organization (Duncan 1976; O'Connor 2008; Tushman and O'Reilly III 1996); generation of R&D (Miller and Morris 1999; Roussel et al. 1991); (O'Connor and DeMartino 2006)
  - From R&D to RID: Le Masson et al. (2006), Garel and Mock (2016)
  - Conceptive research: Felk (2011a, b); capacity for conceptive absorption: Le Masson et al. (2012c, d)
  - Conceptive development: Kokshagina et al. (2012b)

- new ecosystem organizations:
  - On broker-intermediaries: see the literature survey in Agogué et al. (2013a); main references: Sieg et al. (2010), Van Lente et al. (2003); critique of the open innovation in Birkinshaw et al. (2011)
  - On architects: Agogué et al. (2013b, c)
  - On transitions and path creation: Geels (2002, 2005), Geels and Schot (2007)
  - On colleges: Le Masson et al. (2012d), Cogez et al. (2013), Hooge et al. (2016)
  - On the common good (Ostrom 1990) and its creation: Berthet (2013)
  - On expectations: Borup et al. (2006); and managing them in an innovative design situation (Le Masson et al. 2012a)
  - On Co design and social effects: Dubois et al. (2014)

5.5 Workshop 5.1: The KCP Method

An overview of the KCP method was given in the course (Sect. 5.2.4).

With respect to KCPs, a workshop may consist in taking an already addressed case (see, for example, the RATP report available on the subject (Gardey de Soos 2007), the Volvo case (Elmquist and Segrestin 2009), or the Turbomeca case Arnoux 2013), and studying each phase step by step.

Here we have chosen a triple workshop, each part addressing one phase of the KCP method (K, C and P). This allows us to work on the underlying design logic of each phase.

For each phase we explain the reasoning, the objectives (evaluation criteria) and some methodological tips. Note that each phase may be considered an autonomous workshop. This workshop revisits the elements of a three-day training offered by the authors to business executives involved in steering and conducting KCP workshops.

5.5.1 Phase K: Forming a Common K Base with a Strong Partitioning Power

5.5.1.1 General Principle of the Phase

As we saw earlier (Sect. 5.2.4), the purpose of phase K is to pave the way for the emergence of groundbreaking concepts in the subsequent phases. It involves forming a knowledge base that makes defixation possible for KCP participants. However, it is not clear how we should obtain a K base facilitating defixation, i.e. facilitating expansive partitions in C. In particular, it is not enough to "share" knowledge that each participant possesses in order to achieve the result. In fact this sharing may result in strengthening the identity of objects without opening paths for possible breakthroughs. As already mentioned, phase K is a *state of the non-art* and not just a state of the art. Consequently it is a very active phase which sometimes requires acquiring new knowledge.

First we discuss methods to create a K base with a strong partitioning power and then we address the execution of phase K in a KCP process.

5.5.1.2 What Knowledge Should Be Incorporated in Phase K? Methods to Increase the Partitioning and Structuring Power of the K Base

We know that the structure of a partitioning K base follows the splitting condition (non-modularity and non-independence) (Sect. 5.2). But how can we form a K base with such a structure? How can we complete the available knowledge in this sense? In other words: *how can we determine what knowledge to incorporate in phase K?*

Here we indicate some strategies (S1 to S5) which help increase non-modularity and non-independence in a knowledge base. There may be other strategies; these are simply practical examples that have been proven. These strategies do not compete with each other but rather complement each other.

The reader may use these strategies as an exercise to apply to the case of his/her choice.

The reader may also analyze each strategy by asking himself/herself if it leans more towards "non-determinism" or "non-modularity". Answer elements are given at the end of the presentation of each strategy.

S1: "wake up" the dormant partitions of a systematic design.

Consider a well-established systematic design (e.g. the aircraft engine in Chap. 5). Certainly the architectural functions, conceptual models, and choices tend to be traced back to objects of established identity; however, there are often known but neglected alternatives because they are usually not retained. Thus there are as many paths as can be systematically reopened and which can thus indicate the K bases to complete. For example, in the case of aircraft engines:

- at a functional level: a study of uses of energy on an aircraft (beyond mere propulsion: taxiing, cabin pressurization, etc.);
- at a conceptual level: a study of hybrid solutions for propulsion;
- at an embodiment level: a study of architectural alternatives for propulsion. The point is always to complete existing knowledge.

The method tends to *reinforce non-determinism*, as it specifies alternatives where there seemed to be only one path.

S2: activate the tacit facets of the identity of objects: uses, business models, operational ecosystem, and so on.

A systematic design relies upon fixed dimensions. These dimensions are known (see functional analysis in Chap. 2, for example): systematic design is performed based on given *business models, usage,* and *operational ecosystems* already codified in functions. Thus, by learning more on these three dimensions we can activate alternatives to existing systematic design very effectively. This is why studies on uses (particularly deviant ones), alternative business models or detailed knowledge of the operational ecosystem are sources of expansive partitions. There are numerous case studies of uses that encourage research. Refer, for example, to the Telia case (Le Masson et al. 2010b) or the "bolt-cutter" case (see workshop 2, Chap. 3).

The method tends to *strengthen the non-modular character* of the knowledge base: uses, business models, and operational ecosystems were intended to be taken into account in the functions and hence to no longer influence the design process; on the other hand, work in K allows us to identify cases where the interruption goes through original uses and business models.

S3: use generic patterns

Numerous concepts tend to use notions (energy, "fight against", etc.) to which we can associate powerful conceptual models: they constitute "patterns" with a strong partitioning power. For example:

1. "fight against" may consist in eliminating the source (in botanics, fighting against a parasite by eliminating it), preventing attack (preventing the parasite from attacking the plant), preventing the harmful effects of the attack (even if the parasite reaches the plant, it does not attack it in the same way), or treating the harmful effects of the attack.

2. In the energy domain, work on fuel-cell applications has led to an enriched conceptual model of energy (Brun and Polo b). The authors have shown that energy issues made it necessary to address three key questions: the thermodynamic model (efficiency, conversion, etc.), the logistical model (transport, storage, flow –capillary or massive–, exchanges, etc.), and the socio-cultural model (home, access to energy, public access, etc.).

Such patterns can be very helpful: they prevent very quick fixations (examples of fixation: to fight against a parasite necessarily means to kill it; to improve fuel cells means to improve conversion efficiency) by taking into account the dimensions that are often neglected (refer to the case of the energy model usage in Agogué 2013a, b) and they make it possible to find alternatives. Let us specify that it's still about knowledge (we stay in K); these models are nothing but "memory aids".

Note that these integrative conceptual models may be either known "patterns" or the result of work in a phase K: the integrative conceptual models allow the reorganization and synthesis of knowledge reviewed during phase K.

This method can be advantageous both to non-determinism ("to fight against" doesn't necessarily mean "to destroy the aggressor") and to non-modularity (depending on use, energy can take very different forms; use hypotheses may thus involve original design paths).

S4: activate knowledge associated to "good ideas"—departitioning strategies

For departitioning examples refer to the Telia example in (Le Masson et al. 2010a,b), the C-K exercises in Chap. 4, (workshop 4.2) or (Kroll et al. 2013). The logic consists in starting with a "good idea" and gradually abstracting it (in C) until we reach the root concept. This operation also reveals the sometimes implicit knowledge that is used to formulate the idea.

For example, in an exercise like the smart shopping cart, a "good idea" is a "shopping cart on air cushion"; the idea goes back to knowledge on modes of mobility, the difficulty of movement in the store, the comfort of displacement, and so on. Departitioning work is particularly useful in K: K bases touched by the "good idea" often deserve additional research (other types of energy for mobility, causes of movement difficulties, nature of the comfort of displacement, etc.).

The departitioning exercise tends to reveal knowledge bases that are somewhat unexpected for design; in this sense, it contributes to non-modularity as it reveals that knowledge that seemed secondary can be determining. More specifically, the departitioning may also consist in generating alternative ideas to the initial "good idea" and hence the knowledge base also becomes non-deterministic.

S5: formulate "dense" concepts

The "dense" concept tends to create links between multiple knowledge bases. Work on pipes for extraction and transport of oil carried out by Vallourec had started with the "after threading" concept—screwing the pipes being the most sophisticated solution. This concept is somewhat dense: the experts know about threading; the concept does not refer to any simple knowledge "besides" threading. After working for some weeks, the designers reached a more complex formulation of the concept: "multi-functional junctions for platform-module infrastructures". This formulation encourages the acquisition of knowledge on the functions of the junction, on the infrastructures in which it occurs, and on the character of the module or the platform of the junction. Thus it urges participants to expand their exploration beyond knowledge on threading by interesting them in wells, in the installation process, in industrial strategies associated with the commercialization of junctions, and so on.

Dense concepts urge the establishment of links with knowledge that is sometimes distant from the identity of the original object (for example, the work on threading did not necessarily go back to the infrastructure of the entire well). Thus they rather contribute to non-modularity. Non-independence may also result from them if the work in K makes it possible to create alternatives.

5.5.1.3   Objectives and Execution of Phase K in KCP

Having analyzed some strategies to increase the partitioning power of a K base, now we present phase K of a KCP.

Experience has shown us that phase K is critical in a KCP. However, this is a phase that often appears slow, expensive, and demanding for participants. We might be tempted to shorten it. But we would run the risk of starting phase C in a context favorable to fixations. Another risk consists in integrating phase K in an upstream research phase, which leads to a long pursuit of this phase in the hope of finding "solutions" in K, while neglecting the required conceptual effort. One of the challenges of KCP is to show tangible results as from phase K.

Nature of expected results: A phase K whose main objective is to contribute to eliminating fixations; in the table below we explain the objectives according to the four types of fixation seen in this book (Sect. 5.2.1).

Exercise: without looking at the solution provided below, fill out the table by indicating what the contributions of a phase K are on each of the four types of fixation (Solution in Table 5.6).

The main expected effect relates to the lower left quadrant (new knowledge to fight individual fixations). Let us emphasize, for this case, that knowledge does not aim to summarize known objects, as would be the purpose of a "summary" of a topic; instead it aims to *grasp the areas of future development of the object*, and those future areas are not necessarily the same as those of the past.

But a K phase has more than just cognitive defixation objectives in K! In addition to the lower left quadrant, the remaining three cases are also important:

- The K phase also has an impact in C (upper left case); it helps reformulate the concept (making it "denser" and allowing it to spontaneously link more knowledge).
- The K phase also has a significant socialization effect (right column); it makes it possible to incorporate experts in an innovative design process intended to rediscuss established design rules. The K phase must lead experts to present what is known—often more than what novices think—and the limits of validity, which are

Table 5.6 Objectives and evaluation criteria of a K phase. The criteria are formed based on the four fixation effects that the collective-design capabilities must overcome

	Cognitive factors of the fixation effect	Social factors of fixation effects: overcoming the rule breaking resistance
C-expansion (K → C, C → C)	→ **Cover the whole conceptual potential of the initial concept?** *K Phase:* • *stimulating examples,* • *prepare «out-of-the-box » thinking,* • *expansive/integrative conceptual models,* • *formulate dense concepts*	→ **Involve and support people in a rule-breaking process?** *K phase:* • *Let experts present their expertise (and its limits of validity),* • *become aware of false good ideas (trite new ideas, fashionable ideas…),* • *time for sharing synthesis and surprises from K-input (see Ideo prototypes)*
K-expansion (C → K, K → K)	→ **Activate, acquire and produce relevant knowledge?** *K Phase:* • *Multifacets perspective,* • *the object in its operational ecosystem with uses and business model,* • *genealogical perspective,* • *innovation competition analysis,* • *predictable dynamics (of ecosystem, technologies, products,…)*	→ **Manage collective acceptance and legitimacy of rules (re) building?** *K phase:* • *state of the non-art,* • *limits of available knowledge and models,* • *identify and commit stakeholders,* • *become aware of issues and competition (weak signals), of the increase of unknownness,* • *know launched projects (potential resources)*

not always easy for the experts to specify explicitly, much less so for the novices. The state of the art may also involve certain recurrent concepts: the K phase makes it possible to quickly identify the limits of false good ideas and trite points.

• Finally, the K phase prepares people for the need to acquire new and sometimes very important knowledge (lower right); it exposes the limits of trivial solutions and the limits of available expertise; it reveals competitors' most advanced decisions (the K phase must lead to learning about research initiatives already launched by the competition or decisions already made in terms of intellectual property); the K phase must also show a status of the work already started internally in the company, in order to make the best use of available resources; finally, always with a view to using future resources, the K phase aims to mobilize stakeholders by making them share the challenges and the resource needs revealed by the initial work.

Execution: The execution of the K phase consists in identifying a first set of K bases thanks to an initial "control C-K", i.e. a C-K graph that helps to control the overall reasoning. The K phase will be able to use the methods seen above. K bases are presented by experts. Each presentation leads to exchanges with participants, mainly to reinforce the effects of defixation (see above: identification of emerging concepts and reformulation of the main concept, limits of available knowledge, anomalies, missing knowledge, etc.).

During the presentations, the *control C-K* can evolve and lead to the need to present additional K bases.

A synthesis effort occurs in parallel, allowing the gradual formulation of integrating conceptual models.

Reminder: experience with KCPs shows that a K phase must involve at least the following knowledge bases:

a. direct customers: knowledge of markets and products (segmentation), specifications, cost of functions and their value, and so on.
b. users and the entire operational ecosystem: expectations, prices, behavior, deviations, etc.
c. company strategy: knowledge and skills in the company, brand image, identified strategic challenges, existing products and options, market positioning, etc.
d. the competition: products, ongoing projects, research activities, patents, skills, etc.
e. state of the art: patents and existing technical solutions, regulations, past products and genealogy of these products, R&D ecosystem (external laboratories), entrepreneurs, etc.
f. phenomenology: models and usage scenarios, associated physical phenomena, modeling, available observation and analysis instruments, etc.

We pointed out that one of the challenges of a KCP is to obtain tangible results as from the K phase. Below we give examples of results of real cases:

- *"Weak signals."* Example from the Vallourec KCP "after threading": as already mentioned, Vallourec launched a KCP for "after threading". Given that threading is at the heart of Vallourec's work and performance, such a title alludes, for a tire vendor, to launching a project on "after rubber". After threading is a future that the company tends to consider far-fetched. One of the initial results is to discover that certain players are already working on serious alternatives to threading—and "after threading" might become much less "far-fetched": after threading thus becomes a possible future. What's more, the participants also discover that some direct competitors are working on "after threading" activities with some of Vallourec's research partners!
- *Expertise crossing*: KCP Thales cockpit: in a K session on architectures of helicopter cockpits, a specialist of night vision on helicopter helmets that is present in the room discovers that the door pillars of certain models limit the pilot's field of view much more than he thought.
- *Modeling to achieve defixation*: in a project on two-wheeler safety, an expert remembers, in the case of *car* safety, the existence of a decoupled model: over a long period, car safety improves thanks to *independent* work on the vehicle (safety systems, vehicle architecture), on the driver (license, controls, etc.), and on the infrastructure (e.g. dangerous crossings). Used in the case of two-wheelers, the model reveals that for these vehicles there is a fourth variable (equipment: helmets, clothes sets, etc.), that the "vehicle" variable is not enough, and most important, that there are strong interdependences that prevent us from separating the problems: the infrastructure must take into account the vehicles; two-wheeler safety cannot be achieved without taking into account

cars (or other vehicles) which are very often involved in two-wheeler accidents. This little modeling work shows why two-wheeler safety cannot be thought in the same way as that of the car, and which original aspects we risk neglecting by making an analogy between car safety and two-wheeler safety.

- *Provoking examples*: on the same subject of two-wheeler safety, two international examples surprise the participants: the Netherlands, which in the 2000s invent "shared spaces"—spaces that no longer separate the different flows (pedestrians, bicycles, cars) and only impose rules of priority of the "weak" over the "strong" (the pedestrian has priority over the bicycle, which in turn has priority over motorized two-wheelers, which in turn have priority over cars); and Malaysia which at the same time develops infrastructures that allow to completely separate the flows (bicycles and cars no longer share the same lanes at all, even at crossroads—a little like trains and cars which circulate on separate infrastructures). These examples demonstrate the importance of managing the interdependences between different vehicles in the case of two-wheeler safety.
- *Rediscovery of unsuspected internal resources*: KCPs often lead to sharing with all participants the experiences and expertise held by a small number of people: surprising prototypes, exploration of older technical concepts, "wigged" projects, and so on.
- *Modeling efforts directly useful for rule-based design*: work on knowledge makes it possible, first of all, to go back to the simplest development aspects of the objects, namely innovation aspects in rule-based design; it forces us to revisit the conceptual and the generative models of the current object. For example, a KCP on microbusses at RATP (the Parisian transports operator) allowed first of all to update the design "platform" of current bus lines (a model of the vehicle, of the line, its operation, its lifecycle, etc.).

5.5.2 Phase C: Shedding Light on Paths in the Dark Thanks to "Projectors"

5.5.2.1 General Principle of the C Phase

As seen in Sect. 5.2.4, the C phase explores the concept space thanks to knowledge accumulated in the K phase. But this is not a free exploration. Contrary to intuition which would like for the exploration in C to be some kind of unrestrained creativity, the C phase is strictly steered so as to achieve its purpose, which is to overcome fixations. Without steering, ideas—numerous as they may be—tend to explore a limited number of paths, by focusing on variations of the same propositions; the C phase is organized so as to go beyond the classic concepts. Steering is ensured by "projectors", which are approaches deliberately differentiated from the initial concept and which make it possible to guide exploration in C.

First we present the projector construction logic before specifying the details of execution of a C phase.

5.5.2.2   Projector Construction Logic in a C Phase

1. The construction of projectors starts with identifying in the K phase any possible "strains" or expansion lines for the object. These are *emerging spontaneous concepts*, candidates for projector concepts.

These emerging spontaneous concepts may be original notions, "good ideas", or initiatives of competitors or new market players.

They may also be created by identifying the strains on the object: the RATP, working on "walking" as a mobility mode, becomes aware of the evident fact that walking is not a "secondary micro-mode" but a support mode that is essential for the use of transport modes used by the company. A KCP on "night" bus stations (see case study "functional analysis" in Chap. 2 Workshop 2.1 p.45) reveals that in the K phase there are several types of night for public transport: winter night, between 5:00 pm and 9:00 pm, which is already integrated in current transport modes; late night, between 9:00 pm and 2:00 am, also integrated; and the night from 2:00 am to 5:00 am which is the new night, in which the town's configuration changes radically. It is for this third type of night that the station is particularly under strain, given that the town is "closed" and no longer offers conviviality or nearby assistance that is much needed when there are health or safety concerns.

Another example of a concept constructed on a strain on the object: a KCP on smart grids at Areva keeps coming back to the strain between a nuclear power plant zone made for base load operation, and often "fatal" renewable energies (i.e. which depend on external conditions to operate: wind, sun, etc.) and for which, therefore, only a base load operation is possible, with more classic sources supplied as a supplement. This raises a question (in fact, a concept): how can we make nuclear power plants operate at "peak level"?

2. Besides this first review, the construction of projectors depends on a *control C-K* carried out by the steering group, allowing to identify the main expansion principles.

This C-K can itself be controlled: is it strict (i.e. do all Cs have a reference K and vice versa)? Does it allow positioning the common expansion paths (dominant design) and the split paths?

This control C-K helps ensure that the potential expansion paths make it possible to achieve some kind of exhaustiveness in C. In other words, when this work is completed, it shouldn't be possible for a competitor to surprise us with an original offer.

3. *Development of projectors with adornment and wit.*

Projectors structure the exploration of paths around the object. Among these paths we can distinguish those that conserve the identity of the object by enriching it—in that case we are talking about *adornment*; and those that shake it up and question it—in which case we talk about *wit* (these notions were introduced by Hatchuel 2006a).

Adornment adds dimensions to the object. Attribute P adds value to object A and it is new for A. But it does not shake up A. Adornment enriches the object and links it to new K bases. We can also list some classic adornments: sustainable, intelligent, low cost, reassuring, "quick and smart", and so on.

Wit, a term that comes from Spanish rhetoric (see Gracian 1648), about subtlety and the art of genius), designates a new and provoking connection to the object. This time, property P added to the object destabilizes it. For example: "cockpit outside of the airplane" destabilizes the notion of cockpit which is often seen as the piloting space integrated into the aircraft. Wit forces us to extend the object into entirely new categories. Wit has a strong strategic impact: it makes unexpected competition possible, as well as new forms of alliance. Sources of wit: search "beyond the sector", in new technical offers or new business models. Wit counterbalances the fixation in C, the temptation to always search in the main conceptual area. Classic cases of wit:

- dematerialization and despatialization (real/virtual)
- inversion or strong intensification of an adornment
- desocialization: new communities, new partners, new business models.

For practice, in the KCP below the reader can try to justify the evaluation of adornment or wit made on each of the projectors (Fig. 5.31).

4. This way we create several projectors. We must form a set of projectors that will be submitted to the participants. This set must take into account people's fixations. We can distinguish two main categories of contrasting situations: fixation in K or fixation in C.

a. In the first case, expansion in C appears easy to designers… but in fact certain paths are closed because of certain work in K. It is a characteristic situation of weak engineering, of weakly structured industrial systems. In those cases, it is first of all the adornment that fails; projectors look for ways to compensate for this effect and thus give priority to *embellished projectors* The KCPs "operation", "autolib", and "night station" rather fall within this context.

b. In contrast, in the second case, it is the expansion in K that is easy, people are familiar with adornments. But fixation in C prevents them from facing any ruptures. We encounter these situations in industries that perform dominant design, which have strong engineering and research capabilities, in stable disciplines, in stabilized ecosystems. In this case, the set of projectors must *reinforce wit*. The KCPs "smart grids" and "after threading" illustrate this second case.

Hybrid situations are, of course, possible.

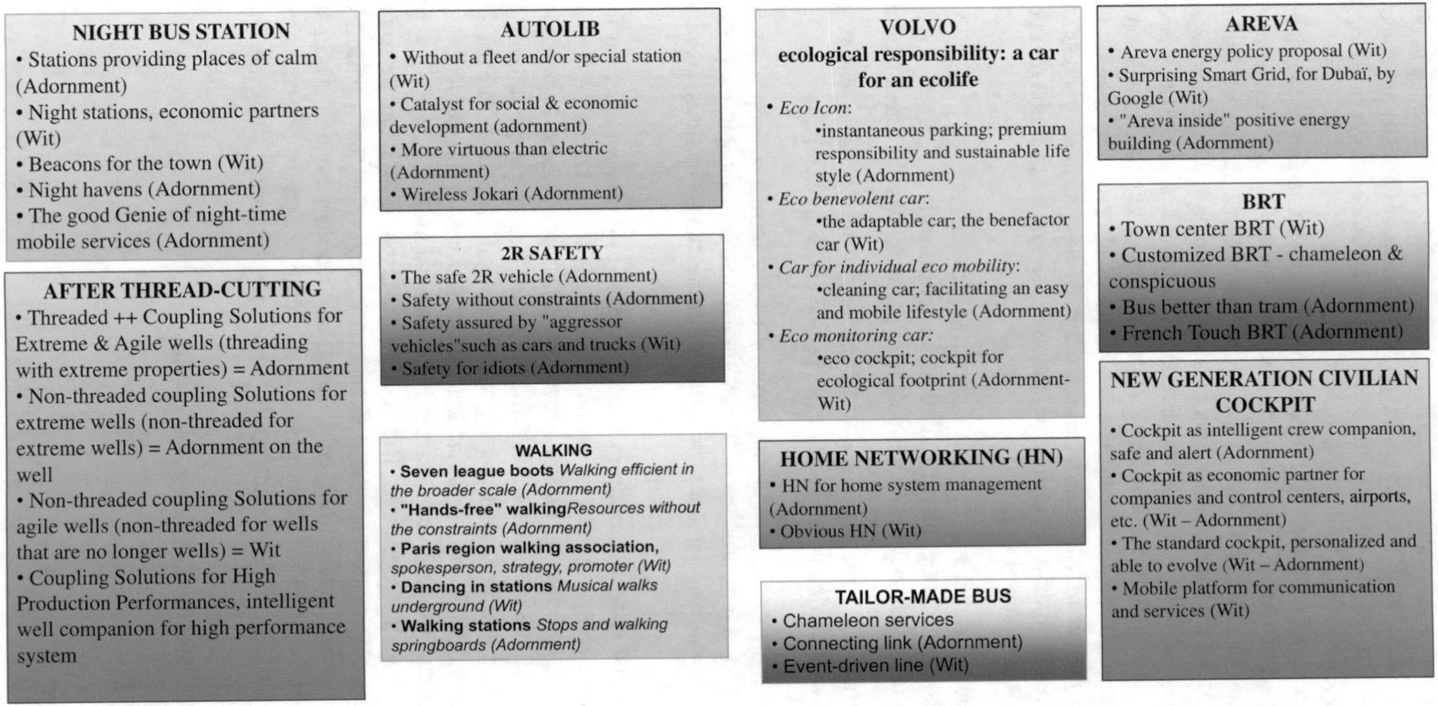

AREVA

- Areva energy policy proposal (Wit)
- Surprising Smart Grid, for Dubaï, by Google (Wit)
- "Areva inside" positive energy building (Adornment)

BRT

- Town center BRT (Wit)
- Customized BRT - chameleon & conspicuous
- Bus better than tram (Adornment)
- French Touch BRT (Adornment)

NEW GENERATION CIVILIAN COCKPIT

- Cockpit as intelligent crew companion, safe and alert (Adornment)
- Cockpit as economic partner for companies and control centers, airports, etc. (Wit – Adornment)
- The standard cockpit, personalized and able to evolve (Wit – Adornment)
- Mobile platform for communication and services (Wit)

VOLVO

ecological responsibility: a car for an ecolife

- *Eco Icon:*
  - instantaneous parking; premium responsibility and sustainable life style (Adornment)
- *Eco benevolent car:*
  - the adaptable car; the benefactor car (Wit)
- *Car for individual eco mobility:*
  - cleaning car; facilitating an easy and mobile lifestyle (Adornment)
- *Eco monitoring car:*
  - eco cockpit; cockpit for ecological footprint (Adornment-Wit)

HOME NETWORKING (HN)

- HN for home system management (Adornment)
- Obvious HN (Wit)

TAILOR-MADE BUS

- Chameleon services
- Connecting link (Adornment)
- Event-driven line (Wit)

AUTOLIB

- Without a fleet and/or special station (Wit)
- Catalyst for social & economic development (adornment)
- More virtuous than electric (Adornment)
- Wireless Jokari (Adornment)

2R SAFETY

- The safe 2R vehicle (Adornment)
- Safety without constraints (Adornment)
- Safety assured by "aggressor vehicles" such as cars and trucks (Wit)
- Safety for idiots (Adornment)

WALKING

- **Seven league boots** *Walking efficient in the broader scale (Adornment)*
- **"Hands-free" walking** *Resources without the constraints (Adornment)*
- **Paris region walking association,** *spokesperson, strategy, promoter (Wit)*
- **Dancing in stations** *Musical walks underground (Wit)*
- **Walking stations** *Stops and walking springboards (Adornment)*

NIGHT BUS STATION

- Stations providing places of calm (Adornment)
- Night stations, economic partners (Wit)
- Beacons for the town (Wit)
- Night havens (Adornment)
- The good Genie of night-time mobile services (Adornment)

AFTER THREAD-CUTTING

- Threaded ++ Coupling Solutions for Extreme & Agile wells (threading with extreme properties) = Adornment
- Non-threaded coupling Solutions for extreme wells (non-threaded for extreme wells) = Adornment on the well
- Non-threaded coupling Solutions for agile wells (non-threaded for wells that are no longer wells) = Wit
- Coupling Solutions for High Production Performances, intelligent well companion for high performance system

Fig. 5.31 Some examples of projectors on KCPs. Caption: Ad = Adornment; Wt = Wit. The reader will be able to justify the characterization of adornment versus wit of the projectors proposed

5.5.2.3 Objective and Execution of a C Phase

Having analyzed the projector strategies in order to collectively explore and structure C, now we present the C phase of a KCP.

1. A phase C whose objective is to contribute to eliminating fixations; we can explain *the objectives according to the four main effects seen in this book* (Sect. 5.2.1).

   Exercise: without looking at the solution provided below, fill out the table by indicating what the contributions of a phase C are on each of the four types of fixation (Solution in Table 5.7).

   As with the K phase, the objectives are diffracted onto the four fixations to oppose. These objectives help specify projector evaluation criteria:

   - activate a knowledge pocket that is far from the dominant design but that constitutes a valuable challenge
   - perform a displacement on the concept tree (gain in originality)
   - strong potential for expansion in C and in K
   - shed light on a critical crossing point

Table 5.7 Objectives and evaluation criteria of a C phase. The criteria are formed based on the four fixation effects that the collective-design capabilities must overcome

	Cognitive factors of the fixation effect	Social factors of fixation effects: overcoming the rule breaking resistance
C-expansion (K → C, C → C)	**→ Cover the whole conceptual potential of the initial concept?** *C Phase:* • *catching the expansion directions of the innovation field,* • *be as exhaustive as possible*	**→ Involve and support people in a rule-breaking process?** *C phase:* • *structure the unknown,* • *organize multiple parallel explorations that will support each other;* • *manage conflicts by investigating all the alternatives, including the conflicting ones (don't select!);* • *give freedom to people (who were impeded in K-phase!)*
K-expansion (C → K, K → K)	**→ Activate, acquire and produce relevant knowledge?** *C Phase:* • *mobilize as much K as possible from the K-phase;* • *share knowledge emerging during the C-phase;* • *identify missing competences*	**→ Manage collective acceptance and legitimacy of rules (re) building?** *C phase:* • *agreement on missing and required knowledge;* • *agreement on critical design paths (and related missing K)* • *all participants are opening their networks for future knowledge acquisition;* • *identify the most mobilizing alternatives*

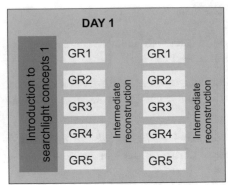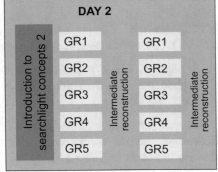

Fig. 5.32 Typical execution of a C phase

We can also specify criteria for a good set of projectors:

- the projectors cover the control C-K significantly in C and in K,
- they respect the good balance between adornment and wit (balance to be defined according to the situation).

2. The *typical execution* of a C phase can be illustrated by the diagram below: we distinguish, over two days, several sequences of work in groups, in parallel, and—between sequences—common restitution points (Fig. 5.32).

Some methodology indications on each phase:

a. *Presentation of projector concepts*: each projector concept can be presented on an illustrated sheet that offers stimulating examples and evokes possible associated knowledge. This sheet may also indicate a systematic-exploration logic: possible concept values (new FRs), possible means of action (new DPs), skills to acquire. The illustration below is the sheet for the projector concept "seven-league boots" for the KCP "the walk" organized by Georges Amar and Blanche Segrestin for RATP (Fig. 5.33).

b. *Group formation and activities*: groups are formed with a view to maximizing the variety of fields of expertise and allowing the hybridization of skills, profiles, and experiences. In each group, a rapporteur is in charge of restitution; to that end he reorganizes and clarifies the exchanges of the group throughout its work. KCP leaders monitor the groups and follow their work progress. They must get an idea of what the group needs to explore (at minimum), they may encourage the use of a certain C-K reasoning, albeit implicit (work on the description of K bases used, strictness in the formulation of concepts); they are attentive and must encourage reformulations and clarifications; they may prevent original ideas from being forgotten by the group.

c. *Intermediate restitutions* are a very important moment for the exploration dynamics. The groups start to discuss among them and influence each other;

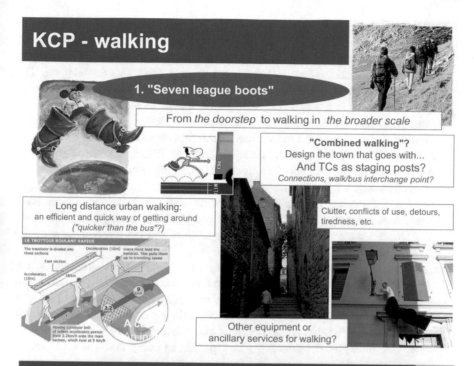

Fig. 5.33 Projector presentation example

certain propositions of one group may solve certain problems or open perspectives for another group. Thus, in the "night station" KCP, a group proposes
the notion of "retractable" station (a station that deploys only from 1:00 am to
5:00 am); this notion stimulates the work several other groups later on. The
groups realize that there is no competition among the different projectors but
that in contrast it is a way to organize collaboration in the unknown.

d. *Going from the first day to the second* demands a lot of work from the facili-
tation team: based on the explorations of the first day, the team must reuse the
control C-K to develop a second series of projector concepts that will impel the
groups to carry out additional explorations. This reformulation work may be
prepared in advance but we must take into account the fixations that could have
revealed themselves on the first day, surprising concepts that could have
appeared and that the steering group had not identified during the preparatory
work. In this stage the KCP leaders have already formed a broad view of the
design strategies that the explorations will help enrich.

e. The seminar concludes with a *short conclusive session* which reports on the
progress and the variety of explorations. In this session, however, we must avoid
making an impulsive summary: there are too many elements to integrate in a
well-constructed design strategy—this is the challenge of the P phase. Variety is
already valuable, though: it often offers a good description of what competitors
can do and thus give the possibility to organize an analysis of the competition
over an extended space.

5.5.3   Phase P: Developing a Design Strategy

5.5.3.1   General Principle of the P Phase

As we saw in Sect. 5.2.4, phase P transforms the propositions developed in phase C
in a program of action. Contrary to intuition, it is not about selecting a proposition.
That would, in fact, be a reasoning error: from a formal point of view, the selection
of the best solution(s) among n, at a constant budget, is based upon the hypothesis
that the alternatives are independent. Yet, by nature, conceptual propositions all
come from the same origin and, most often, share numerous resources in K. Always
from a formal point of view, it is thus about analyzing *combinations of interdependent alternatives*. It is, actually, what we have been able to study in the
risk-management strategies in innovative design (See Sect. 5.2.3). Phase P, therefore, consists in developing *an innovative design strategy taking into account the
interdependences between propositions*.

First we present the logic of the design strategy before specifying the details of
execution of a P phase.

5.5.3.2   Notion of Design Strategy

In rule-based design, we saw (Chap. 2) that the value of a design activity is that of a singular project and it depends to a large extent on the part (or the volume) of the market acquired by the company, V, and on the margin M. An actualization over the course of time helps us find the net present value (NPV) .

In innovative design, the value is related to the design of a new identity on the market, I, which will be able to capture an unknown margin M and the volume V of a market that remains to be defined. This identity on the market can be constructed in several stages, during which I, V, and M may increase.

In traditional logic, this identity I is fixed (by dominant design) and the company forms a compromise between V and M (see figure below). In innovative design, *the design of I* makes it possible to overcome the compromise V versus M (increase M holding V constant or increase V holding M constant or increase them both!) (Fig. 5.34).

In innovative design, designing new identities on the market demands (or makes possible) more complex and richer design strategies than in rule-based design (for a more detailed explanation see Sect. 5.2):

- On *risks*: Risk logic not only takes into account the technical risks of the project or market risks but must also include the *risks of things no to do*, the risks of things done incorrectly (reinforce fixations, cause speculation bubble effects), the risks of doing too much of something too fast (very high risk investments), the risks of something done by one person (ignorance of time lines and of learning progressiveness, negligence with respect to the contributions of other designers, etc.). In very general terms, these risks impose, in fact, approaches that are rigorous, progressive, prudential, and collective.
- On *the value*: Valuation logic not only takes into account the value of the singular project but must also include multiple valuations: valuation done as soon as possible on products that exists or that are being developed, valuation on the entire ecosystem (with partners, suppliers, advisors, the R&D ecosystem and, more generally, design partners such as design or engineering schools), valuation on market capitalization, recruitment, brand image, and so on.

Fig. 5.34 Displace the margin/volume equilibrium by designing new identities on the market. V = market volume, M = margin, I = new identity on the market

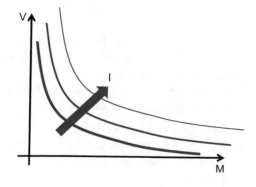

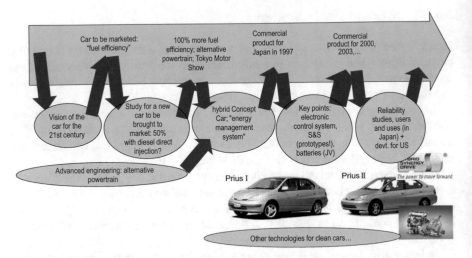

Fig. 5.35 Example of gradual creation of an identity on the market: Toyota and hybrid cars

An example of risk management and valuation in innovative design is given by the design of Prius and other hybrid vehicles by Toyota in the late 1990s. The reader may refer to (Magnusson and Berggren 2001) and to the examination instructions of the "Product Design and Innovation" course of 2005, included in the appendix. The article (and the instructions given for the exam) leads to an innovative design process of the "value management" and "design spaces" type with several stages (see diagram below). These stages allow a systematic and quick valuation (quick gain in image, in prospective knowledge, etc.) and risk management (gradual elimination of market risks, setup of a network of research partners, creation and gradual extension of a strong identity on the market, etc.) (Fig. 5.35).

5.5.3.3 Objectives and Execution of a P Phase

1. A phase P, like the other two phases, whose objective is to contribute to eliminating fixations; we can explain *the objectives according to the four main effects seen in this book* (Sect. 5.2.1).

Exercise: without looking at the solution provided below, fill out the table by indicating what the contributions of a phase P are on each of the four types of fixation (See solution in Table 5.8).

As with phases C and K, the objectives are diffracted onto the four fixations to oppose. In particular, we may emphasize that a P phase:

- certainly aims to define projects to guide innovative design in the initial design field
- but that these projects are in fact the first steps towards a design strategy, and it is the coherence (in terms of risk and value) of this strategy that adds quality to phase P.

Table 5.8 Objectives and evaluation criteria of a P phase. The criteria are formed based on the four fixation effects that the collective-design capabilities must overcome

	Cognitive factors of the fixation effect	Social factors of fixation effects: overcoming the rule breaking resistance
C-expansion (K → C, C → C)	→ **Cover the whole conceptual potential of the initial concept?** *P Phase:* • *More than "one good idea": a strategy to cover the whole tree, with multiple, contrasted and complementary projects*	→ **Involve and support people in a rule-breaking process?** *P phase:* • *Not necessarily "compatible" with predefined strategy of the firm, organization and brands: take the opportunity to rediscuss firm strategy, brand and organization*
K-expansion (C → K, K → K)	→ **Activate, acquire and produce relevant knowledge?** *P Phase:* • *Not only "feasible" ideas, with "limited investment in K": take into account K production and acquisition*	→ **Manage collective acceptance and legitimacy of rules (re) building?** *P phase:* • *Not necessarily limited to the competences of the initial team: redefine (extend) the team, play with the ecosystem*

- the objective is therefore not the success of a singular project but the implementation and the gradual exploration of new identities on the market
- by ensuring the commitment and gradual involvement of the company and its ecosystem (current and... future stakeholders).

2. *General process of a phase P*:

The general process (see diagram below), starting from knowledge and concepts accumulated in the previous phases, consists in:

(1) identifying new possible identities on the market, as well as the associated ruptures, which may be cognitive (in C and in K) but also organizational (new processes, committees, new company identity, etc.).
(2) Then it is about identifying the associated design strategies: starting point, time horizon, actions, martingales, etc. We must associate a cohesion logic to these strategies, i.e. a logic of commitment and involvement of partners: this concerns the company's internal partners (based on the strategy, the organization, possible internal shareholders), customers, users, advisors, and certifiers, as well as external design partners (suppliers, R&D ecosystem, and design ecosystem).
(3) The work finally leads to a set of contrasting and interdependent projects.

This phase P entails:

1. significant work on concepts (reformulation of the initial C_0, restructuring of Cs and Ks accumulated in a control C-K, etc.)

2. work to position C-K with respect to challenges, which are identified mostly thanks to the work on competition/innovation and to the strategic guidelines developed in the K phase
3. a constant effort for restitution and socialization of the work: the P phase entails regular meetings of the steering committee and opening the work group to relevant stakeholders (Fig. 5.36).

3. *Methods and example of identification of new identities on the market and associated ruptures.*

Three types of rupture are expected:

1. rupture in C: change in the identity of objects, components, service, business model, uses, etc.
2. rupture in K: new knowledge on competing innovation initiatives (competition/ innovation analysis, based both on the initiatives of known competitors and on innovation initiatives by other players), identification of critical knowledge that is often deficient (uses, value, new technologies, etc.), knowledge on new partners (new alliances, new networks to incorporate, etc.)
3. organizational rupture: review of internal or external organizations. This includes: integration of new skills (design in certain technical universes, R&D in certain business universes, etc.), new types of logic to organize exploration within the company (implementation of a high-level innovation company, a logic for inter-projects exploration platforms, new types of portfolio management, new processes to involve stakeholders in breakthrough projects with round-table logic and internal business angels), organization of links with the external environment (new method for facilitating research networks, ecosystems, connection with users and advisors, etc.).

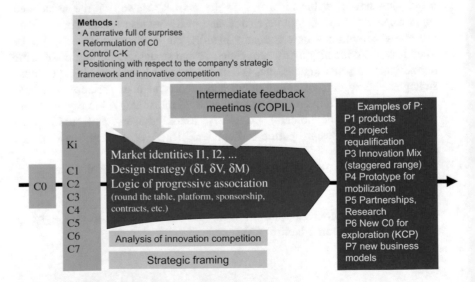

Fig. 5.36 Execution and methods in phase P

These ruptures result from explorations of previous phases in C and K. We have to do their mapping. First a synthesis C-K is made, with reformulation of the initial C_0, identifying ruptures in K (in value and in robustness) and ruptures in C (identification of expansive partitions and rupture in design rules, mostly related to the market), and clarifying the structures in place (within the company but also the structures of the ecosystem and of the competition).

Example: identification of ruptures in the P phase of the "self-service car rental" KCP. This KCP was launched in the late 2008 by a consortium of partners (including the RATP) to prepare an innovative bid to the call for tenders by the Paris City Hall on the future Autolib (self-service car rental). In May 2009 (a few months before the bid submission to the call for tenders and after phases K and C of the KCP), the steering group, coached by the authors, was drawing up the mapping of ruptures. To that end, the group started by reproducing in C-K the design reasoning foreseen before the start of the KCP (see Fig. 5.37):

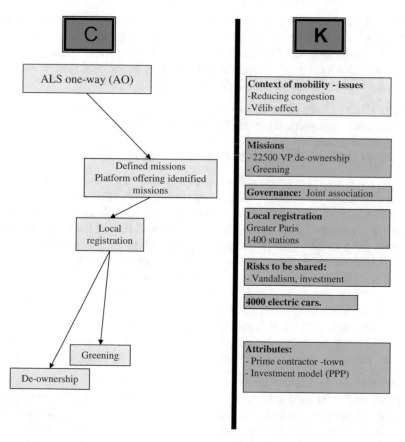

Fig. 5.37 C-K graph representing the state of knowledge and concepts before the "self-service car rental"

- In K: to simplify we distinguish context knowledge (upper section), the knowledge related to the value (FR, middle sections), and the knowledge related to the means of action (DP, lower section).

  - Context: the initiative follows on from the success of Vélib and aims to address the challenges of reducing traffic jams in the city.
  - Value: a self-service car rental is proposed here, which, in time, would result in the "non-ownership" of 22,500 private vehicles (i.e. 22,500 vehicles not purchased by Parisians, with respect to the current situation), with a view to respecting the environment and achieving sustainability, with 4,000 electric vehicles and 1,400 stations installed in Paris and its environs, based on a risk-sharing model that includes vandalism risks (after the vandalism observed on Vélib).
  - Means of action: the project owner is the city, and funding could be based on a PPP (public-private partnership) investment model. These elements constitute initial knowledge.

- In C, a specific project is quickly designed: a "one-way" self-service car (the vehicle is not necessarily returned to the point from where it was rented), with a set of predefined mobility missions (rental of a car for a specific amount of time), with a local registration in stations marked on the road (Vélib model) with a view to de-ownership and greening (Fig. 5.37).

After phases C and K we obtain the graph of Fig. 5.38. Note in K an extension on the context (see particularly the "ownership" challenges), on the values and on the means of action.

Thus we can easily identify the ruptures in C and in K.

- In K: the work in the previous phases helps identify seven critical and, at the same time, deficient knowledge pockets (see graph). This is rupture knowledge. For example: knowing the risk-sharing models, working with manufacturers in order to better understand the constraints associated with the car itself, getting to know better the possible structure schemes of third parties (e.g. communities), etc.
- In C: with respect to the original direction, we note a reformulation of the concept and several new alternatives: paths allowing open missions (autolib as a laboratory for new means of mobility, both in technical and in social terms), paths exploring alternative local registration types (autolib models for private fleets, for private stations, etc.), and paths exploring types of logic for subsequent new types of temporary ownership for people that don't have access to full ownership. All these paths are ruptures with respect to the original direction (Fig. 5.38).

4. *Methods for developing projects of a design strategy.*
   Based on these ruptures we must develop a design strategy. Said strategy entails a reformulated root concept, a set of structured alternatives (promise, values, etc.) expressing a design basis, the critical missing knowledge, martingales identifying the first moves and the manner in which those first moves can help play subsequent ones, related organizational mechanisms (platform, committees, contracts, etc.) and a commitment logic to describe the involvement of partners in the design.

Context of mobility - issues:
-*Hard to access (ownership) means of transport* (Tunis): living space, access to mobility, gaps in public transport system, shared cars
- *De-ownership of individual vehicle* (Sao Paolo): dense, structured social networks, creditworthy customers
- Exploration of new forms of mobility: new uses (neo-teleworking, services at/in/towards urban mobility)
- Urban mobility system: private car & public transport & X-lib (car, bike, etc.) & walking & etc.

Car/microcar ("voiturette")
- Expensive/cheap
- Risk sharing/ low risk

Risks to be shared:
- Financial exposure
- Vandalism
- Low take-up

Contexts:
- mobility requirements
- existing means: dense
- user capabilities
- Objective: greening, for everybody, vehicle for the specific objectives of the prime contractor

Urban booster
- Economic & social devt.
- Energy/mobility bills

Urban media
- Social networks
- Information systems 2.0 (open hard/soft)

Urban dynamism
- Centre (Paris "bobos" [bourgeois-bohemians])
- Outskirts (Outer Paris commuters)
- Neighborhood (Social integration)

AutolibLAB
- R&D Labs (β-cars, etc.)
- AppLibs (AppStore-like)
- SmartGrid (power cylinder)

Missions: defined by whom and for whom?
- De-ownership
- Ownership
- Greening

Local registration
- open (public places) or gradually open
- closed: private fleet and stations

Governance
All-embracing urban mobility system

Attributes:
- **Financial exposure** : expensive/cheap vehicle/
- **Third parties** : under operator, confidence, payer, supplier
- **Performance model** (associated ecosystem): usage/ main operator /external ecosystem
- **Services** including "lights and beacons"
- **Evolutionary** or not
- **Prime contractor** (town,…) / operator (PF, without vehicle or stations) / third parties (investor) / users / others
- **Investment model**

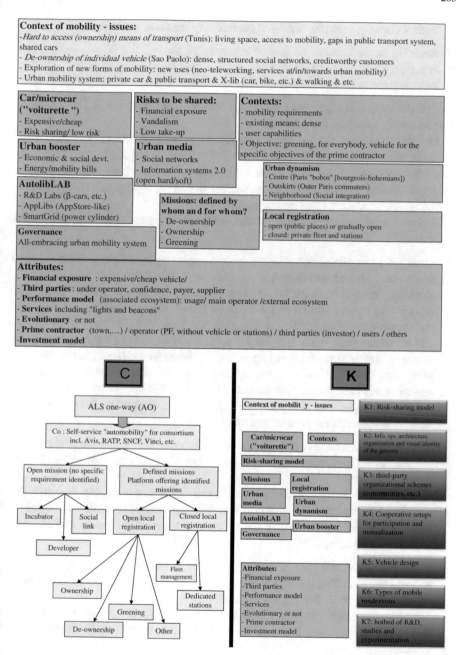

Fig. 5.38 C-K graph representing the state of knowledge and concepts after phases C and K and after a synthesis at the start of the P phase (KCP "self-service car rental"). On the top: K bases constructed in phases C and K. On the bottom: C-K graph. In K of the C-K graph: on the *left*, synthesis of K bases constructed in phases C and K; on the *right*, K bases necessary but deficient

Such a strategy must be deployed particularly on three main axes:

- ensuring an exploration logic with repeated steps
- ensuring a "quick and smart" response, not just "quick and dirty"!
- ensuring commitment

Classic decisional strategies tend to overlook the last two axes and to reduce the first to one or two flagship projects.

During the process, the steering group reports to the KCP players but it must also try to expand the field of thought by gradually including the relevant people and functions to continue the exploration. The results obtained are submitted to an enlarged steering group, which includes, for example, the upper management of R&D, of the product, of the strategy, etc.

As an example, below we present the design strategy considered in the autolib KCP (see Fig. 5.39).

1. The C-K synthesis revealed 5 major conceptual paths (slightly different from those seen above): autolibLAB (autolib as a laboratory for new means of mobility), autolib "first car" ("temporary ownership" path), autolib "better than my car" (path very similar to the main path but favoring the original services), private autolib ("closed" local registration path), and autolib "light" (this too is a variant of the main path with major ruptures with respect to the means); the synthesis has also revealed, as seen above, seven knowledge pockets to complete. The challenge is to cover these 5 concepts and these 7 pockets with projects that correspond to the three axes of a P phase (exploration logic, "quick and smart" logic, and commitment logic). These three axes are reformulated here to indicate three types of research with respect to the call for tenders:

   a. launch *demonstrators* quickly, particularly to make the call for tenders more flexible so that it leaves several paths open ("quick and smart" and commitment logic)
   b. *prepare the future "project plateau"*, mainly by strengthening the skills that will be necessary for the development of the project (exploration logic)
   c. reinforce *the capabilities of the entire ecosystem* (exploration and commitment logic) (Fig. 5.39).

2. Six projects are constructed (see Fig. 5.40). Three of them are demonstrator projects (P1, P2, P3), two are about skills (P4 and P5), and one is rather about the ecosystem (P6). See the six projects below (Fig. 5.40).

3. Thus we can show (see Fig. 5.41) that the six projects "cover" well the 7 K bases, the 5 concepts, and the 4 types of research discussed above. Some comments:

   a. Project P4 is a "risk-elimination" project: if such a project becomes possible, it opens up a whole new realm which questions the other paths;
   b. Project P5 is the one that is mostly geared towards the creation of new expertise;

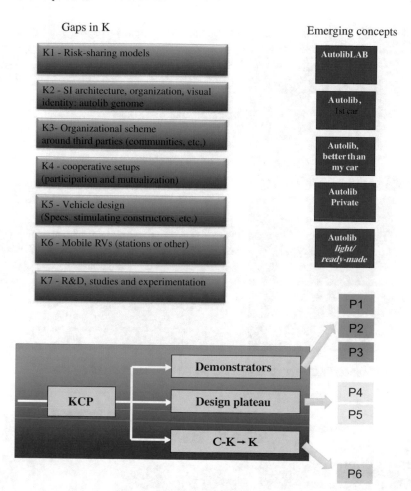

Fig. 5.39 KCP autolib, phase P, development of a design strategy: deficient Ks, emerging concepts and types of project to cover

c. Projects P1, P2, and P3 aim mostly towards "opening" perspectives with respect to car mobility (avoid being trapped in a "vélib with cars" adding up "impossibilities"). This "opening" is mostly intended for stakeholders and therefore takes the form of demonstrators. These demonstrators are targeted both towards technical knowledge and towards knowledge on ecosystem players (Fig. 5.41).

5. *Projects derived from a phase P*

Phase P also leads to a project portfolio and organizational mechanisms in charge of managing said portfolio. Among the most common projects we find:

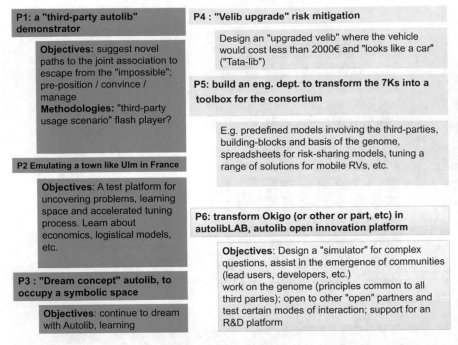

P1: a "third-party autolib" demonstrator

Objectives: suggest novel paths to the joint association to escape from the "impossible"; pre-position / convince / manage
Methodologies: "third-party usage scenario" flash player?

P2 Emulating a town like Ulm in France

Objectives: A test platform for uncovering problems, learning space and accelerated tuning process. Learn about economics, logistical models, etc.

P3 : "Dream concept" autolib, to occupy a symbolic space

Objectives: continue to dream with Autolib, learning

P4 : "Velib upgrade" risk mitigation

Design an "upgraded velib" where the vehicle would cost less than 2000€ and "looks like a car" ("Tata-lib")

P5: build an eng. dept. to transform the 7Ks into a toolbox for the consortium

E.g. predefined models involving the third-parties, building-blocks and basis of the genome, spreadsheets for risk-sharing models, tuning a range of solutions for mobile RVs, etc.

P6: transform Okigo (or other or part, etc) in autolibLAB, autolib open innovation platform

Objectives: Design a "simulator" for complex questions, assist in the emergence of communities (lead users, developers, etc.) work on the genome (principles common to all third parties); open to other "open" partners and test certain modes of interaction; support for an R&D platform

Fig. 5.40 "Autolib" KCP: the six exploration paths upon completion of phase P

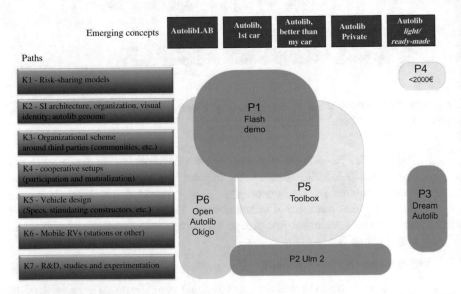

Fig. 5.41 "Autolib" KCP: six paths to cover 5 concepts, 7 pockets of deficient knowledge, and 3 exploration challenges (demonstrators, skills plateau, mobilization of the ecosystem)

- Options or variants of ongoing projects or products
- Patent-filing projects
- Research projects on topics that seem critical
- "Platform" projects: simulators, benches, virtual mockups, serious games, etc. which often constitute an infrastructure integrating multiple initiatives resulting from a KCP. In the Thales case, one of the major results of the KCP on the cockpit of the future was the creation of an original simulator that allows making a demonstration of certain innovation paths and hence makes it possible to integrate new propositions, paving the way. This type of simulator is also a medium of exchange with other designers (suppliers) or users (pilots, in the Thales case).
- "Demonstrator" projects: often light (flash animation, for example), these demonstrators are used for sharing original conceptual paths and for including designers that so far are little involved.
- New organizational structures, such as high-level innovation committees (see Vallourec KCP) or other prototyping centers (see Turbomeca KCP).
- The implementation of processes that help activate knowledge bases in the innovation field: competitive intelligence, patent landscaping, participation in conferences and trade shows, etc.
- New forms of interaction with the ecosystem: renewal of the network of partners, of the research-project portfolio with external laboratories, approaching new players, etc.

Then the evaluation of the project portfolio brings out the general principles we saw in Chap. 5. Thus we consider:

- the profitability of each project (static efficiency criterion, see Chap. 3)
- risk management made possible by this portfolio: reduction of uncertainty, exploration of original alternatives, creation of generic technologies
- Expected learning outcomes (formation of new resources)

In lieu of a conclusion: some additional questions:

1. *Compare the KCP process with the IDEO process (see case study 4.2 in Chap. 4).*
2. *Show how a KCP process allows us to work in the three dimensions of an innovative design system: reasoning, performance, organization.*
3. *For readers that are professionals: choose a well-known company (i.e. a company whose fixation effects you are familiar with), choose an innovation field (C0), and propose the first elements for a KCP in this C0.*

    a. *K bases to review for a K phase in the C0*
    b. *Control C-K*
    c. *First projectors suggestion (justify their adornment or wit character and propose an adornment/wit balance according to the context)*
    d. *Design strategy with some typical projects.*

5.6 Case Study 5.1: Edison, from Inventive Genius to Creator of an Innovative Firm: Edison's Invention Factory

The purpose of this case study is to analyze the design abilities of an original but controversial enterprise, that of Thomas A. Edison (1847–1931).

5.6.1 Why Edison?

Edison is a legend in pioneering innovation work. Edison remains, especially in the United States, an exemplary figure of invention. He showed that innovation was open to "common men": Edison was self-taught and used to be a paper boy when he was young. Drucker praised Edison as a model of "discipline in innovation" (Drucker 1985b), with the capacity to search for opportunities and with a certain obstinacy to attain them. Many authors have insisted on Edison's experimentation capacity: from 1913 Arthur D. Little places Edison's laboratory among the biggest research laboratories in the United States and writes about him that he conducts a "specialized intense research which knows no rest until everything has been tried" (Little 1913). Much more recently, Thomke evokes Edison's example to describe an ability for fast experimentation (Thomke 2003).

However, Edison is also presented as an atypical, extreme case, even as a "fender". For Bertin, who wrote in 1938 an exhibit on the organization of professional collective research in France (Bertin 1938), Edison cannot be an example, as he is an unlikely combination of three figures: inventor, implementer, and guide/director; yet we cannot "expect that God wants there to be scientists and directors of this caliber, because they are relatively limited in number", which makes it necessary to "organize" applied research. In other words Edison is not a model of organization. For Locke, Edison is one of the last "practical men" who will disappear with the arrival of scientists and engineers that graduate from specialized schools (Locke 1984). He would, therefore, be an old-fashioned model. Edison's laboratory is not really a research laboratory; however, it does not resemble a business either. Thus, certain authors have criticized Edison's business sense: for Drucker, Edison is not a good model for business development; Edison "so totally mismanaged the businesses he started that he had to be removed from every one of them to save it" (Drucker 1985a).

Edison's enterprise, therefore, does not have the characteristic traits of an industrial organization and it seems to be far from being able to embody a business model. A typical case of unmanaged innovation? Do the vocabulary and the models developed in the previous chapters allow us to overcome this paradox and to demonstrate interesting forms of organization which would have been concealed by the myth?[8]

We study successively these three dimensions: performance, design reasoning (at least the available elements), and organization.

The reader will be able to answer the following three questions:

1. *How can Edison's performance (in innovative design) be evaluated?*
2. *How can innovative-design reasoning and methods in Edison's enterprise be described?*
3. *What types of organization are implemented in Edison's enterprise?*

The answers are given throughout the text. Probing questions appear in the conclusion.

5.6.2   Some Elements to Evaluate Innovative Design Performance

We can assess the growth of certain businesses based on their size, their turnover, etc., but when it comes to design we have seen that growth, even before we express it in these accounting terms, translates to the ability to propose new concepts and develop new knowledge in order to materialize those concepts. Yet, even if it is difficult to assess the accounting aspect of numerous businesses founded by Edison, we can only be impressed by the conceptual innovativeness. Edison filed 1368 different patents, a number that was for a long time the highest in the world. Although Edison is famous for developing an electrical lighting system by incandescence, he has also developed products in the most varying industrial sectors:

- Electricity (since 1885), from the power plant to the filament of the incandescent lamp; then electric motors with applications on the nascent tramways (since 1888) and on tests on electric vehicles (1899), which propelled Edison onto the production of batteries for railways and industrial plants and then onto the research for a universal electric motor.
- Sound, from the recording of music in a fully assembled studio on the third floor of the West Orange laboratory to the design and manufacture of phonographs, cylinders, and discs (the support), both for professional applications (dictating machine) and for recreation (peep show!).
- Cinema with kinetographs, from the manufacture of cameras to the recording of films in a studio, the Black Maria, built in 1893 in West Orange, and the "kinetoscope", a new hobby consisting in paying to see a short film individually (see pictures below).

---

[8]This reexamination has benefited greatly from the remarkable editing work of the "Edison papers", carried out in the last twenty years and which was accompanied by several publications, among which we cite in particular (Israel 1998; Millard 1990).

- Edison also carries out titanic work in the 1890s on a new type of grinding and iron-ore extraction on a very large scale (which Henry Ford said was his inspiration for the development of the Model T assembly line) which will lead to the birth of a cement manufacture factory (Portland Cement, 1990–1914), based on previously developed techniques.

This creation ability is also demonstrated by the creation of new words. We can randomly cite the following: the facsimile telegraph (1868), the magnetograph (1869), the quadruplex telegraph, the inductorium (a "shocking machine" to treat rheumatism—about a hundred devices were sold), the acoustic telegraph (1875) (precursor of the telephone), the etheric telegraph (1875), the famous phonograph, the phonoplex (a system that allowed small relay stations of a telegraph to communicate with each other without disturbing communication between the large ones), the aerophone (compressed air to amplify a recording), the phonomotor (which transformed a sound wave into a rotational motion), the megaphone and the aurophone (for the hard of hearing), the mimeograph (1880s), the dictating machine (1886) (a system for recording and replaying messages for professional use), the talking doll (1888), the coin slot phonograph (1890), and the kinetoscope (1890s), among others.

This creation of words is also due to the fact that many of Edison's inventions were starting points for new industrial sectors which have become familiar to us. With Edison we are typically in a framework where the identity of objects to be designed is not known. Edison embodies this capacity to conceive new object identities.

5.6.3   What Reasoning and Design Methods Were Used by Edison?

We don't find the big models of classical management in Edison's companies: neither research-lab management nor technical-office management nor big-company management. We do have, however, extensive documentation which gives us a deeper understanding of the reasoning and methods of design used.

5.6.3.1   Production and Reuse of Knowledge

The story of Edison is the story of implementation and sophistication of knowledge processes involved in design. At first Edison reluctantly follows a more traditional company model based on a product (e.g. the concept of the telegraph for the gold and stock exchange quotation) but he quickly abandons that project (in fact in its year of establishment, 1869) to focus more and more on the production of knowledge. In 1874, after disengaging himself from another project designed based on the automatic telegraph, he actually resells the manufacturing part to focus on the experimental part.

This production of knowledge becomes more refined after his trip to Great Britain (1873): "although ultimately unsuccessful from a commercial stand point, Edison's trip to England proved extremely important to his career as an inventor. His encounter with the sophisticated British electrical community and with new problems of electrical transmission would lead Edison to a growing appreciation of how much he did not know about the electrical and chemical phenomena involved in cable and automatic telegraphy" (Israel 1998). He develops an experimental program and equips his workshop with an electrochemical laboratory (by bringing precision measurement devices from Great Britain). After establishing his laboratory in Menlo Park in 1876 he soon heads an organization of about sixty people, all working on tests and experiments.

This experimentation power is not uncontrolled and does not overlook the economic criteria of knowledge production: it is not a blind empiricism of which Edison was often accused; on the contrary there is a prior systematic use of existing knowledge (thanks to the library, which is always at the heart of Edison company buildings) and, depending on the results, that preliminary state-of-the-art knowledge will be revisited and deepened. Even better: this experimentation is often based on establishing *models* that drive and steer experiment plans: thus, in the research for a filament for incandescent bulbs, Edison starts by proposing a new law: "the heat released by a body is proportional to its radiative surface, not to its resistance" (Hughes 1983). In other words, a high-resistance lamp will not necessarily consume more than a low-resistance lamp if its radiative surface is reduced. He therefore looks towards a very thin filament wound in a spiral, which makes it possible to reduce the radiative surface and increase the resistance of the set, and he takes an interest in insulation layers that prevent short-circuits in the spiral.

Edison also gets organized to familiarize himself with the market: the launch of new products—with different levels of success—allow him to better understand market developments and the emergence of new values. Originally active in the business-to-business market, Edison will gradually start investing in the consumer market (mainly with the talking doll or with certain types of phonograph) and bring out new value spaces such as entertainment, a market that had just started to emerge before Edison and which will later become successful as we know it.

This intense production of knowledge is complemented by an extraordinary mechanism of reuse of produced knowledge. Experiment books are signed by the experimenters and classified by topic, all experiments are recorded in a centralized manner, and each experiment is monitored all the while placing the greatest importance on the "result": it is always about looking for *unexpected* results "that might lead to an invention" (Israel 1998, p. 199). This logic of knowledge reuse is particularly present in the field of materials, about which Edison clearly thinks that knowledge obtained in a certain line of research can be reused in another line of research: that is the case for research on filament materials (platinum, then bamboo charcoal, etc.), wax for phonographs or insulators for electric cables. Edison also notes that chemical substances that are useful for industrial applications (hard rubber, celluloid, glass, soap, paper, etc.) have barely been studied and need systematic experimentation. More generally, historians that have studied Edison have underlined the extent to which Edison's success was linked to that capacity of establishing fruitful relationships between dissimilar fields; Thomas Hughes underlined Edison's capacity to find metaphors "that allowed him to draw on what he knew to suggest order in what he did not know" (Hughes 1983). Therefore the reuse of produced knowledge relies not only on the capacity to memorize and use "ready-made" models but also on a strong capacity of modeling, i.e. creating conceptual models that circulate more easily through very dissimilar design spaces.

5.6.3.2 Design Reasoning

What design reasoning is associated with this production of knowledge?

We should discuss here whether Edison's success is simply due to a large number of trials: we cannot exclude this hypothesis a priori—the high number of patents filed by Edison over the years could correspond to a Poisson statistical process! What are the elements that lead us to think that this hypothesis is restrictive and does not grasp Edison's way of thinking?

Studies on Edison's companies and written work help us understand better the reasoning applied in the organizations that he established. For each of the products we find precursors and prior research and knowledge. Is the phonograph a spontaneous creation? No; interestingly it stems from work on the acoustic telegraph and the recording of telegraph messages: for telegraph creators who are interested, like Edison, in the nascent telephone, the telephone is a type of telegraph and we should produce, similar to the telegraph, a written recording! Edison becomes interested in X-rays, discovered by Röntgen, because during the same period he is trying to understand electromagnetic phenomena with a view to standardizing dynamos and other electrical devices. He is interested in the electric car because he is looking for possible uses of electric energy produced by all the new electric power plants— Edison had quickly identified the challenge of distributing equally the load of power plants. Why go into iron-ore grinding? First of all because he was looking for a way to lower the price of platinum which had been a serious candidate for incandescent-bulb filaments. It is precisely that work on iron-ore grinding that led to the work at Portland Cement, which made Edison one of the biggest producers of that product in the 1910s (Fig. 5.42).

There exist, therefore, important links between research efforts carried out by Edison. As we examine genealogies more closely, we are surprised to find out that there is a reasoned order that interlinks the designs of different products. The succession of innovations reveals developments in the nature of products and techniques which correspond to forms of learning about the product and about the market.

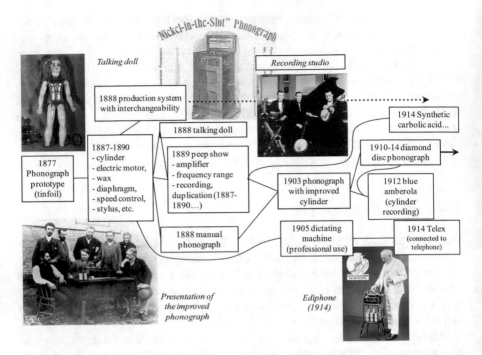

Fig. 5.42 Edison's lineage of products, starting with the phonograph

Let us take, for example, the products that descend from the phonograph (see Fig. 5.42): this line has an impact on innumerable areas! Thus, in the field of electricity, the phonograph influences ongoing research and leads to the development of low-power electric motor and then, in 1907, to the development of the universal electric motor. In the field of sound, the phonograph has a long linage: the first prototype is born in 1877; ten years later many innovations are added (cylinder, electric motor, wax, diaphragm, speed control, stylus, etc.) to create the first actual product. In 1888 Edison uses this knowledge to produce the first talking doll, and a year later, having realized the colossal impact of recorded sound in the recreation field, he launches the "coin slot amusement machine" with an amplifier, recording and copying of the recorded cylinders, etc. In the same year he also launches a manual phonograph. He also becomes interested in the developments of professional fields and designs the first dictaphones for offices and administrations which were booming at the time. For those products Edison also developed a production system based on interchangeability of parts (1888). In 1903 an improved phonograph comes out, in 1910 a phonograph that uses a disc and no longer a cylinder, etc. (see diagram below).

We can also reconstruct the genealogy of products descending from the "visual phonograph" which will lead to the birth of cinema, of the kinetoscope, to the creation of recording studios, etc. (see diagram below, based on Millard 1990) (Fig. 5.43).

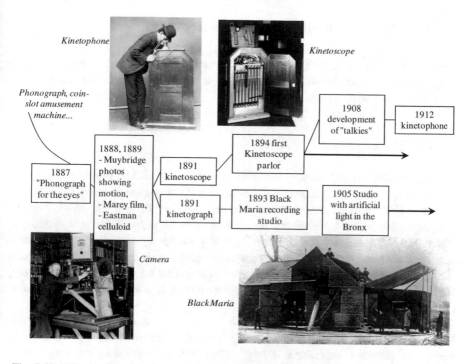

Kinetophone

Kinetoscope

Phonograph, coin-slot amusement machine...

| 1908 development of "talkies" | | 1912 kinetophone |

| 1894 first Kinetoscope parlor |

| 1888, 1889 - Muybridge photos showing motion, - Marey film, - Eastman celluloid |

| 1887 "Phonograph for the eyes" |

| 1891 kinetoscope |

| 1891 kinetograph |

| 1893 Black Maria recording studio |

| 1905 Studio with artificial light in the Bronx |

Camera

Black Maria

Fig. 5.43 Edison's line of products, starting with the "visual phonograph"

Thus Edison puts in place a true *innovative design strategy*, tracing paths of research and learning, using at the same time techniques (development of new technologies), markets (development of new markets with complex systems of licensed sellers), the industrial system (implementation of complete systems, from the recording of sound to its paid distribution!), and customer values (exploring new markets for group entertainment with sound and image).

We also recognize the characteristics of a *risk-management strategy of the unknown*: maintain the abilities to play several moves on one line of products or many lines in parallel: avoid expensive research and learning that would put the company at risk; ability to launch breakthrough studies (crazy-concept logic) but also the ability to explore the market in a cost-effective way, either by pursuing medium-term agreements with large companies (Western Union and later one Edison General Electric) or by updating particularly niche products such as the coin slot phonograph and more generally products for the entertainment market. This prudential management certainly contributed to the extraordinary longevity and the exceptional growth of organizations conceived by Edison. It is those organizations that we will now focus on.

5.6.4   Organization of the Invention Factory

Contrary to what the legend might lead us to think, Edison is not alone. Edison's story is in fact also the story of invention of original forms of organization of collective innovative design.

5.6.4.1   The Beginnings

Edison's solitude is real when he is starting out. He will quickly join the community of precision mechanics, designers of devices for the telegraph industry (1860s). So he works in the service department at Western Union of Boston, one of the biggest telegraph companies. He attempts to create his own business, then he is employed as a superintendent at Gold and Stock Reporting Company. Up until this point, it is the somewhat chaotic journey of a relatively independent engineer, who explores different ways to collaborate with companies (as a contractor or salaried employee) or is trying out a classic form of entrepreneurship (creating a firm based on a concept-product).

A new type of cooperative appears with Pope, Edison & Co. This enterprise will play the role of consulting office for big telegraph companies and offers a great variety of services: instrument testing, patent filing, preparatory drawings, purchasing agent for telegraph and electric devices, design of special devices, construction of experimental devices, precision testing, etc.

Knowledge production and client relations are given structure. However, the company will quickly decline. After working as a "company inventor" for Western Union, Edison will found a new company with great ambitions for knowledge production, especially after his trip to Great Britain (1873). He will reinforce, as we have seen, experimental abilities. Thus we go from a company that resembles a machine shop to a true laboratory that can produce knowledge in an intensive and controlled manner. At the time, Edison is surrounded by some assistants as well as some experienced designers. He works, for example, with Bachelor and hires Robert Spice, chemistry professor and expert on acoustics. On the other hand, the company's commercial relations are stabilized: it has rather close ties with large companies (such as the powerful Western Union) which regularly assign it new demands. After working as an electrical engineer at Atlantic & Pacific, Edison reviews the organization and its scope of activity again: he abandons manufacturing to focus exclusively to the experimental part.

5.6.4.2   Menlo Park (1876)

This embryonic model comes into its own with the creation of Menlo Park, "the invention factory" (1876). Edison is 29 years old at the time. Menlo Park represents a change in scale: almost sixty employees work there, the buildings include a small library, a kind of "show room" used to showcase inventions, a fully equipped machine shop, and a vast laboratory with all the necessary instruments and substances. It is a mechanism intended exclusively for invention, which makes it possible to create, test, and quickly modify prototypes and thus to accelerate the flow of innovation. This experimentation power is rigorously managed: on one hand Menlo Park follows a research program discussed regularly with industrial funders; on the other hand, this sophisticated mechanism helps Edison start to explore new value spaces such as electric lighting. Such a mechanism also requires a certain division of tasks: Edison is no longer an inventor who experiments with some close associates; he is truly the director of a large-scale R&D laboratory. Little by little, two types of division of tasks will start to form:

- on one hand, divisions of tasks related to the production of knowledge: experimentation spaces (laboratories, workshops, even factories later on, etc.) and fields of expertise. Edison surrounds himself with competent experts in chemistry, glasswork, mathematics, and physics, and as the electric system is developed, he will hire more experts;
- on the other hand, divisions of tasks related to lines of products. Two types of lines can be distinguished:

  - those that are emerging: it is at Menlo Park that the first incandescent electric lighting system and the first phonograph will be designed; subsequently research is conducted on electric transport (car, train, etc.) or even on engines.
  - those that, being more mature, need a design-office type of support: that was the case for the telegraph at first, and then for all devices associated with electric lighting (bulbs, dynamos, construction of electric power plants, cabling, etc.) which are commercialized by companies established in Menlo Park (Edison Lamp Company, Electric Tubes Company, Machine Works, Isolated Company, Thomas A. Edison Construction Department, etc.).

5.6.4.3 West Orange

Structuring becomes even more important when Edison creates the West Orange laboratories (1887, Edison is forty years old). As the laboratories emerge, Edison writes: "I will have the best equipped & largest laboratory extant and the facilities incomparably superior to any other for rapid and cheap development of an invention, & working it up into commercial shape with models patterns special machinery—in fact there is no similar institution in existence" (cited by Israel 1998). The machine shop can fit 50 people; dozens of experimenters "carry stock of almost every conceivable material of every size and with the latest machinery." Another machine shop is entirely destined for precision mechanics. There is a physics laboratory (galvanometer room) made without ferrous metals in order to avoid electromagnetic interference, a chemistry laboratory and an annex specially equipped for handling dangerous products, a metalworking laboratory, a personal experimentation room for Edison, a library with 10,000 books and with subscriptions to all scientific magazines of that period, stocks of materials, and the widest range of devices and components. A total of more than 100 people work there from the very first years. There are approximately thirty talented experimenters (young scientists that have come to get trained near Edison), about forty mechanics working on the machines, 3 to 5 modelers, the same number of designers, a firefighter, two blacksmiths, some carpenters, and a shop assistant.

The organization also becomes more complex. Heading the laboratories are seasoned experts who master a theoretical corpus but who also possess a great practical sense; nevertheless, Edison encourages not so much specialization but rather a wide range of knowledge: experimenters must be able—according to the long tradition of the machine shop—to create any device on demand. In this tradition, and following the growth of the "invention factory", an important novelty appears: we no longer find individual mechanics-inventors-entrepreneurs; instead, now we find small teams in charge of exploring subjects assigned by Edison in the form of written instructions which constitute a brief and leave extensive room for initiatives to the experimenters that have thoroughly understood Edison's methods. Edison himself explains this organization as follows: "to organize a gang of one good experimenter and two or three assistants, to appropriate a definite sum yearly to keep it going… have every patent sent to them and let them experiment continuously." It is therefore an organizational mechanism which couples product and skills locally; it may include exploring a new field or researching a new alternative within an existing line of products, or developing certain components for a complex system. This last situation is demonstrated by new generations of the phonograph, where several aspects are explored in parallel by different assistants: the motor and the battery, recording cylinder waxes (700 blends will be tested in 5 months, which —combined with theoretical work and encyclopedic knowledge on the subject— will turn Aylworth into the greatest international expert on the subject for future decades), experimental recording of voice and music, means to copy recordings, and finally the part that Edison reserves for himself: the phonograph itself.[9]

Moreover, assistants are now in charge of entire lines of products. Israel points out that Edison was barely involved personally in the entertainment business; we can therefore conclude that this business was developed in big part by assistants. As regards cinema, Dickson leads the design work for the kinetoscope: work on the lens and on a microscopic objective for taking photographs, work on the emulsions, complete construction of the recording studio, pilot work on filmmaking, research

---

[9]Note that we are talking about value spaces and not only a decomposition of objects in smaller parts, insofar as each aspect may lead to new products or services for the entire sector which is emerging at the time.

on talking films, and even construction of new experimental laboratories to complement this work. Kennelly heads the work on standardization of products in the lighting sector. He is director of the galvanometer room and makes significant contributions in the electrical engineering field. We can also evoke the case of Insull who, after working as Edison's assistant, becomes in charge of design at Machine Works, contributes to launching work on insulation of cables for underground placement, and will become shop manager at Edison General Electric. Thus we see an emergence of leaders in different lines of products, who are very competent both in the business aspect and in techniques.

Some years later, from 1916, the organization will be structured based on a precise organization chart which distinguishes various roles:

- Engineering services in charge of developing prototypes and products, which constitute the Development Department. They are in fact the support services in the development of well-established lines.
- Product engineers who explore new concepts and ideas, based on a divisionalization principle which is very unusual at such early stages in the life of ideas; in some cases, the department's activity is rather exploratory (such as for cinema) and structured as a special section dedicated to innovation. These are the successors of small research teams.
- Laboratories and test workshops with an area reserved for Edison's experiments; this activity closely resembles research activity insofar as it explores new domains of knowledge. It always relies upon a very well equipped library, on impressive stocks of the most varied raw materials and products of all kinds.

The reader will easily recognize an RID organization (Fig. 5.44).

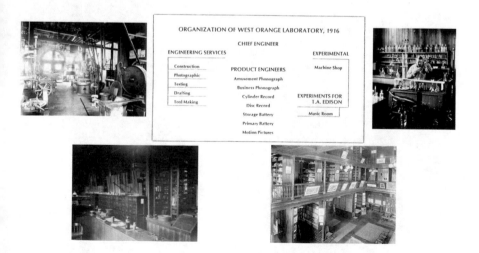

Fig. 5.44 West Orange: an organization for repeated innovation

In lieu of a conclusion, the reader will find below some elements to remember and discuss on this case:

(1) Edison's enterprises show forms of expansion that focus mostly on the identity of objects.

To take this idea further, discuss the expansion/resources relation implemented at Edison's enterprises. Conclude on the performance of innovative design.

(2) The products launched by Edison show that the design reasoning that underpinned their emergence surpassed significantly the simple variation on a known identity, on an established dominant design. Edison's lines of products are far from a simple "industrial sector", a simple "family of products", or a simple variation on a "core competence"; and yet they constitute coherent entities, which are diligently managed by designers.

To take this idea further, discus the structure of knowledge: can we identify elements that indicate a knowledge structure that respects the splitting condition?

(3) As regards organization, we have seen how Edison's organizational model itself had gone through a complex embryogeny, with several years passing before clear RID logic concepts were implemented (at Menlo park (1876) and especially at West Orange (1887), with Edison having more than twenty years of working life behind him).

To take this idea further, analyze the different forms of risk management implemented by Edison.

5.7 Case Study 5.2: Organization of the Innovative Design of Thales Avionics (Author: Denis Bonnet)

Below we explain in detail the organization of the innovative design at Thales Avionics (this case is based on Denis Bonnet's presentation at the Innovative Design Theory and Methodology department on January 8, 2013, supported by a longitudinal study on several years of activity of the department and on several projects that it carried out).

The organization implemented has made it possible to achieve significant success on several aspects. Many elements related to products remain confidential to date but certain innovations, derived from programs such as Cockpit Designer, ODICIS (One DIsplay for a Cockpit Interactive Solution), or Avionics 2020, have recently known great media success (Bourget 2011 and 2013, Janus de la Prospective 2012, Red Dot Concept Award 2013, etc.).

The questions are asked throughout the case study.

5.7.1 Origins

Thales Avionics is an entity of Thales, in charge of designing, developing and selling cockpits, both for military and civil aircraft.

Several years ago Thales Avionics developed an organization for innovative design within the Innovation and Human Factors department of the Cockpit center of competence.

The origins date back to a KCP project on the innovative cockpit (led by Guillaume Lapeyronnie) and several experiments on innovative projects conducted by using the C-K theory. Those first experiments yielded two major results: the implementation of a "cockpit design center" (a prototyping space dedicated to the exploration of innovative-cockpit concepts) and the launching of work on the single-screen cockpit. Those first experiments made it possible to incorporate the KCP method as a "routinely" used method for new concepts to emerge. However, they also showed that the KCP method is not sufficient, that its efficacy depends on the quality of the knowledge base used, and that its outputs are not easily incorporated by other design stakeholders. That is how the need emerged for a process that is better integrated within a department, with four objectives:

- to be able to incorporate the innovative concept into the more classic processes of TLR (i.e. Technology Readiness Level logic, which is widely used in the aviation industry)
- to better train teams on value reasoning in innovative design
- to better manage relationships with external factors, mainly by avoiding misunderstandings which may arise when concepts are materialized into prototypes
- to take into greater account the risks associated to breakthrough concepts (risks related to costs, to certification, to performance, etc.).

5.7.2 Global Approach

The global approach follows the TRL milestone structure (see Fig. 5.45; source: Denis Bonnet) (TRL stands for Technology Readiness Levels—it is a method developed by NASA for estimating technology maturity during an acquisition process).

How does this approach differ from the traditional development "funnel", especially from the aforementioned phases of conception of said funnel?

We note several specific features: we are not talking about initial "ideas" but of themes or, in the C-K language, innovation field concepts. It is worth mentioning that these concepts are validated by marketing.

The work on the concepts does not consist in selecting ideas but instead in organizing an innovative-design process with divergence, exploration of knowledge, and refining of concepts, particularly due to the use of prototyping in the very early phases. Said divergence does not lead to an idea-selection process but to a *recombination* and development of a multi-project design strategy that is validated with the product policy.

We describe four aspects of the organization: the innovative-design process in the very early phases (TRL 1–3), the logic of demonstrators supporting this process, the organization of the workspace, and the process leading to the creation of a new rules system for rule-based design, a process that we have called design "adjustment" (TRL 4-5).

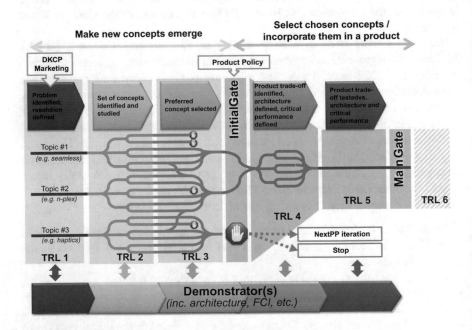

Fig. 5.45 The innovative design approach at Thales Avionics (*source* Denis Bonnet)

5.7.3   The Innovative Design Process

The initial concepts are shared with the marketing teams and are linked to strategic elements and known values. The teams are responsible for carrying out a C-K reasoning. To that effect they have been extensively trained on the C-K theory and on its implementation (more than 50% of the staff have taken the courses at the School of Mines); they have also been trained on the most important skills in the field (human factors, man-systems interfaces, etc.). Innovative-design "referents" have been put in place. Based on their personal profiles, the team members can lean towards the concept, the knowledge, or the prototyping aspect of a project. The C-K work leads to proposals that are systematically explored by means of prototypes ("demo or die" logic).

Innovative-design work is not isolated. Instead it is regularly presented to "major clients" (including internal ones); the timing chosen (the TRLs) is that used at a sector level: the initial concepts are worked on jointly with the other functions of the firm (starting from recurring problems, embryonic innovation fields, often after a reinterpretation and conceptualization of the questions asked initially). A particular effort of communication and training on the approach was carried out: sharing a common reference base, understanding the nature of the process outputs, understanding the usual traps (dominant design, fixation), accepting failure, and valuing variety/originality. Concept trees are regularly presented to partners (development, client, marketing, strategy, and other roles), often in simplified form and with examples (and not only abstract proposals). When the process leads to the identification of pockets of knowledge to explore, this exploration is conducted–to the extent possible–with the help of experts that are closest to the associated domains. The exchange is done around prototypes, to the extent possible.

5.7.4   Demonstrators and Prototypes

Innovative-design activity relies on a structured and constant demonstration approach (one demonstrator every six months). The demonstrator incorporates innovations in a "coherent cockpit".

What are the objectives of demonstrators?

A demonstrator has five main objectives:

1. serve as a catalyst for innovation by accelerating readiness
2. validate identified value elements for clients and users
3. bring out new ideas when in contact with users (stimulation logic)
4. evaluate proposed solutions/concepts in a representative operational context and verify that they satisfy the identified value elements
5. allow the transition to development.

This integrated demonstration logic (multiple innovations in one demonstrator) contributes to the acceptance of failure: the few local failures (innovations that do not reach the expected value elements) are offset by innovations that hit the market. Thus the perception of the demonstrator always proves to be positive overall.

5.7.5 The Work Space—Innovation Hub

The teams meet in a dedicated space, where they may also incorporate clients, end users, and external teams (which have very specific knowledge and consist mainly of designers). So this work space is a showcase for technical and technological innovations developed by the department.

The space is carefully structured around a central discussion area with spaces for informal talks, brainstorming and creativity, and presentations. Around the work space there are several other areas:

- specialized laboratories (electronics, optics, software benches)
- prototyping environments
- thematic open spaces (according to the concepts being explored).

The space fits 30–45 people.

5.7.6 The Process of "Regulating" Design (TRL 4-5)

This is the phase of transition to the rule-based design, which consists in developing a set of new and relevant rules. This phase goes mainly through a new technology or product definition.

How does this phase differ from a "rule-based" development project?

This phase has three objectives:

- Define the product based on the innovations retained: this has to do with identifying a good combination of innovations while ensuring their coherence based on the points of view of engineering (performance, operation safety, maintainability, certification, etc.) and removing the most critical risks.
- Define the complete life cycle of the future product: development, manufacture, maintenance, recycling, etc.
- Stabilize the economic equation, making sure that the necessary last adjustments will respect the value/cost balance (in a broad sense: perceived, operational, technical, etc.).

When it comes to technology, we will mainly try to identify the functional specifications, then proceed to prototypes that are all the more representative, and validate the technology to the extent possible with respect to the different environments where it is to be integrated.

When it comes to a product or a system, the TRL phases serve as milestones.

- Identification (TRL 4): transposition of TRL 3 concepts to a products/systems variant, exploration of alternatives of gathering retained concepts, identification of the best compromise of gathering concepts based on the different points of view of engineering.
- Readiness (TRL 4-5): identification of the functional specifications, definition of usage scenarios, identification of the context(s) of the object, logical architecture, physical architecture, evaluation.

Thus we get closer to the phases of rule-based design but this "first passage" aims primarily to constitute a new base of associated rules, particularly a development reference base.

Study the manner in which function I implemented by Thales indicates both a cohesion and a coordination concern.

5.8   Case Study 5.3: Conceptive Research for Conceptual Absorptive Capacity: The Non-CMOS Image Sensors by STMicroelectronics

This case study illustrates in a more precise manner a critical capacity for conceptual research: conceptual absorptive capacity.

In the 1990s Cohen and Levinthal showed that the value of industrial research was not only linked to prior related knowledge but also depended greatly on the knowledge that it allowed the firm to acquire in external laboratories according to the problems with which it was faced (Cohen and Levinthal 1989, 1990). In effect, R&D constituted an *absorptive capacity*. The characteristics proposed for describing this capacity were the following: recognizing the value of information, assimilating it, and applying it.

In Chap. 5 we showed that this absorptive capacity covered, in fact, two different capacities: an epistemic absorptive capacity and a conceptual absorptive capacity (Le Masson et al. 2012c).

The epistemic absorptive capacity can be characterized by the ability to recognize the value of external knowledge, assimilate that knowledge, and use it (Lane et al. 2006). As regards innovative design, absorptive capacity cannot be built upon this principle as "recognition" is not possible: it has no pre-defined value; on the contrary, the objective is to revise its main attributes. Does that mean that absorptive capacity is not necessary in innovative design? Intuitively we feel that the capability to absorb external knowledge is important in innovative design. In fact, the demands and the difficulties of innovative design make it possible to describe the expected properties of absorptive capacity in the case of innovative design: contributing to rule-breaking, helping to reorganize knowledge on well defined concepts, and generating–if needed–the knowledge necessary to certain breakthrough concepts.

Conceptual absorptive capacity thus satisfies these three objectives. We demonstrate this point by means of the following case study on conceptual research on "non-CMOS" image sensors (Felk 2011a, b; Le Masson et al. 2012c, d). We use the C-K theory to illustrate the reasoning of the case study.

First we try to illustrate the reasoning with respect to image sensors without breakage (in a rule-based design). We find the expected pattern of an epistemic absorptive capacity: the concept alludes to key roles (CMOS materials, CMOS processes, etc.) within the firm, which themselves shall mobilize external knowledge (clients, suppliers, laboratories, etc.). The value is determined in C and the associated questions are addressed to the professional roles in K, who recognize and assimilate new knowledge external to the firm, knowledge that is used to respond to the question initially put forth.

Note that this first C-K graph demonstrates the hereditary structure of the object (see in this chapter for details on this notion) (Fig. 5.46).

What happens in the case of innovative design? We have identified three main phases:

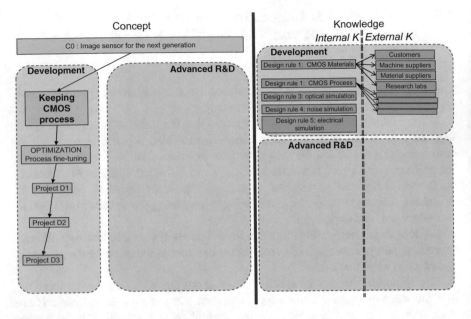

Fig. 5.46 Epistemic absorptive capacity illustrated within the C-K framework

1. The first stage is about rule breaking: research identifies a particularly blocking rule (the CMOS process) and decides to break the heredity from the point of application of that rule. Therefore it introduces the "non-CMOS image sensor" concept which dictates entering a new knowledge space (Fig. 5.47).

2. The second stage consists in recreating a cognitive reference associated to the breakthrough concept. To that effect the researcher creates conceptual models associated to the breakthrough concept. Here the challenge is to find a silicon (Si) layer capable of playing the critical role of a photodiode (photon-to-electron transformer). These conceptual models make it possible to activate knowledge. They are *hooks* for the knowledge already existing in the ecosystem but which has been inaccessible so far. This phase requires an internal knowledge production effort (see knowledge pockets in light-colored characters on gray background in Fig. 5.48).

3. The third stage consists in stimulating the production of knowledge in the ecosystem. The logic of the cognitive reference makes it possible to identify the concepts that cannot be used together with available knowledge (or are being produced by other knowledge). Thus the conceptual absorptive capacity consists in organizing the production of external knowledge based on those concepts. This is the milieu stimulation aspect (Fig. 5.49).

As an exercise we can ask ourselves questions on the role of prior knowledge in the aforementioned reasoning. Indeed the notion of epistemic absorptive capacity builds upon prior knowledge (see figures below: it is prior knowledge which allows

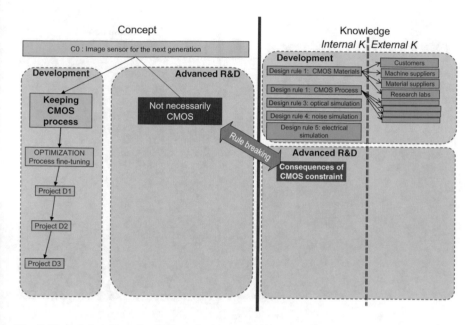

Fig. 5.47 Rule breaking, first aspect of conceptual absorptive capacity

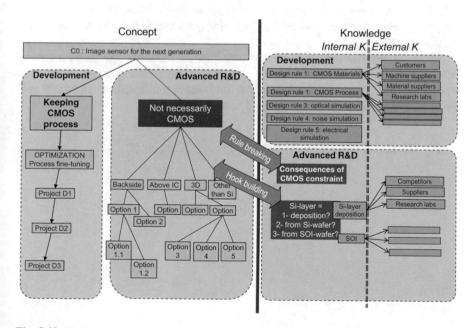

Fig. 5.48 Hook building, second stage of conceptual absorptive capacity

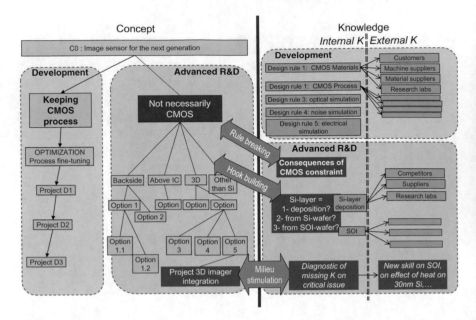

Fig. 5.49 Milieu stimulation, third aspect of conceptual absorptive capacity

links to be made with the external environment). In case of a breakthrough concept, this prior knowledge often present the following question: is it useful or useless and even stagnant?

Based on the above we can demonstrate that prior knowledge plays a critical role: it makes it possible to identify which rules to break, recreate conceptual models which will serve as hooks, and identify actors in the ecosystem who would likely be interested in the production of new knowledge.

5.9   Case study 5.4: Building with Hemp,—Taming Technological Bubbles by Managing Generative Expectations

This case study is derived from (Le Masson et al. 2012. It illustrates how a college of actors (the association "building with hemp") was able to organize collective exploration by shaping expectations (i.e. shared concepts). To guide the reader:

Question 1: Compare the behavior of an actor confronted to uncertainty and an actor confronted to the unknown (see the two models: anticipative expectation vs. generative expectation)

Question 2: how does the college leverage the positive effects of expectations and tame their negative effects ?

A recent body of literature underlined a fascinating phenomenology of expectations, in cases of changing identity of objects (see in particular the special issue of TASM on expectations in 2006). Such situations tend to create hype and disappointments that can be assimilated to technology-driven speculative bubbles that repeat over time and sometimes slowly converge towards economic growth. *Can these cycles be avoided? Is it possible to manage expectations in cases of innovative design?*

5.9.1   Contrasting Two Models of Expectations Management: Anticipative Expectations Management Versus Generative Expectations Management

The literature has helped uncover a phenomenology of expectations (Geels and Raven 2006; Robinson and Propp 2008). Expectations have been described as a "natural" phenomenon associated with radical systems innovation (Smith et al. 2010). As underlined by Borup et al. (Borup et al. 2006), expectations can guide activities, provide structure and legitimacy, attract interest and foster investment. They give definition to roles, clarify duties, offer a common perception of what to expect and how to prepare for opportunities and risks. They help to mobilize resources at all levels and to build bridges across boundaries, between communities or groups, between different levels or scales (micro-level of projects, meso-level of companies, macro-level of institutions) and different times.

However, these phenomena are not necessarily positive and also raise critical issues. They are supposed to create hype—and hence involvement, implication, investment, etc.—but, as underlined by several authors (Borup et al. 2006; Callon 1993; Nowotny and Felt 1997), they may also generate disappointment by "creating lasting damage to the credibility of industry, professional groups and investment markets" (Brown 2003). "Expectations are accompanied by serious costs in terms of reputations, misallocated resources and investment" (Borup et al. 2006). Some authors showed that expectations follow cycles linking expectations and requirements and that these cycles can converge (requirements are met) or diverge (when requirements are unmet, i.e. expectations are not fulfilled, new promises have to be created) (Geels and Raven 2006; Van Lente 1993; Van Lente and Rip 1998). Expectations support alignment towards a single shared vision, but this alignment can prevent exploration and divergence (Van Lente and Rip 1998) and create 'irreversibilities' (van Merkerk and Robinson 2006). Expectations are supposed to involve actors, but cognitive models of expectations and uncertainty (Sung and Hopkins 2006) have underlined that some expectations tend to involve non-experts, who are less sensitive to uncertainties, contrary to experts who are more knowledgeable about uncertainties and therefore believe less in the anticipated future… but might be the most relevant people to deal with the uncertainties!

These issues might be explained by the fact that there are two very different types of expectations management, associated with two different types of expectations, clearly distinguished in the literature.

5.9.1.1 Anticipative Expectations

The first model is very well described in Propp and Moors' synthesis (Propp and Moors 2009). The authors give the following definition: "An expectation is an anticipation of the kind of future that may be 'on its way' from within the present. Experience may tell us in some instances which outcome is likely, but the results have not occurred yet and are uncertain." Expectations can be managed so that they follow that pattern: "Actors simultaneously talk up the deterministic momentum of current developments and suppress uncertainty and alternative futures, hoping for alignment of other actors—and the resources they have or can distribute—around these expectations" (Propp and Moors 2009). We find several examples of these kinds of expectations in the literature. For instance, Geels and Raven 2006 showed that in the 1970s biogas development was based on the expectation of "cheap alternative energy generation" based on existing knowledge on digestion. van Merkerk and Robinson (2006) gives the following example: "In 1993, Harrison and Manz revealed a large breakthrough in the journal Science with a successful miniaturization of the analytical technique of capillary electrophoresis. They articulated their expectations as follows (p. 897): 'The application of micromachining techniques to the miniaturization of chemical analysis is very promising and should lead to the development of analytical laboratories on a chip.' Typical advantages of chip-based analysis systems are speed, less sample needed and possibly portable." In both cases it is expected that there is a high probability that a new piece of knowledge A (in the example above, micromachining techniques applied to chemical analysis) will give a future F (in the example above, "analytical laboratories on a chip") that meets performance criteria (or more general speaking sources of users value, S) (speed, portability, small sample,…). The anticipative expectation is the probability to realize F given the assets As, $P(F/As)$, to get the value of F given the sources of value Ss, $V(F/Ss)$.

We can underline that these anticipative expectations are *constructive* in the sense of Borup et al.: there is not necessarily a "real" future, this future can be imagined and figured out by people at a certain moment of the innovation process. We call them "anticipative" because it confronts people to a predictable representation of the future (value and probability) and lead them to decide to "take" this future F or not (anticipate = ante-cipare, take in advance according to the latin root).

How is this type of expectation managed? The "expectation raiser" has to *design* the good sentence "$A,S \rightarrow F$ with a high value $V(F/S)$ and high probability of success $P(F/A)$". An anticipative expectation is well-formed when it meets certain criteria: the value of the future state $V(F/S)$ has to be desirable and clear; the probability of reaching F also has to be high and warranted by the initial asset (A). F can be partly undefined but this "degree of design freedom" shouldn't imply strong changes in As and Ss that would change $P(F/A)$ and $V(F/S)$. The expectation raiser tries to prove that the relationship between A, S and F is almost certain (almost deterministic). *Ideally, anticipative expectations are design-free.* Anticipative expectations provoke an alignment of the actors towards F, based on A and S. Alignment means three simultaneous effects regarding collective innovation:

- it leads to identify a network of partners
- it supports their coordination (division of labour) to realize F
- it supports their cohesion (common interest), based on sharing the value of the future F

It represents a kind of "miracle" in collective action.

5.9.1.2   Generative Expectations

Borup et al. suggested that there is another model of expectations, which they called "second-order expectations". In some cases, only some assets and some sources are known and they are insufficient to promise a predictable future with a stable probability and value. Nonetheless, there is an innovation field in which some actors propose to work with others with a view to make emerge new assets, sources of value and futures. In that sense, this proposal is a "second-order expectation,": it is the promise, of *creating* first-order expectations. What is expected is the creation of the unexpected, or the 'unexpectable', at a certain moment when well-formed "first-order" expectations cannot be formulated yet. Strictly speaking, rather than generating a single stabilized deterministic link between assets As, sources of value Ss and one valuable future F, expectations tend to support the *emergence* of new assets (As), new sources of value (Ss) and new futures, Fs. Hence, such expectations should help design many possible futures! This is why we call our model a model of ***generative expectations management***. This means that in case of generative expectations an expectation F based on As and Ss will be so unknown that it is impossible to define $P(F/A)$ and $V(F/S)$. Or better: F claims that the exploration (required for its design) will generate As, Ss that will change F itself, $P(F/A)$ and $V(F/S)$ in unexpectable ways. Generative expectations are asset-free and design-intensive, whereas anticipative expectations are asset-intensive and design-free.

For the reader: show that when generative expectations are managed as anticipative ones, it leads to technological bubbles—answer below.

When a situation of generative expectation is managed in the same way as anticipative expectations, this tends to create hype and a technological speculative bubble. Initially, the expectation raiser formulates promises to attract investors. He identifies a credible A, credible source of value S, and a credible F, with an apparently high probability of reaching F on the basis of A to get high value on the basis of S (anticipative, first-order expectation). In this case, the logic of anticipative expectation does not lead to the exploration of a variety of As, Ss and Fs, but will tend to *reduce* explorations around A, S and F (Propp and Moors 2009). This promise tends to be "low-hanging fruits" for non-designers (Sung and Hopkins 2006) who will ask for simple "first-order" expectations, reinforcing the drift. Since second-order expectations are related to radical innovation with high levels of uncertainty and unknownness, disappointment is very likely to follow on. This explains why, in cases of higher uncertainties, this kind of expectation provokes disappointment (Geels and Raven 2006; Van Lente 1993; Van Lente and Rip 1998) and emerging irreversibilities (Robinson and Propp 2008).

Note that the bubble is not caused by a distance between belief and reality (as is the case for classical financial bubbles) but a distance between belief and the outcomes of the design process. This is not even an overestimation of the design process (as if we thought we could get F but finally we can't) but a misunderstanding (or an underestimation) of its generative potential (the design process is non linear, expanding, and it generates future that couldn't be anticipated at the beginning).

To conclude, when second-order (generative) expectations are managed as anticipative ones, they tend to create fragile collaborations, with non-relevant stakeholders (non-designers), resulting in cycles with slow or even no convergence.

Hence generative expectation requires a specific form of management. To characterize generative expectation management one should answer two series of questions:

(1) The group or the network of actors associated to generative expectations: who is attracted? To do what? For which reasons?
(2) The means to shape and manage generative expectations over time in a dynamic way?

5.9.2   Some Elements on the Research Method

Case relevance: why building with hemp? Over the last two decades, a whole industry has emerged, revolving around the use of a natural fibre, hemp, as a new construction material. What was initially no more than a set of loosely connected actors gradually became an industry, with a host of coordinated actors: hemp producers, hemp transformers, lime producers, architects, entrepreneurs, masons, prescribers, state agencies, insurance companies, research labs, etc. It is now based on new products, services, competencies and publicly recognized R&D programs. The process was a clear success: with neither large financial resources nor the intervention of a prominent, powerful actor, it was more the result of the active cooperation of heterogeneous actors. The actors had limited resources; industrial hemp was a marginal crop grown in small areas of land, with small R&D investments and a public image associated with the use of illicit substances, even though industrial hemp is different from the varieties cultivated for such purposes.

Expectations helped to overcome these obstacles. They created the 'bundle' of actors and to support their action, avoiding the usual pathologies identified in the literature.

Case analysis technique: One of the key issues in the analysis consists in identifying the expectations and their nature. We conducted interviews with the main actors of the "Building with Hemp" case, we also had access to technical reports. The case study was done in 2010. Building on C-K theory, we analyzed the *various* expectations of the actors over time, in *mapping all the identified C and K*, and the type of evaluation made by the actors. For each expectation, one identifies the future F, the related assets As and sources of value Ss and we analysed (based on the interviews of the main actors in the field) whether the relationship between A, S and F was sufficiently well-defined so that the probability of success $P(F/A)$ and the associated value $V(F/S)$ are high and stable (anticipative expectation) or whether the proposal linking A, S and F clearly require to discover new As (to be able to define a probability $P(F/A)$) and/or new Ss (to be able to define a value $V(F/S)$).

In our particular case we realized C-K graphs for each "building with hemp meetings". The assets (As) and sources of value (Ss) are mapped in K-space and each imagined, yet partially unknown future F appears as a proposition in C, which is related to some knowledge in K but yet is not true in K. In the general case such a proposition is a second-order expectation since it requires further explorations (new As and Ss) to refine the future to such an extent that, for the refined, new future, it is possible to define a reliable value and a reliable probability. At this point the refined future is a first-order expectation. Hence this gives an analytical tool to identify first order and second order expectations (see Fig. 5.50).

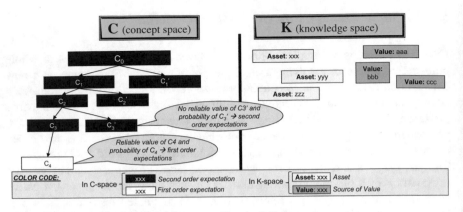

Fig. 5.50 first order and second order expectation in C-K theory

5.9.3 Smart Expectation Management in "Building with Hemp"

The cooperative La Chanvrière de l'Aube (LCDA) transforms raw hemp into by-products (fibre, hemp chaff [*chènevotte*], fruit, oil, etc.) with commercial value. Historically, only the hemp fibre, which was sold to cigarette paper manufacturers, was really profitable. However, this business decreased steadily and LCDA tried to find new markets for hemp products. We studied the history of one of these explorations—building with hemp—from it's beginning in 1986 until 2008, focusing on how LCDA employees were able to manage expectations and finally create a new industry.

From the collected data, we can clearly distinguish four main moments in the history of the exploration. Phase 0 (1986–1993): the limits of system innovation based only on anticipative expectations; phase 1 (1993–1998): raising generative expectations; phase 2 (1998–2005): realizing the newly generated anticipative expectations while rejuvenating generative expectations; phase 3 (2005–…): balancing newly generated anticipative expectations and generative expectations.

5.9.3.1 Phase 0: Innovation Based on Anticipative Expectations (1986–1993)

In 1986, Mr. Rasetti, a mason, asked LCDA to supply him with some hemp chaff for building purposes. With Mr. Rasetti, LCDA developed its first product for building, an aggregate of hemp chaff for light cement, Canobiote®. Competitive products were also launched (Isochanvre®, Canosmose®). In the following years, LCDA provided several masons with hemp chaff, which they used in cement to obtain a daub-like concrete. As its properties are close to those of daub, the historians and architects involved in the project considered that hemp cement was an appropriate material for

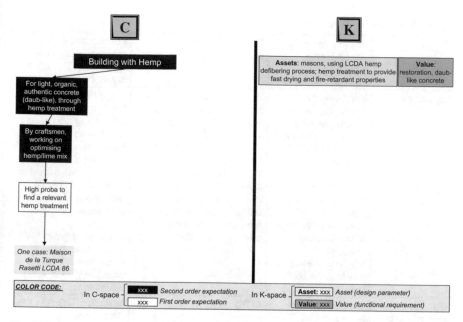

Fig. 5.51 design reasoning on hemp building (1986–1993)

the restoration. One example at that time was the restoration of the *Maison de la Turque* in Nogent sur Seine. However, beyond local successes, there was little market growth and no reliable solutions were found to use hemp in construction.

"Building with hemp" was associated with one clearly defined future, i.e. hemp for renovation, for light organic, daub-like concrete, for masons, etc. (see Fig. 5.51), with a well-identified value, based on the competences of masons and hemp providers. Knowledge production was restricted to experiments on fibre treatments, with the hope to finally realize F with a supposedly high probability. This is clearly an anticipative expectation. It had the classical effect: mobilize people who believe and see value in the expectation (masons) to realize the promise in a coordinated way (masons build; hemp providers provide material) and share the associated value (new business for masons, new market for hemp provider) (see Table 5.9).

5.9.3.2 Phase 1: Raising Generative Expectations (1993–1998)

In 1993, LCDA decided to invest more heavily in the creation of new applications for hemp, in particular in construction, "*because of several years of very low growth*" (to quote Bernard Boyeux, head of new applications for hemp in building at LCDA). "*[W]e had the intuition that the initial path was too narrow and that we should find new alternatives, not necessarily based on hemp treatments for special masonry.*" The past experiments had mainly resulted in unsuccessful trials and difficulties in addressing problems such as the choice of the binder, the type of cementing process,

Table 5.9 Case study synthesis

	Phase 0: reference (no generative expectation management)	Phase 1: raising generative expectations	Phase 2: realizing anticipative expectations, rejuvenating generative expectations
Expectation	F = "hemp for renovation, for light organic, daub-like concrete, for masons" A = masons and hemp provider competences S = daub like concrete is needed in restoration P(F/A) = probability to find an hemp treatment V(F/S) = value of the new business for hemp in restoration	Fs: "substitute for concrete not based on the existing tack coats", "hemp concrete for new uses",... As: hemp-cement properties, tack coats,... Ss: construction likes eco-friendly, green, structural properties,... Design degrees of freedom: "new uses?" "Not the existing tack coats?"	Well-formed anticipative one: "professional rules for building with hemp, with known tack coat, known hemp, for known applications...": The future F is based on clear A and S and doesn't require critical A or S exploration (high P and V). Rejuvenated generative expectations: "for do-it-yourself", "for other users", with other processes"...
1. The group of actors: Who is attracted?	Some masons and hemp provider	Designers	Designers (including free riders or opponents of anticipative expectations)
To do what?	To realize the hemp renovation daub-like concrete	To generate well-formed anticipative expectations	To realize the first newly generated anticipative expectation *and* to generate new anticipative expectation
For which reasons?	Efficient work and value division to get the future market	Based on the fact that there are more chances to create new anticipative expectations in collaboration	Collaboration favors the realization of anticipative expectations *and* the creation of new anticipative expectations from the generative one
2. Means of the expectation management	Shared vision of a well-defined future, trial and error process	Design second order expectations and share them with potential partners by avoiding confusion with anticipative expectations	Confirm that the generative expectation did generate well-formed anticipative expectation; (high probability, high value). Rejuvenate generative expectation

etc. LCDA decided to embark on broader explorations. In this perspective, Bernard Boyeux started to contact lime producers and managed to launch a development program with one of them, on a new tack coat. He also got in touch with new actors —architects, material researchers—to convince them to work on hemp. For instance, a public research program was launched jointly by LCDA and ENTPE (a technical university specialized in construction techniques) on hemp cement characterization (thermal, acoustic and mechanical properties). As shown in the C-K graph below (Fig. 5.52), these explorations led to provide new assets and new sources of values

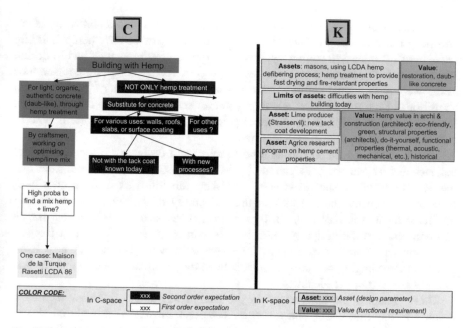

Fig. 5.52 Phase 1 (1993–1998): raising generative expectations

and help to design new futures. Moreover these futures were explicitly so unknown that their probability of success and value couldn't be evaluated. They specified open research questions to support the design of new assets (e.g.: "not with the tack coat known today") as well as new values ("for other uses?").

At the end of 1997, LCDA organized a workshop with a wide attendance base (all the hemp builders of the time), to clarify the potential of "building with hemp". Several experts made presentations on previously unknown aspects of building with hemp, revealing the limits of past experiments.

At the Building with Hemp meeting, LCDA did not provide one well-defined vision of the future but proposed new, partially designed paths with several, explicitly open questions. New pieces of knowledge were also assembled by means of the presentations made by researchers, architects, masons, etc. These preliminary results opened more questions than they provided answers. As L. Goudet, a mason-entrepreneur explained: *"We had a roundtable discussion where everybody explained how they had experimented with hemp building and that there were no problems. When it came to my turn, I said: 'I don't understand, you all build with hemp but nobody has mentioned the problems. I have problems, for example that the mortar doesn't dry, I don't know how to mix it, I don't know how to fix it on the wall, the setting is uncertain, etc.' Then we went round the table again and it became clear that we all had the same problems."* The meeting made people aware of the *absence of any reliable techniques.* LCDA raised expectations at this meeting, not by validating the technique but by opening and keeping 'open' multiple alternatives

During this meeting, LCDA confronted the participants with a simple decision: whether or not it was worthwhile for them to take part in the "hemp building movement". Rather than having to decide to "fund" a business, they had to decide whether or not to commit to an open design process. As one expectation raiser once said: "we say to people coming: 'don't wonder whether you buy [building with hemp] or not but ask yourself whether you *can see a potential to collaborate* on [building with hemp]'"; if one "sees a potential", this means that one is able to contribute to the generative expectations.

This type of expectations hence tend to reject non designers: if the stakeholder believes that he can not influence the probability of success, he will consider that the probability of the success is very low (there are still a lot of open questions, there were so many alternatives that the probability of the "winning" one being profitable for him is very low,…). By contrast, if the stakeholder believed that he *can* contribute to the design process, he will participate in the hope that the collaboration will lead to create alternatives (yet unknown). The unknowness itself results in encouraging participants who believe they can contribute to the creation of alternatives (not yet identified).

These elements are summarized in the analytical framework below (see table below). In this first phase, generative expectations are well-identified and clearly distinguished from anticipative one. The expectation raisers aim at involving designers (more than funders), to work to generate well-formed anticipative expectations; they want to convince these designers by showing that working together, they have more chance to generate anticipative expectations than working alone (they don't try to convince them by sharing the value of the future, precisely because the probability of this value can not be defined). These goals are achieved through two main means:

(1) design generative expectations by enriching assets, and sources of values and by designing alternatives which are explicitly so incomplete that (i) their further design would require to create (discover) new assets and/or new sources of values and (ii) consequently no reliable probability of success and value can be computed.
(2) Show these partially unknown futures to potential contributors, by underlining that (i) there are not anticipative expectations; (ii) it might be fruitful to design together.

This first phase ended in 1998 with the creation of an association for building with hemp designed to act as a research and innovation platform for all the participants.

5.9.3.3   Phase 2 (1997–2005) First Outcomes: Realizing the Newly Generated Anticipative Expectation While Rejuvenating Generative Expectations

After the creation of the "building with hemp" association, the participants started to explore new paths and design complementary experiments. LCDA and Strasservil developed their new tack coat. Research labs started to work on hemp cement properties. Entrepreneurs developed different processes. Masons worked on how to use hemp concrete in several geographical locations and for various uses. The core team met every two months between 1998 and 2000 to set in place the association's status, organization, projects and strategy. During this period, the process of path creation was maintained. New forms of business were discussed, such as selling hemp concrete building blocks with a complete construction "system". New properties for hemp concrete were investigated in the meantime: as a porous system, hemp concrete enables new types of air circulation in buildings; as a water absorbing product, it can help regulate humidity; and it can also offer new forms of acoustic comfort. This corresponded to new assets, new sources of value and new futures but these futures were still so incomplete that probability of success and value could not be reliably computed and their design would lead to the creation of new As and sources of value.

In parallel, well-formed anticipative expectations emerged. It became increasingly clear that it was indispensable to obtain decennial liability coverage from insurance companies. This required stable, standardized, validated materials (hemp, lime, aggregates, etc.) as well as routinized practices, implemented by masons qualified to use hemp. This consisted in drafting a set of "professional rules" to be validated by the building profession and to be followed to obtain decennial liability insurance. To write these rules, the actors had to select (and validate) some types of products (and hence exclude others) and select (and validate) some practices (and exclude others) to finally stabilize a division of labour (between hemp producers, hemp transformers, lime producers, aggregate producers, masons, architects, etc.) and unavoidably, the division of value. "getting professional rules on building with hemp based on the known hemp, coats, practices…" emerged as a clear anticipative expectation (assets, value for the customers/users, well-defined future, with a reasonable probability of success and value). But their design automatically meant that potential "winners" and "losers" were identified, the latter being potentially hostile, which could decrease the probability of success of this project.

The second major meeting on "building with hemp" took place in this context in 2001. 250 people took part, far more than at the first one. In particular, new construction firms and institutions attended the conference (craftsmen, the French Federation of Building (FFB), CSTB (Scientific and Technical Centre for Building), ADEME (French Environment and Energy Management Agency). The participants were informed of the progress made since the last meeting and gradually became familiar with the innovation field as a whole (synthesized in Fig. 5.53).

Interestingly, the focus of the conference was *not* on the professional rules, i.e. not on the anticipative expectations. Participants were simply asked whether or not they would go on building with hemp in general, which encompassed anticipative *and* generative expectations. This led to the paradoxical consequence that the association launched the professional rule project while opponents of the project were members of the association.

This can be logically explained by the co-existence of generative and anticipative expectations. Just as in the first phase, generative expectations are likely to

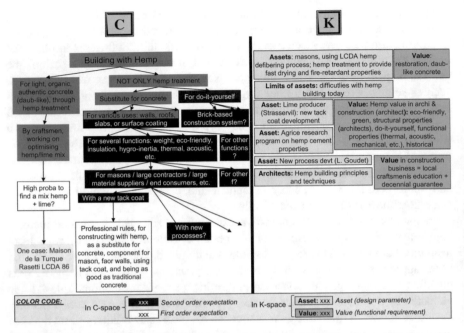

Fig. 5.53 Phase 2 (1997–2005): realizing the newly generated anticipative expectation while rejuvenating generative expectations

attract designers. Designers will come even if, in the perspective of anticipative expectation alone, they wouldn't have participated. For instance the anticipative expectation might correspond to some complementary assets (Teece 1986); if the collaboration addresses only anticipative expectation, the owner of complementary assets should not participate (since whatever his collaboration, the other will pay for the development and he will profit from it without paying the development costs). When generative expectations are associated to anticipative one, the emergence of alternatives is always possible so that it might become critical for the complementary asset owner to participate. The same is true for a loser in the anticipative expectation: if the collaboration addresses only the anticipative expectation, then he won't participate and will even fight the project. If generative expectations are associated to generative ones, then it is more reasonable for him to participate to contribute to the development of other solutions.

Finally this means that *generative expectations reinforce anticipative ones*. Since free rider will contribute and potential opponents won't fight against its development, it contributes to make anticipate expectations more "well-formed" in the sense that it reinforces the probability of success.

Phase 2 is summarized in the table below. It shows both anticipative and generative expectations. The expectation raiser aims at keeping in the game all possible contributors to building with hemp (i.e. including opponents or free-riders of anticipative expectations); it aims at both realizing the first newly generated anticipative expectation *and* generating new anticipative expectation on the basis of rejuvenated expectations. These effects are obtained through several means: (1) confirm that generative expectation led to anticipative expectations; (2) ensure that the latter are well-formed (high probability, high value); (3) rejuvenate generative expectation.

The first professional rules were established in early 2007. New research projects were also launched during that period, in particular one launched in 2004 to study prefabricated hemp concrete building blocks with associated systems design principles.

5.9.3.4   Phase 3 (2005–2010): Towards a Balanced Growth Path? Balancing Newly Generated Anticipative Expectations and Generative Expectations

After 2005, LCDA and its partners followed two main paths in terms of the design process. On the one hand, following on from the professional rules project, they began to design related services (communications on the professional rules; education and training for masons, based on the rules; generalization of the rules in the European Union; increased standardization, etc.). One striking action consisted in involving new partners from the construction materials business. The "building with hemp" association contacted the competitors of Strasservil (at the time called Lhoist-BCB), global firms such as Lafarge, Calcia and other concrete, lime and cement producers. These new partners provided their experience of the construction market and gave increased legitimacy to the association, which could no longer be viewed as an agent for one particular lime producer. In research, the new national Prebat research program (on energy efficient construction) allowed agronomic researchers to gain a better understanding of hemp in concrete. All these were anticipative expectations.

On the other hand, new explorations were prolonged and launched: a Eureka European program continued to study construction principles and building blocks for hemp concrete; at the same time, alternative properties were explored for building with hemp (acoustics, hygrometric comfort, etc.); and connections with other fibres were put on the research agenda, for instance by studying complementarities in hemp, straw and wood. Once again generative expectations were rejuvenated (Fig. 5.54).

The third meeting on "building with hemp" took place in 2006. Hemp was recognized as a matter of national interest. Instead of taking place at LCDA's premises in Bar-sur-Aube, near Troyes (Champagne region), it was held at the Ministry for the Environment in Paris under the patronage of national representatives. Representatives from the ministries of equipment and of agriculture, standardization agencies and social housing agencies also attended.

From the point of view of LCDA and its partners this event was a great opportunity but they were concerned about the risk of it creating bubbles of hype and expectations that the actors would not be able to meet. Confronted with the

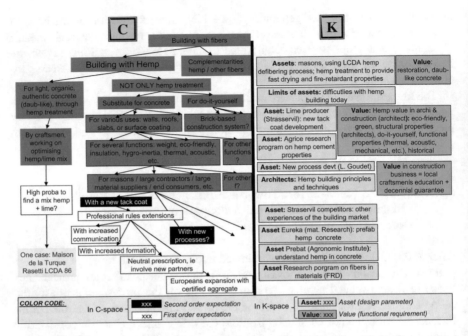

Fig. 5.54 Phase 3 (2005–…): towards a balanced growth path? Balancing newly generated anticipative expectations and generative expectations

growing fad for building with hemp, the head of LCDA, Benoit Savourat, tamed expectations (see for instance his interview in *Ouest-France*, July 2009), by stressing that there were still many uncertainties in the business and many unknown factors to be dealt with. He underlined that large-scale industrial and commercial applications were still far off and that R&D efforts were still required.

This phase confirms the pattern already observed in phase 2: coexistence of anticipative and generative expectations; to attract designers to realize the anticipative expectations but also to generate new anticipative expectations, on the basis of more favourable conditions to design; this mobilization is obtained by confirming the capacity to generate well-formed anticipative expectations and by rejuvenating the generative expectations.

5.9.4 Main Results and Implications

The story is still in the making. Rising expectations are moving in tandem with continuous growth. New farmers have started to cultivate hemp and important investments have been made to develop industrial hemp processes. How far can this innovation lead? This is still an open question. Based on this empirical study, we can now return to our model of generative expectation management.

We should first underline that we were practically able to distinguish between anticipative and generative expectations, depending on the design degree of freedom in the expectations. The future is either sufficiently defined to get reasonably reliable probability of success and value; or it precisely requires that new assets and sources of values are designed before speaking of probability of success and value.

This simple distinction on the types of expectations leads to a simple distinction on their respective management. Anticipative expectation requires a management that stabilizes and monitors a well-formed expectation. Generative expectation management consists in creating the conditions for other to generate again and again multiple well-formed anticipative expectations by expanding assets and sources of value. This general criteria leads to two contrasted models of expectations management: anticipative expectations and generative expectations (see Table 5.10)

R1: Who is attracted, to do what, for which reasons?

Whereas an anticipative expectation aims at attracting funders and suppliers of pieces of the expected future to realize the defined future, generative expectation management aims at attracting designers able to imagine futures that cannot be described in advanced. In the former case, the "expectators" are convinced by the value of the future; in the latter case, they are convinced by the more favourable conditions made to generate new futures.

LCDA designed an innovation field that made the stakeholder able to design, underlining missing knowledge and unknown dimensions of the proposals. It hence helped stakeholders to organize their own agenda of research questions and their design process. As a researcher said: "By working with the association, I can identify relevant research questions far more easily. My colleagues who don't participate are often reproducing trials and results already obtained some years ago".

R2: What are the means to manage expectations over time?

We have identified three critical means for managing generative expectations:

3. uncover generative expectation by revealing the open questions, the "design degree of freedom". The generative expectation manager will insist on what is not known, what remained to be explored.
4. give convincing proofs of the generative power: make sure that anticipative expectations are generated and that they are well-formed.
5. balance the generation of anticipative expectations (from generative expectations) with the rejuvenation of generative expectations. The generative expectation manager actively contributes to formulate new generative expectations.

The management of generative expectations is marked by a dual generation process (generate anticipative expectations/rejuvenate generative expectations). This dual generation corresponds to a **mutual conditioning,** with generative expectations supporting the realization of anticipative ones and vice versa. Expectation management consists in using realization in one path to create new paths in another direction and conversely to create new paths to support the realization of another path. For instance, opening paths on building blocks and sophisticated properties of hemp concrete helped to go on with the professional rule project; opening new paths on building with other fibres helped to increase the

Table 5.10 A model of generative expectation management

	Anticipative expectation management	Generative expectation management
Expectation	A proposition P_{1stOE} with properties: – there is an asset A – there are sources of values S – there is a future F – This future F based on S has a high value – this future F based on As is very probable	A proposition P_{2ndOE} with properties: – there are multiple futures Fs – these future Fs and the related assets As and sources of values Ss, are not sufficiently defined to speak about their value V (Bs/S) and the probability P (Bs/As) (or equivalently: there are enough degrees of freedom on the futures Fs to create new As and Ss. – P_{2ndOE} defines an action space to generate P_{1stOE}
1. Who is attracted?	Funders attracted by V(F/S)	Designers attracted by the collective action space
To do what?	To realize F	Create anticipative expectations, i.e. design As, Ss, Fs such as P (F/A) and V(B/S) are high
For which reasons?	Coordinate resources to realize F based on A, take part in V(B/S)	Participate to a collective action that is more generative (more and better anticipative expectations)
2. Means of expectation management	(1) Find and keep over time the relevant A and S and F that create the right alignment (right people, right work division, right value division) (2) Monitor the convergence of P (F/A) towards 1 over time	(3) Uncover generative expectation and its related action space (4) Prove the generative power by generating well formed anticipative expectations (5) Balance the generation of anticipative expectations with the rejuvenation of generative expectations

business based on professional rules. Conversely, realizations of initial promises convinced stakeholders to go on with emerging promises.

When expectation management focuses on the realization of an anticipative expectation, it is likely to get stuck in political games and conflicts of interests decreasing the probability of success and the value; when it focuses purely on the creation of new anticipative expectations, there is a risk of it getting stuck in speculative hype. Generative expectations create new horizons for the likely losers of the first business created, while well-formed anticipative expectations convince them of the likelihood that the next promise will come to fruition; conversely, anticipative expectations create resources for the winners of the first business created and path creation encourage them to participate further and to reinvest these (knowledge) resources in the next business creation.

References

Agogué M (2012a) L'émergence des collectifs de conception inter-industries. Le cas de la Lunar Society dans l'Angleterre du XVIII siècle. *Gérer et comprendre* 109:55–65.

Agogué M (2012b) *Modéliser l'effet des biais cognitifs sur les dynamiques industrielles: innovation orpheline et architecte de l'inconnu.* MINES ParisTech, Paris.

Agogué M (2013) *L'innovation orpheline. Lutter contre les biais cognitifs dans les dynamiques industrielles.* Economie et Gestion. Presses de l'Ecole des Mines, Paris.

Agogué M, Arnoux F, Brown I, Hooge S (2012a) *Introduction à la conception innovante— Eléments théoriques et pratiques de la théorie C-K.* Economie et Gestion. Presses de l'Ecole des Mines, Paris.

Agogué M, Berthet E, Fredberg T, Le Masson P, Segrestin B, Stötzel M, Wiener M, Ystrom A (2013a) A contingency approach of open innovation intermediaries—the management principles of the "intermediary of the unknown". Paper presented at the *European Academy of Management*, Istambul,

Agogué M, Berthet E, Fredberg T, Le Masson P, Segrestin B, Stötzel M, Wiener M, Ystrom A (2016) Explicating the role of innovation intermediaries in the "unknown": A contingency approach. *Journal of Strategy and Management* 9 (2).

Agogué M, Cassotti M, Kazakçi A (2011) The Impact of Examples on Creative Design: Explaining Fixation and Stimulation Effects. Paper presented at the *International Conference on Engineering Design, ICED'11*, Technical University of Denmark,

Agogué M, Comtet G, Le Masson P, Menudet J-F, Picard R (2013b) Managing innovative design within the health ecosystem: the Living Lab as an architect of the unknown. *Revue Management et Avenir Santé* 1 (1):17–32.

Agogué M, Kazakçi A (2014) 10 years of C-K theory: a survey on the academic and industrial impacts of a design theory. In: Chakrabarti A, Blessing L (eds) *An Anthology of Theories and Models of Design. Philosophy, APproaches and Empirical Explorations.* Bangalore, pp 219–235. doi:10.1007/978-1-4471-6338-1

Agogué M, Kazakçi A, Hatchuel A, Le Masson P, Weil B, Poirel N, Cassotti M (2014a) The impact of type of examples on originality: Explaining fixation and stimulation effects. *Journal of Creative Behavior* 48 (1):1–12.

Agogué M, Le Masson P, Robinson DKR (2012b) Orphan Innovation, or when path-creation goes stale: missing entrepreneurs or missing innovation? *Technology Analysis & Strategic Management* 24 (6):603–616.

Agogué M, Poirel N, Pineau A, Houdé O, Cassotti M (2014b) The impact of age and training on creativity: a design-theory approach to study fixation effects. *Thinking Skills and Creativity* 11:33–41.

Agogué M, Yström A, Le Masson P (2013c) Rethinking the Role of Intermediaries as an architect o f collective exploration and creation fo knowledge in open innovation. *International Journal of Innovation Management* 17 (2):24.

Amsterdamer Y, Molin H (2011) Les Experts-Leaders face à l'innovation de rupture: des méthodes de construction et de gestion de l'expertise. Ingénierie de la Conception. MINES ParisTech, Paris.

Arnoux F (2013) *Intégrer des capacités d'innovation radicale: Le cas des mutations des systèmes d'énergie aéronautiques.* MINES ParisTech, Paris.

Barthelemy M, Guémy A (2012) Méthodes de conception innovante pour accompagner les processus d'entrepreneuriat interne en photonique. Option Ingénierie de la Conception. MINES ParisTech, Paris.

Becker B, Gassmann O. (2006) "Gaining leverage effects from knowledge modes within corporate incubators." *R&D Management*, 36 (1):1–16.

Ben Abbes A (2007) Cadrer des projets collaboratifs en structurant des champs d'innovation. Ecole des Mines de Paris, Paris.

Bergek A., Jacobsson S, Carlsson B, Lindmark S, Rickne A. (2008) "Analyzing the functional dynamics of technological innovation systems: A scheme of analysis." *Research Policy*, 37 (3):407–429.

Berthet E (2013) *Contribution à une théorie de la conception des agro-écosystèmes. Fonds écologique et inconnu communs.* MINES ParisTech) AgroParisTech, Paris.

Berthet E, Bretagnolle V, Segrestin B (2012) Introduction of semi-perennial forage crops in an intensive cereal plain to restore biodiversity: a need for collective management *Journal for Sustainable Agriculture.*

Bertin A (1938) Etude sur le mécanisme des recherches scientifiques et en particulier examen des possibilités de recherches collectives sur le plan professionnel. Paris.

Bessant J, Rush H (1995) "Building bridges for innovation: the role of consultants in technology transfer." *Research Policy*, 24 (1):97–114.

Birkinshaw J, Bouquet C, Barsoux J-L (2011) The five Myths of Innovation. *MIT Sloan Management Review* 52 (2):43–50.

Borup M, Brown N, Konrad K, Van Lente H (2006) The sociology of expectations in science and technology. *Technology Analysis & Strategic Management* 18 (3/4):285–298.

Bresnahan TF, Trajtenberg M (1995) General Purpose Technologies: Engines of Growth? *Journal of Econometrics* 65 (1):83–108.

Brogard C, Joanny D (2010) Stratégies d'innovation pour préparer les moteurs d'avion vert de 2025. Rapports de l'option Ingénierie de la Conception. MINES ParisTech, Paris.

Brown I, Le Masson P, Weil B (2011) Enabling and controlling third-party developers—a study of Apple's iPhone App design environment. Paper presented at the *International Product Development Management Conference*, Delft,

Brown N (2003) Hope against hype: accountability in biopasts, presents and future. *Science studies* 16 (2):3–21.

Brun E, Polo K (2008) Comment construire une offre énergétique basée sur les plateformes Piles à combustible (PAC) AXANE? Rapports de l'option Ingénierie de la Conception. MINES ParisTech, Paris.

Callon M (1993) Variety and irreversibility in networks of technique conception and adoption. In: Foray D, Freeman C (eds) *Technology and the Wealth of Nations—The Dynamics of Constructed Advantage.* Pinter, London,

Carreel C (2011a) Conception de nouveaux pansements grâce à une exploration structurée du diagnostic de la plaie—Urgo Médical. Rapports de l'option Ingénierie de la Conception. MINES ParisTech, Paris.

Carreel C (2011b) Entrepreuriat et théories de la conception. Rapports de l'option Ingénierie de la Conception. Chalmers University & MINES ParisTech, Paris.

Chapel V (1997) *La croissance par l'innovation intensive: de la dynamique d'apprentissage à la révélation d'un modèle industriel, le cas Téfal.* Thèse de doctorat en Ingénierie et Gestion, Ecole des Mines de Paris, Paris.

Christensen CM (1997) *The Innovator's Dilemma. When New Technologies Cause Great Firms to Fail.* The Management of Innovation and Change. Harvard Business School Press, Boston, MA.

Cogez P, Felk Y, Le Masson P, Weil B (2011) Absorptive Capacity for Radical Innovation: a Case Study in the Semiconductor Industry. In: *IEEE International Technology Management Conference*, San Jose, California, 2011.

Cogez P, Kokshagina O, Le Masson P, Weil B (2013) Industry-Wide Technology Road Mapping in Double Unknown—The Case of the Semiconductor Industry. In: *IEEE International Technology Management Conference*, San Jose, CA, 2013.

Cohen WM, Levinthal DA (1989) Innovation and Learning: The Two Faces of R & D. *The Economic Journal* 99 (397):569–596.

Cohen WM, Levinthal DA (1990) Absorptive Capacity: A New Perspective on Learning and Innovation. *Administrative Science Quarterly* 35 (1):128–152.

Cohen WM, Levinthal DA (1994) Fortune Favors the Prepared Firm. *Management Science* 40 (2):227–251.

Coombs R, Narandren P, Richards A (1996) A literature-based innovation output indicator. *Research Policy* 25:403–413.

Dantas Machado Rosa C, Raade K (2006) Profitability of venture Capital investment in Europe and in the United States. European Economy Economic Papers. European Commission, Directorate-General for Economic and Financial Affairs,

Davila A, Foster G, Jia N (2010) Building Sustainable High-Growth Startup Companies: MANAGEMENT SYSTEMS AS AN ACCELERATOR. *California Management Review* 52 (3):79–105.

de Metz I (2010) Dispositifs d'étude des compé-tences du consommateur dans la con-ception de produits innovants. Ingénierie de la Conception. MINES ParisTech, Paris.

Defour M, Delaveau C, Dupas A (2010) *Avionique. Des technologies innovantes au services des plus belles réussites aéronautiques.* Gallimard Loisirs, Paris.

Drucker P (1985a) *Innovation and Entrepreneurship, Practice and Principles.* reprint 2002 edn. Butterworth-Heinemann, Oxford.

Drucker PF (1985b) The Discipline of Innovation. *Harvard Business Review.*

Dubois L.-E, Le Masson P, Weil B, Cohendet P (2014) "From organizing for innovation to innovating for organization: how co-design fosters change in organizations." 21st International Product Development Management Conference (Thomas Hustad best paper award), Limerick, Ireland, 23.

Duncan RB (1976) The ambidextrous organization: designing dual structures for innovation. In: Kilmann LR, Pondy LR, Slevin DP (eds) *The management of organizational Design: Strategy Implementation,* vol 1. North Holland, New York, pp 167–188.

El Qaoumi K (2012) Compétitivité des entreprises et conception. Mémoire du Master MTI. MINES ParisTech—Paris Dauphine Université, Paris.

El Qaoumi K (2016) *L'expansion fonctionnelle, nouvelle mesure de l'innovation—Étude empirique et modèle post-lancastérien de la transformation des biens de consommation.* PSL Research University—MINES ParisTech, Paris.

El Qaoumi K, Le Masson P, Weil B, Ün A (2016) Testing Evolutionary Theory of Household Consumption Behavior in the case of Novelty—Product characteristics approach. *Journal of Evolutionary Economcs* (accepted).

Elmquist M, Le Masson P (2009) The value of a 'failed' R&D project: an emerging evaluation framework for building innovative capabilities. *R&D Management* 39 (2):136–152.

Elmquist M, Segrestin B (2007) Towards a new logic for Front End Management: from drug discovery to drug design in pharmaceutical R&D. *Journal of Creativity and Innovation Management* 16 (2):106–120.

Elmquist M, Segrestin B (2009) Sustainable development through innovative design: lessons from the KCP method experimented with an automotive firm. *International Journal of Automotive Technology and Management* 9 (2):229–244.

Felk Y (2011) *Evaluation et pilotage des activités de recherche pour la rupture dans la R&D centrale de STMicroelectronics: réviser les classiques du management de la recherche industrielle.* MINES ParisTech, Paris.

Felk Y, Le Masson P, Weil B, Cogez P (2009) Absorptive or desorptive capacity? Managing advanced R&D in semi-conductors for radical innovation. In: *International Product Development Manageent Conference,* Enschede, the Netherlands, 2009.

Felk Y, Le Masson P, Weil B, Hatchuel A (2011) Designing patent portfolio for disruptive innovation—a new methodology based on C-K theory. In: *International Conference on Engineering Design, ICED'11,* Copenhagen, Technical University of Denmark, 2011. p 12.

Finke RA (1990) *Creative Imagery: Discoveries and inventions in visualization*. Erlbaum, Hillsdale, NJ.

Freitas Salgueiredo C (2013) Modeling biological inspiration for innovative design. Paper presented at the *i3 Conference*, Paris,

Freitas Salgueiredo C, Hatchuel A (2016) Beyond analogy: A model of bio-inspiration for creative design. *AI EDAM* 30 (Special Issue 02):159–170.

Gapihan O, Le Mestre S (2008) Méthodes de conception et compétences des consommateurs—le cas de produits de la grande distribution. Ingénierie de la Conception. MINES ParisTech, Paris.

Gardey de Soos P (ed) (2007) *Conception innovante à la RATP: la méthode KCP. Cinq cas pratiques de conception innovante collective*. Les rapports de la prospective, n°146. RATP, Paris.

Garel G, Mock E (2016) *The Innovation Factory - Taking the Plunge!*, CRC Press, Boca Raton, FL. 166 p

Gawer A, Cusumano M (2002) *Platform leadership: How Intel, Microsoft, and Cisco Drive Industry Innovation*. Harvard Business School Press, Boston, MA.

Geels FW (2002) Technological transitions as evolutionary reconfiguration processes: a multi-level perspective and a case-study. *Research Policy* 31 (8–9):1257–1274.

Geels FW (2005) The dynamics of transitions in socio-technical systems: A multi-level analysis of the transition pathway from horse-drawn carriages to automobiles (1860–1930). *Technology Analysis & Strategic Management* 17 (4):445–476.

Geels FW, Raven R (2006) Non-linearity and Expectations in Niche-Development Trajectories: Ups and Downs in Dutch Biogas Development (1973–2003). *Technology Analysis & Strategic Management* 18 (3/4):375-392.

Geels FW, Schot J (2007) Typology of sociotechnical transition pathways. *Research Policy* 36 (3):399–417.

Gillier T, Piat G, Roussel B, Truchot P (2010) Managing Innovation Fields in a Cross-Industry Exploratory Partnership with C–K Design Theory. *Journal of product innovation management* 27 (6):883–896.

Gracian B (1648) *Agudeza y arte de ingenio*. Iuan Nogues, Huesca.

Hassen A (2012) Modeling and simulation of risk management strategies in double unknown. Introducing the concepts of temperature and density of the economical environment. MINES ParisTech, Chair of Design Theory and Methods for Innovation, Paris.

Hatchuel A (2006) Quelle analytique de la conception ? Parure et pointe en design. In: Flamand B (ed) *Le design. Essais sur des théories et des pratiques*. Editions du Regard, Paris, pp 147–160.

Hatchuel A, Le Masson P, Weil B (2009) Design Theory and Collective Creativity: a Theoretical Framework to Evaluate KCP Process. In: *International Conference on Engineering Design, ICED'09*, 24–27 August 2009, Stanford CA, 2009.

Hatchuel A, Le Masson P, Weil B (2011) Teaching Innovative Design Reasoning: How C-K Theory Can Help to Overcome Fixation Effect. *Artificial Intelligence for Engineering Design, Analysis and Manufacturing* 25 (1):77–92.

Hekkert MP, Suurs RAA, Negro SO, Kuhlmann S, Smits REHM (2007) "Functions of innovation systems: A new approach for analysing technological change." *Technological Forecasting and Social Change*, 74 (4):413–432.

Hesham Mohamed R (2012) Measuring firms'competencies using an IPC-bsed instrument. MINES ParisTech, Paris.

Hooge S, Kokshagina O, Le Masson P, Levillain K, Weil B, Fabreguette V, Popiolek N (2016) "Gambling versus Designing: Organizing for the Design of the Probability Space in the Energy Sector." *Creativity and Innovation Management*, 25 (4):pp. 464–483.

Hughes TP (1983) *Networks of power: electrification in Western Societies, 1880–1930*. John Hopkins University Press, Baltimore.

Israel P (1998) *Edison: A Life of Invention*. John Wiley ans Sons, New York.

Itten J (1961) *The art of color*. John Wiley & Sons, Inc., New York, NY.

Itten J (1975) *Design and Form, the Basic Course at the Bauhaus and Later*. Revised edition (first edition 1963) edn. John Wiley and Sons, Inc., London.

Jansson DG, Smith SM (1991) Design Fixation. *Design Studies* 12 (1):3–11.

Jech T (2002) *Set Theory*. Springer Monographs in Mathematics, 3rd millenium edition, revised and expanded edn. Springer, Berlin.

Kandel E, Leshchinskii D, Yuklea H (2011) VC Funds: Aging Brings Myopia. *Journal of Financial and quantitative analysis* 46 (2):431–457.

Kandinsky W (1975) *Cours du Bauhaus (1929)*. Traduit de l'allemand d'après des notes manuscrites par Suzanne et Jean Leppien edn. Denoël, Paris.

Klee P (1922). *Beiträge zur bildnerischen Formlehre ('contribution to a pictorial theory of form', part of Klee 1921-2 lectures at the Bauhaus)*, Weimar p.

Klee P (1966) On modern Art, Faber and Faber Limited, London. p., English translation of the conferences given in 1924–1926.

Kleinknecht A, Bain D (eds) (1993) *New concepts in inno–va–tion out-put measure-ment*. Mac-millan, London.

Klerkx L, Leeuwis C (2008) "Balancing multiple interests: Embedding innovation intermediation in the agricultural knowledge infrastructure." *Technovation*, 28 (6):364–378.

Kline SJ, Rosenberg N (1986) An Overview of Innovation. In: Landau R, Rosenberg N (eds) *The Positive Sum Strategy, Harnessing Technology for Economic Growth*. National Academy Press, Washington, pp 275–305.

Koh EY (2013) Engineering design and intellectual property: where do they meet? *Research in Engineering Design* 24 (4):325–329.

Kokshagina O (2014) *Risk Management in Double Unknown: Theory, Model and Organization for the Design of Generic Technologies*. MINES ParisTech, Paris.

Kokshagina O, Cogez P, Le Masson P, Weil B (2012a) The Impact of Sector Dynamic Rules on the Choice of Management Strategy in Highly Uncertain Market and Technology Situation. Paper presented at the *Informs*, Phoenix,

Kokshagina O, Le Masson P, Weil B (2013) How design theories enable the design of generic technologies: notion of generic concepts and Genericity building operators Paper presented at the *International Conference on Engineering Design, ICED'13*, Séoul, Korea,

Kokshagina O, Le Masson P, Weil B, Cogez P (2012b) Platform emergence in double unknown: common challenge strategy. Paper presented at the *R&D Management Conference*, Grenoble, France,

Kokshagina O, Le Masson P, Weil B, Cogez P (2014) Innovative field exploration & associated patent portfolio design models. Paper presented at the *IDMME 2014*, Toulouse, France,

Kokshagina O., Le Masson P., Weil B. (2016) Should we manage the process of inventing? Designing for Patentability, Research in Engineering Design (online)

Koza JR, Keane MA, Streeter MJ, Adams TP, Jones LW (2004) Invention and creativity in automated design by means of genetic programming. *Artif Intell Eng Des Anal Manuf* 18 (3):245–269.

Kroll E, Le Masson P, Weil B (2013) Modeling parameter analysis design moves with C-K theory. Paper presented at the *International Conference on Engineering Design, ICED'13*, Séoul, Korea,

Lancaster KJ (1966) Change and innovation in the technology of consumption. *American Economic Review* 56:14–23.

Lane PJ, Koka BR, Pathak S (2006) The reification of absorptive capacity: a critical review and rejuvenation of the construct. *Academy of Management Review* 31 (4):833–863.

Lefebvre P (2013) Organizing Deliberate Innovation in Knowledge Clusters: From Accidental Brokering to Purposeful Brokering Processes. *International Journal of Technology and Management* 63 (3/4):212–243.

Leguay A, Rousseleau D (2012) Conception des couples {technologie; marché} dans le cas de la recherche avancée chez STMicroelectronics. Rapports de l'option Ingénierie de la Conception. MINES ParisTech, Paris.

Le Masson P, Aggeri F, Barbier M, Caron P (2012a) The sustainable fibres of generative expectation management: The "building with hemp" case study. In: Barbier M, Elzen B (eds) *System Innovations, Knowledge Regimes, and Design Practices towards Transitions for Sustainable Agriculture.* INRA Editions, Paris, pp 226–251.

Le Masson P, Cogez P, Felk Y, Weil B (2012b) Absorptive Capacity for Radical Innovation: A Case Study in the Semiconductor Industry. *Advances in Technology and Innovation Management* 1.

Le Masson P, Cogez P, Felk Y, Weil B (2012c) Revisiting Absorptive Capacity with a Design Perspective. *International Journal of Knowledge Management Studies* 5 (1/2):10–44.

Le Masson P, Weil B, Hatchuel A, Cogez P (2012d) Why aren't they locked in waiting games? Unlocking rules and the ecology of concepts in the semiconductor industry. . *Technology Analysis & Strategic Management* 24 (6):617–630.

Le Masson P, Hatchuel A, Weil B (2010a) Modeling Novelty-Driven Industrial Dynamics with Design Functions: understanding the role of learning from the unknown. In: *13th International Schumpeter Society*, Aalborg, Denmark, 2010a. p 28.

Le Masson P, Hatchuel A, Weil B (2011) The Interplay Between Creativity issues and Design Theories: a new perspective for Design Management Studies? *Creativity and Innovation Management* 20 (4):217–237.

Le Masson P, Hatchuel A, Weil B (2013a) Teaching at Bauhaus: improving design capacities of creative people? From modular to generic creativity in desing-driven innovation. Paper presented at the *10th European Academy of Design*, Gothenburg,

Le Masson P, Weil B (2016) Fayol, Guillaume, Chevenard—la Science, l'Industrie et l'exploration de l'inconnu: logique et gouvernance d'une recherche conceptive. *Entreprises et Histoire* 83:79–107.

Le Masson P, Weil B, Hatchuel A (2006) *Les processus d'innovation. Conception innovante et croissance des entreprises.* Stratégie et management. Hermès, Paris.

Le Masson P, Weil B, Hatchuel A (2010b) *Strategic Management of Innovation and Design.* Cambridge University Press, Cambridge.

Le Masson P, Weil B, Kokshagina O (2013b) A new perspective for risk management: a study of the design of generic technology with a matroid model in C-K theory. In: Taura T, Nagai Y (eds) *Nara Workshop*, Nara, Japan, 2013b. p 15.

Lenfle S, Le Masson P, Weil B (2016) "When project management meets design theory: revisiting the Manhattan and Polaris projects to characterize "radical innovation" and its managerial implications." *Creativity and Innovation Management*, 25 (3): pp. 378–395.

Little AD (1913) Industrial Research in America. *Journal of Industrial and Engineering Chemistry* 5 (10, oct. 1913):pp. 793–801.

Locke RR (1984) *The End of the Practical Man.* Industrial Development and the Social Fabric. JAI Press Inc, Greenwich, Connecticut.

Magnusson T, Berggren C (2001) Environmental innovation in auto development. Managing technological uncertainty within strict limits. *International Journal of Vehicle Design* 26 (2/3):101–115.

Miller WL, Morris L (1999) *Fourth Generation R&D, Managing Knowledge, Technology, and Innovation.* John Wiley & Sons, Inc, New-York.

Millard A (1990) *Edison and the business of innovation.* Johns Hopkins Studies in the History of Technology. The Johns Hopkins University Press, Baltimore and London.

Nambisan S, Sawhney M (2007) "A Buyer's Guide to the Innovation Bazaar." *Harvard Business Review*, June 2007, pp 109–118.

Nowotny H, Felt U (1997) *After the Breakthrough—The Emergence of Hight-Temperature Superconductivity as a Research Field.* Cambridge University Press, Cambridge, UK.

O'Connor GC (2008) Major Innovation as a Dynamic Capability: A Systems Approach. *Journal of product innovation management* 25 (4):313–330.

O'Connor GC, DeMartino R (2006) Organizing for Radical Innovation: An Exploratory Study of the Structural Aspect of RI Management Systems in Large Established Firms. *Journal of product innovation management* 23 (6):475–497.

OECD (2005) Oslo Manual—the measurement of scientific and technological activities—proposed guidelines for collecting and interpreting Innovation data. 3rd edition edn. OECD,

Osborn AF (1953) *Applied Imagination: Principles and Procedures of Creative Problem Solving*. First edition edn. Charles Scribner's sons, New York.

Ostrom E (1990) *Governing the Commons: The Evolution of Institutions for Collective Action*. Cambridge University Press, New York.

Ozman M (2013) *The Network Approach in Innovation Studies: An appraisal of its theoretical and empirical foundnations and identification of critical issues for future research*. Paris Est Marne la Vallée, Paris.

Paulus PB (2000) Groups, Teams, and Creativity: The Creative Potential of Idea-generating Groups. *Applied Psychology: An International review* 49 (2):237–262.

Propp T, Moors E (2009) Strategic policy impacts of the uptake of genomics-related expectations: the case of the Netherlands. *Science and Public Policy*.

Reich LS (1985) *The Making of American Industrial Research, Science and Business at GE and Bell, 1876–1926*. Study in economic history and policy, the United States in the twentieth century. Cambridge University Press, Cambridge.

Robinson DKR, Propp T (2008) Multi-path mapping for alignment strategies in emerging science and technologies. *Technological Forecasting and Social Change* 75 (4):517–538.

Roussel PA, Saad KN, Erickson TJ (1991) *Third Generation R&D, managing the link to corporate strategy*. Harvard Business School Press, Boston, Massachusetts.

Schaller RR (2004) *Technological Innovation in the Semiconductor Industry: A Case Study of the International Roadmap for Semiconductors (ITRS)*. George Mason University, Fairfax, VA.

Schofield RE (1957) The Industrial Orientation of Science in the Lunar Society of Birmingham. *Isis* 48:408–415.

Schofield RE (1963) *The Lunar Society of Birmingham, A Social History of Provincial Science and Industry in Eithteenth-Century England*. Clarendon Press, Oxford.

Shinn T (2004) New sources of Radical Innovation. Research-technologies, Transversality and Distributed Learning in a Post Industrial Order. Paris.

Shohet S, Prevezer M (1996) "UK biotechnology: institutional linkages, technology transfer and the role of intermediaries." *R&D Management*, 26 (3):283–298.

Sieg JH, Wallin MW, von Krogh G (2010) Managerial challenges in open innovation: a study of innovation intermediation in the chemical industry. *R&D Management* 40 (3):281–291.

Sinclair B (1974) *Philadelphia's philosopher mechanics—a history of the Franklin Institute 1824–1865*. History of Technology. The Johns Hopkins University Press, Baltimore.

Smith SM, Ward TB, Schumacher JS (1993) Constraining effects of examples in a creative generation task. *Memory and Cognition* 21 (6):837–845.

Smith A, Voss J-P, Grin J (2010) Innovation studies and sustainability transitions: The allure of the multi-level perspective and its challenges. *Research Policy* 39 (4):435–448.

Stewart DD, Stasser G (1995) Expert role assignment and information sampling during collective recall and and decision-making. *Journal of Personality and Social Psychology* 69:619–628.

Sung J, Hopkins M (2006) Towards a method for evaluating technological expectations: Revealing uncertainty in gene silencing technology discourse. *Technology Analysis & Strategic Management* 18 (3/4):345–359.

Talke K, Salomo S, Wieringa JE, Lutz A (2009) What about Design Newness? Investigating the Relevance of a nNeglected Dimension of Product Innovativeness. *Journal of Product Innovation Management* 26:601–615.

Teece DJ (1986) Profiting from technological innovation: Implications for integration, collaboration, licensing and public policy. *Research Policy* 15 (6):285–305.

Thomke SH (2003) *Experimentation Matters. Unlocking the Potential of New Technologies for Innovation*. Harvard Business School Press, Boston.

Thursby JG, Jensen R, Thursby MC (2001)"Objectives, Characteristics and Outcomes of University Licensing: A Survey of Major U.S. Universities." *The Journal of Technology Transfer*, 26 (1):59–72.

Turpin T, Garrett-Jone S, Rankin N (1996) "Bricoleurs and boundary riders: managing basic research and innovation knowledge networks." *R&D Management*, 26 (3): pp 267–282.

Tushman ML, O'Reilly III CA (1996) Ambidextrous Organizations: Managing Evolutionary and Revolutionary Change. *California Management Review* 38 (4):8–30.

Ün A (2011) On the Expansion of Lancasterian Characteristics Space and Learning Patterns. Chair of Design Theory and Methods for Innovation, Paris.

Van Lente H, Hekkert M, Smits R, Van Waveren B (2003) Roles of systemic intermediaries in transition processes. *International Journal of Innovation Management* 7 (3):247–279.

Van Lente H (1993) *Promising Technology: The Dynamics of Expectations in Technological Development*. Eburon, Delft.

Van Lente H, Rip A (1998) Expectations in technological developments: an example of prospective structures to be fiiled in by agency. In: Disco C, van der Meulen BJR (eds) *Getting New Technologies Together*. Walter de Gruyter, Berlin, pp 195–220.

Van Merkerk RO, Robinson DKR (2006) Characterizing the emergence of a technological field: Expectations, agendas and networks in Lab-on-a-chip technologies. *Technology Analysis & Strategic Management* 18 (3/4):411–428.

Veryzer RW (1998) Discontinuous Innovation and the New Product Development Process. *Journal of product innovation management* 15:304–321.

Ward TB (1994) Structured Imagination: The Role of Category Structure in Exemplar Generation. *Cognitive Psychology* 27 (1):1–40.

Ward TB, Smith SM, Finke RA (1999) Creative Cognition. In: Sternberg RJ (ed) *Handbook of Creativity*. Cambridge University Press, Cambridge, pp 189–212.

Appendix
Past Examination Questions in the Course 'Product Design and Innovation' at MINES, ParisTech (2004–2011)

Hints for correct answers are given for the examinations 2004–2011.

© Springer International Publishing AG 2017
P. Le Masson et al., *Design Theory*,
DOI 10.1007/978-3-319-50277-9

Appendix A
Knowledge Control 2004—Product Design and Innovation

A.1. Course Question (1/3 of Total Marks)

1. What are the basic principles of systematic design?
2. What is a C-K operator called? Give examples.
3. State the first axiom of systematic design. Discuss its relevance to design.
4. Define the research function in a business. What is its relationship with innovation?
5. Give two examples, with justification, of products whose identity has recently been altered (excluding examples given in the course)

A.2. Workshop Summary (1/3 of Total Marks)

1. Systematic design stage: give an example of a conceptual model of the vacuum cleaner.
2. Innovative design stage: reasoned presentation of the exploration of a field of innovation (free topic; you can revisit the topics encountered in the workshop).

A.3. Case Study (1/3 of Total Marks)

After studying the accompanying text due to Peter Rice {Rice, 1994 #333}(Fiat) develop the following points:

1. Diagnosis of the design process at Fiat; in particular, is Systematic Design involved (justify to what extent)? Do we see the underlying structure of the languages of systematic design? Is this process conducive to innovation, and if so, to what extent and for what reasons?
2. Diagnosis of the innovative design process put forward by P. Rice; in particular, does the implementation of the procedure allow us to think otherwise, and why? On what types of language is this procedure based? Analyze where and why it fails?
3. Given the examples in the course on the automobile, how would you analyze the two situations described by Peter Rice? Given the procedures and organizations as described do the difficulties encountered still seem relevant today?

© Springer International Publishing AG 2017
P. Le Masson et al., *Design Theory*,
DOI 10.1007/978-3-319-50277-9

4. On the basis of this analysis, what are the main structural characteristics of the "automobile" from the point of view of its design? Have these characteristics changed since the time described by P. Rice? Using other features characteristic of today's automobile, can you complete the process by taking these into account in the design of this type of object today?

Appendix B
Knowledge Control 2005—Product Design and Innovation

B.1. Course Question (1/3 of Total Marks)

1. In C-K theory, what is a concept? What is a piece of knowledge?
2. In the sense of C-K theory, when can a design be assumed to have been achieved?
3. What dimensions must be analyzed to characterize an industrial design regime?
4. State the major characteristics in the particular case of a regime based on systematic design.
5. What are the assessment criteria for the exploration of an innovative field?

B.2. Workshop Summary (1/3 of Total Marks)

1. Systematic design stage: using the case of the vacuum cleaner, give an example of a function (provide justification that this is indeed a function); give an example of a conceptual model associated with this function (provide justification that this is indeed a conceptual model).
2. Innovative design stage: reasoned presentation of the exploration of an innovative field (free topic: you may make use of the topics you encountered in the workshop).

B.3. Case Study (1/3 of Total Marks)

Article to read: **Magnusson, T. and Berggren, C. (2001)**. "Environmental innovation in auto development. Managing technological uncertainty within strict limits." *International Journal of Vehicle Design*, 26, (2/3), 101–115.

This article analyses the design of the Prius 1 which came out in 1997 in Japan, where 70,000 models have been sold. From 2000 Toyota marketed a version outside Japan, the Prius 2, with the Prius 3 coming out in 2003. This car was voted car of the year for 2004, and was a great success. In 2005 Toyota launched a hybrid SUV (RX400h) which met the new emissions regulations covering CO_2/km.

© Springer International Publishing AG 2017
P. Le Masson et al., *Design Theory*,
DOI 10.1007/978-3-319-50277-9

After studying the accompanying article, develop the following points:

1. The issues related to innovation for companies today:

   - The authors mention contemporary changes in NPD (New Product Development) in the automobile industry. Illustrate these changes using recent examples in this sector.
   - Can you see other sectors in which these changes can also be seen? Illustrate your answer.
   - Can you see other forms of innovation that the authors may not have mentioned and which might be more visible in other sectors?

2. The industrial design regime for the automobile:

   - On the basis of the discussions of the participants and from the text, qualify the industrial design regime in the automobile industry.
   - Is this regime stable or not? If not, give a few symptoms of the crisis.

3. The design process in the case of the Toyota hybrid: using the information provided in the article, characterize Toyota's design reasoning.
   Questions to guide the analysis:

   - What concepts are successively dealt with? Use is made of what items of knowledge?
   - What do you think of this reasoning? Is this a systematic design reasoning? Does this involve developing a new engine technology?
   - What are the objectives sought by Toyota (according to the article, and according to you)? How would you characterize the value of the design process used by Toyota (i.e. what are the results of the design process)?

4. Organization of development and prototyping:

   - What types of prototype are employed in the development of a standard product? (how many? What is their purpose? etc.). How would you characterize them in terms of design reasoning?
   - What prototypes are mentioned in the article? What do you think about this number of prototypes? What points do they focus on? According to you, what is their purpose? How would you characterize them in terms of design reasoning?
   - How would you characterize the forms of organization mentioned in the article? Is this systematic design? What do you think? Compare with other exploratory situations encountered in the course (cf. Ideo).

5. How would you finally analyze Toyota's design strategy for the hybrid? (you can use some of the elements concerning Toyota's latest developments on the hybrid).

Appendix C
Knowledge Control 2006—Product Design and Innovation

C.1. Course Question (1/3 of Total Marks)

1. In the theory of systematic design, what is a function?
2. What is the difference between a rule-based design regime and an innovative design regime? You may draw on the three dimensions of a design regime.
3. What are the various platform logics?
4. What assessment criteria are used to identify an innovative firm?

C.2. Summary of Workshops (1/3 of Total Marks): Supermarket Shopping Cart

1. Rule-based design stage: describe the supermarket shopping cart using the language of systematic design. What innovative leads are opened up by these languages?
2. Innovative design stage: propose a design strategy in the innovative field "smart shopping cart for the elderly".
   Questions to guide the work:

   - Using C-K formalism, endeavor to present a set of structured alternatives;
   - What program of action can you put forward for a specialist shopping cart firm that might want to implement this strategy?

C.3. Case Study (1/3 of Total Marks)

Article to read: **Jenkins, R.V. (1983)**. "George Eastman et les débuts de la photographie populaire" [*George Eastman and the beginnings of popular photography*]. *Culture technique*, No.10, June 1983, `USA', pp. 75–87.

This article analyzes the birth of the photographic industry aimed at the general public.

Additional items: articles in Le Monde 2004 and 2006.

© Springer International Publishing AG 2017
P. Le Masson et al., *Design Theory*,
DOI 10.1007/978-3-319-50277-9

After studying the accompanying documents, develop the following points:

1. Reconstruct the design reasoning:
   Questions to guide the analysis:

   - What concepts are successively dealt with?
   - Of what items of knowledge is use made? Where do they come from?
   - Can you detect expansive partitions over the course of the reasoning process?

2. Assess the design strategy
   Questions to guide the analysis:

   - Is this a "disruptive innovation"?
   - Set out the product lines; characterize them according to their explored value spaces, specific resources (skills and technologies) and conditions of deployment (market, business models, financing, income, etc.)
   - Set out the dynamics of the relationships between lines.

3. Describe the organization of a "start-up" making use of the ideas in the course.
   Questions to guide the analysis:

   - What are the various prototypes, tests, demonstrators, products, etc.? What were the support organizations?
   - What are the links between these successive "attempts"? Is there an organization responsible for these links?

4. Design regimes:

   - What can you say about the transition from the innovative design described in the article to rule-based design (traditional photography for the general public)? What parts of the article describe the emerging "dominant design" (technical object, skill, major players in the sector)?
   - Based on articles in Le Monde, discuss the contemporary changes in this rule-based design. Do you agree with Eric Leser, author of the 2004 article, in considering this to be essentially a change in technology?
   - What did these changes mean for a firm such as Kodak?

Appendix D
Knowledge Control 2007—Product Design and Innovation

D.1. Course Question (1/3 of Total Marks)

1. What are the four dimensions of the canonical model?
2. What are the fundamental principles of systematic design?
3. What are the criteria required to say that an industry is in "dominant design"?
4. In C-K theory, what is an "expansive partition"?
5. What are the assessment criteria for the exploration of an innovative field?

D.2. Summary of Workshops (1/3 of Total Marks)

1. In the case of the bolt cutter:

   - Carry out a functional analysis of the bolt cutter.
   - Mob, the company making the bolt cutter, greatly valued the suggestion made by Archilab. Mob wanted to continue the work and requested a structured set of possible innovations based on the concept of a "bolt cutter minimizing user energy input". To prepare this work:

     a. Provide a conceptual model (provide the justification that this is a conceptual model)
     b. In the Mob brief, give a set of innovative paths generated by this conceptual model.

2. Using C-K formalism, endeavor to present a set of structured alternatives based on the concept of the "smart cart for shopping with children".

D.3. Case Study (1/3 of Total Marks)

Article to read: **Gaudillière, J.-P. (2005)**. "Hormones, régimes d'innovation et stratégies d'entreprise: les exemples de Schering et Bayer." [*Hormones, innovative regimes and business strategies: examples from Schering and Bayer*]. *Entreprises et histoire*, 36, pp. 84–102.

This article analyses two hormone treatment design modes in the 1930s. You can start reading from the 2nd paragraph on page 87.

Using the accompanying text, develop the following points:

1. Reconstruct the design reasoning process for the two companies. *Questions to guide the analysis:*

   - Recall the contemporary model for designing medicines in the contemporary pharmaceutical industry (see lecture given by A. Ceccaldi).
   - Bayer: using the information in the text (Prolan and DES in particular, pp. 96–100), reconstruct the design reasoning model. Set out:

     - The types of initial concept
     - The types of operator used, the nature of the knowledge produced and possible partitions.
     - Compare this reasoning with the contemporary design reasoning for pharmaceuticals.

   - Schering: using the information in the text (Progynon and DHA in particular, pp. 91–96), reconstruct the design reasoning model. Set out:

     - The types of initial concept.
     - The types of operator used, the nature of the knowledge produced and possible partitions.
     - Compare this reasoning with the contemporary design reasoning for pharmaceuticals.
     - Compare this reasoning with the design reasoning applied to the case of Nanobiotix. In particular, how do the two companies deal with the question of interaction between treatment and organization. On the basis of these four cases (contemporary model of pharmacy, Bayer, Schering and Nanobiotix), construct a rapid C-K tree for the concept: "controlling the interaction between the pharmaceutical product and the organization".

2. Compare the design organizations. *Questions to guide the analysis:*

   - Bayer:

     - The role of research? Who carries it out? (see the case of DES)
     - The role of clinical trials? Who conducts them?

   - The same questions for Schering.

3. Assessment of design performance.
   Economic data are lacking and it is difficult to assess the company's financial performance. However, it is possible to assess the type of expansion sought by each company and the resources allocated. To compare the two regimes, state for Bayer and Schering:

   - The nature of the conceptual expansions sought.
   - The effort made in the production of knowledge by each of the companies.

Conclusion

Hints for III (2007)

<u>Article to read:</u> **Gaudillière, J.-P. (2005)**. "Hormones, régimes d'innovation et stratégies d'entreprise: les exemples de Schering et Bayer." [*Hormones, innovative regimes and business strategies: examples from Schering and Bayer*] *Entreprises et histoire*, 36, pp. 84–102.

This article analyses two hormonal treatment design modes in the 1930s. You can start reading from page 87, 2nd paragraph.

Using the accompanying text, develop the following points:

1. Reconstruct the design reasoning process for the two companies:

Questions to guide the analysis:

- Recall the contemporary model for design medicines in the contemporary pharmaceutical industry (see lecture given by A. Ceccaldi).
- Bayer: using the information in the text (Prolan and DES in particular, pp. 96–100), reconstruct the design reasoning model. State:

  – The types of initial concept

    synthetic hormone substitute (DES = estrogen)

  – The types of operator used, the nature of the knowledge produced and possible partitions

    $C \rightarrow K$: Candidates for substitute discovered externally;
    $C \rightarrow K$: validation of this candidate in two dimensions: non toxic and same applications
    No partitions, validation only

  – Compare this reasoning with the contemporary design reasoning for pharmaceutical products.

    Very close; the screening phase barely appears here; logic of validation

- Schering: using the information in the text (Progynon and DHA in particular, pp. 91–96), reconstruct the design reasoning model. State:

  – The types of initial concept

    Preparation of natural hormone molecules

  – The types of operator used, the nature of the knowledge produced and possible partitions.

    $C \rightarrow K \rightarrow C$: "natural" preparation = tissue or semi-synthesis extraction;
    K for semi-synthesis = manufacturable since it works in nature;
    $K \rightarrow C$: strategy for finding precursors

*K→C: intermediate precursor between cholesterol and testosterone =
DHA, which can be synthesized*
*C→K: validation of this candidate: non toxic since it is natural;
application to be conceived but if natural then high potential*

- Compare this reasoning with the contemporary design reasoning for pharmaceutical products.

  Somewhat different: applications sought at the end!
  procedures worked on rather than products

- Compare this reasoning with the design reasoning used for Nanobiotix. In particular, how do the two companies deal with the question of interaction between treatment and organization. On the basis of these four cases (contemporary model of pharmacy, Bayer, Schering and Nanobiotix), construct a rapid C-K tree for the concept: "controlling the interaction between the pharmaceutical product and the organization".

  1st partition: K = either validate all interactions by clinical trials
  Or validate the interactions without trials:
  2nd partition in the "without trials" case: either minimize the interactions (Nanobiotix)
  or use only risk-free interactions (Schering)

2. Design organizations
   Questions to guide the analysis:

- Bayer:

  - The role of research? Who carries it out? (see the case of DHE)
    Screening = UK lab; no toxicity, internal
  - The role of clinical trials? Who conducts them?

    Dosage, thanks to a network of correspondents. No work on the application

- Same questions for Schering. Types of initial concept

  Synthesis = Berlin lab; no toxicity, internal
  Clinical trials = dosage on the one hand but also new applications

3. Assessment of design organizations.

Economic data are lacking and it is difficult to assess the company's financial performance. However, it is possible to assess the type of expansion sought by each company and the resources allocated. To compare the two regimes, state for Bayer and Schering:

–   The nature of the conceptual expansions sought.
    S = *expansion of applications*
    B = *synthesis that has the same applications as the natural molecule*
–   The types of knowledge produced
    S = *expansion on synthesis and application*
    B = *minimize K*

Conclusion

1. *Schering regime appears more expansive;*
2. *Bayer regime depends on Schering results for applications*
3. *Bayer regime is more robust as regards the assumption of non-toxicity*
4. *Schering appears as an innovative design company. However, while an innovative design company can also function under a rule-based design regime (see the Saint Gobain example in this course), Schering appears not to use rule-based design. This could possibly prove fatal for future business growth.*

Appendix E
Knowledge Control 2008—Product Design and Innovation

E.1. Course Question (1/3 of Total Marks)

1. Give an order of magnitude for the entry ticket for an automobile program (Renault Megane), the number of engineers and technicians in an automobile engineering department, and the cost of a project for producing a new molecule in the pharmaceutical industry.
2. In the public transport sector, give a rule-base design type of innovative path and an innovative-design type of innovative path; justify your choices.
3. In the theory of systematic design, what is a function? Give an example of a function in the case of the bolt-cutter.
4. State the first axiom of the axiomatic theory due to Nam P. Suh.
5. In C-K theory, what is a concept? What is an item of knowledge?

E.2. Summary of Workshops (1/3 of Total Marks)

E.2.1. Exercise in the Economic Assessment of a Platform

A company that designs and manufactures equipment for electrical networks (e.g. transformers, circuit-breakers, etc.) (e.g. Schneider Electric) has organized the design of its products in platforms. At the request of its major customers, the company initiates the development of a circuit-breaker variant for niche applications. The project manager is faced with the following alternatives:

a. To develop this as a standalone product for a total cost of 100k€ over one year (year 0). Over years 1 to 6 the product thus developed will bring in 20k€ per year for the company (this figure is the difference between the annual turnover and the annual direct costs of production and distribution).
b. To develop this product using an existing platform. The cost is then 40k€ but the product has a lower performance and brings in only 15k€ per year.

1. Using the economic project assessment tools you have encountered in the course, what advice would you give to the project leader?
2. The platform manager suggests the following variant:

© Springer International Publishing AG 2017
P. Le Masson et al., *Design Theory*,
DOI 10.1007/978-3-319-50277-9

c. develop the product using the existing platform while developing a module that could be re-used for future products on the platform. This development would then cost 80k€ and the product would bring in the expected 20k€.

Does this alternative alter the project leader's decision? Why?

3. As arbiter between the exchanges of information between the platform manager and the project leader, the business unit manager requests that the dynamic returns be taken into account. Calculate a dynamic returns indicator for each of the three alternatives. On what value-added condition of the new module does alternative c) become worthwhile? Justify your answer.

(hint for numerical applications: $\sum_{i=1}^{5} \frac{1}{(1+0.10)^i} = 3.79$)

E.2.2. The "Shopping Cart" Example

1. Analysis of the shopping cart using the languages of systematic design:

    a. give three functions of the shopping cart. Justify your answer.
    b. give (at least) one conceptual model describing the role of the user in a supermarket.
    c. Using this knowledge base, suggest two expansive partitions on the "smart shopping cart" concept (justify the expansive character). Towards what knowledge bases might these expansive partitions lead?

2. Using C-K theory, provide a structured set of alternatives based on the concept of "a smart shopping cart in an economic crisis situation".

E.3. Case Study (1/3 of Total Marks)

Article to read: **Gehry, Frank O. (2004).** "Reflections on Designing and Architectural Practice." Extract from "Managing as designing", edited by R. J. Boland and F. Collopy, Stanford Business Books, Stanford, CA, pp. 19–35.

Bibliographical information about Frank Gehry: born in 1929, with a degree in architecture from the university of Los Angeles, Frank Gehry was the architect for numerous works, including in particular the Guggenheim museum in Bilbao (1997); MIT Stata Center in Cambridge (MA) (2003) and the future Louis Vuitton creative foundation in Paris (2010). He created his agency, "Frank O. Gehry and Associates Inc" in Los Angeles in 1962. He won the Pritzker Prize for architecture in 1989.

The accompanying text is the written version of Frank Gehry's lecture at a seminar convened on the occasion of the inauguration of the "Peter B. Lewis" building at the Weatherhead School of Management, Case Western Reserve University, Cleveland, Ohio.

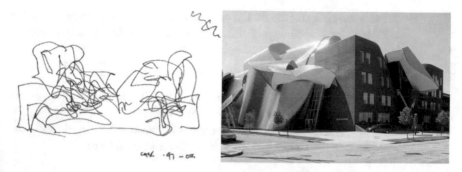

One of Frank Gehry''s first sketches for the Lewis Building—photo of the building after construction

Using the accompanying text, develop the following points:

1. Frank Gehry tries to involve the clients in the design. Discuss this process.
   Questions to guide the analysis:

   - Using C-K theory, analyze the design process to which Frank Gehry invites his clients (in particular, see the design of the MIT building): identify the concepts and knowledge used in the process. Comment on the expansive character of this process.
   - What are the differences between customer relations in this process and customer relations as prescribed by systematic design?
   - What is a "successful" design in the Gehry sense? (in particular, see the discussion on the notion of "functional" design at the end of the chapter).
   - As far as the users are concerned, what, in your opinion, are the conditions for the success of such a design process?

2. The operation of Gehry Partners. What controls the growth of this company?
   Questions to guide the analysis:

   - Comment on the logic of human resources management.

     - Compare this with the dominant logic of the sector (as presented by Frank Gehry).
     - In your opinion, why does Gehry present this as a determining factor for the growth of his company (you may draw on other examples of companies encountered in the course). In terms of innovative design, what "expansions" are allowed by managing human resources in such a way?

   - Comment on the relationship Franck Gehry and his company have with CAD:

     - in what domain has CAD become necessary? Why, in your opinion, does the expertise assembled by Frank Gehry and Partners in this domain appear better than that of the reference companies on this topic?

- – What is the effect on the design capabilities of Gehry and partners? Is this a systematic design logic? What additional knowledge is required to answer this question with complete rigor?

- Comment on the logic of the company's cohesion:

  - – What is the circle of partners, and how has it evolved? What are the "common interests" and how have they evolved?
  - – How has Franck Gehry prepared for his succession? What are the possible futures for Gehry and Partners without Franck Gehry?

Hints for III (2008)

1. Frank Gehry likes to involve clients in the design. Discuss this process.
   Questions to guide the analysis:

   - Using C-K theory, analyze the design process to which Frank Gehry invites his clients. (in particular, see the case of the MIT building): identify the concepts and items of knowledge used in the process. Comment on the expansive character of this process.
   C1: satisfying the budget/without; C2 = the concept of togetherness without necessarily being together; K1 = show what can be done with the budget; K2 = show different forms of collective habitat (colonial style, orang-utan, etc.)

   - How does this process differ from the client relationship as prescribed by systematic design?
   Function = no client learning process!
   - What is a "successful" design in the Gehry sense? (in particular, see the discussion on the notion of "functional" design at the end of the chapter).
   Two issues: contribute to the history of architecture (new expansions, see also Mies van der Rohe's discussion of the corner) AND construct buildings
   - As far as the users are concerned, what, in your opinion, are the conditions for success of such a design process?
   The users must participate (obviously) AND there must be very talented designers to ask the users (AND also: don't expect too much, as Gehry says: 'when we get there, the client hasn't shifted much' :)

2. The working of Gehry Partners. What are the factors determining the growth of this company?

   Questions to guide the analysis:

   - Comment on the logic behind the management of human resources.

     - Compare it with the dominant logic of the sector (as presented by Frank Gehry).
     - In your opinion, why does Gehry present this as a determining factor for the growth of his company (you may draw on other examples of companies encountered in the course). In terms of innovative design, what "expansions" are allowed by managing human resources in such a way?
     The model of the unpaid intern = min costs → multiple projects (quantity); by paying the intern, the issue is: (1) to put him to good use; (2) to use him again; (3) to bill the agency → create value-skill co-evolution effects (i.e. minimum cost vs. maximum coupling)

- Comment on the relationship Franck Gehry and his company have with CAD:

  – in what domain has CAD become necessary? Why, in your opinion, does the expertise assembled by Frank Gehry and Partners in this domain appear better than that of the reference companies on this topic?
  – What is the effect on the design capabilities of Gehry and partners? Is this a systematic design logic? What additional knowledge is required to answer this question with complete rigor?

    CAD = creation of original resources, creating a rule base for accelerating design on certain points; a rule base more effective than the engineering department since it is constantly being pressed by the "innovative design"; HOWEVER, it is not the "rule-based design" since Gehry gives the impression that he is ready to re-discuss the identity of objects; but this is what would need to be analyzed for Frank Gehry's various works.

- Comment on the logic of the company's cohesion:

  – What is the circle of partners, and how has it evolved? What are the "common interests" and how have they evolved?
  – How has Franck Gehry prepared for his succession? What are the possible futures for Gehry and Partners without Franck Gehry?

    Gradual enlargement of the circle of "partners", where the issue is the ability of each to express themselves. Preparing succession = find and train partners with the ability to make architecture evolve. Possible futures: in all cases, Frank Gehry Mk 2 → nothing? use just K bases of CAD type and expert skills? Or also adopt the forms and languages proposed by Frank Gehry to do like Mies van der Rohe with Peter Behrens: carry on with expansion?

Appendix F
Knowledge Control 2009—Product Design and Innovation

F.1. Course Question (1/3 of Total Marks)

1. What are the criteria used to assess a rule-based design project?
2. In C-K theory, what is an expansive partition?
3. In systematic theory, what is a function? Using the case of the bus station, give an example of a function.
4. State the first axiom of axiomatic design. Discuss its relevance for design.
5. Within the framework of platform-based design, a platform director proposes to initiate a project P for a new product. Product development lasts one year (year 0). The service life of the product is also one year (year 1). Development costs are 100k€ (in year 0). The product is expected to bring in 110k€ over year 1, the year in which it is sold. The project leader believes that the skills consequently developed will increase the turnover for other products sold in year 1 by 20%. Before project P, this turnover was estimated at 50k€.

   - What is the static return for the project?
   - What is the dynamic return?

   You may take an actualization rate of 10%.

F.2. Summary of Workshops (1/3 of Total Marks)

F.2.1. Analysis of the Design Regime in an Industrial Sector

Vincent Ventenat (Décathlon) presented the case of a composite tennis racket made from carbon fiber and/or flax fiber. Alexandre Ceccaldi presented the case of new cancer therapies using nanoparticles activated by X-rays. Georges Amar presented the case of the Rapid Transit Bus (also called, in French, the *Bus à Haut Niveau de Service*, BHNS) for the town of Curitiba. Nils Saclier presented the case of "Tom-Tom Navigation for All" built into Renault vehicles.

© Springer International Publishing AG 2017
P. Le Masson et al., *Design Theory*,
DOI 10.1007/978-3-319-50277-9

You may choose *one of these cases* on which you must answer the following questions:

1. What aspects of these innovations appear to come under rule-based design in the sectors concerned? Justify your answer.
2. What aspects of these innovations indicate a crisis of identity for the objects in the sectors concerned? Justify your answer.

F.2.2. The "Shopping Cart" Example

A team working on the smart shopping cart has produced the following C-K diagram.

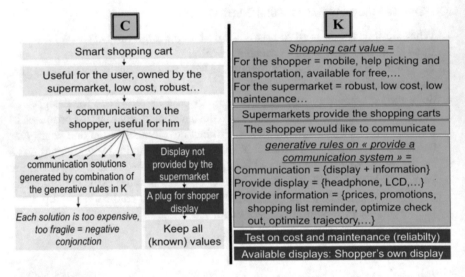

"Smart shopping cart" C-K diagram. Key: text in *white* on a *gray background* represent (in K space) expansions of knowledge and expansive partitions in C space

1. Why did the team consider that the proposition "the screen is not provided by the supermarket" gives rise to an expansive partition for the smart shopping cart?

   The same question for "A shopping display electrical socket".
2. Assess this work using V2OR criteria (variety, value, originality, robustness)
3. Using the suggested K base, make a new expansive partition on the proposed tree.
4. Using C-K theory, present a structured set of alternatives based on the concept "smart shopping cart for sustainable development". Do not exceed four levels in the concept tree.

F.3. Case Study (1/3 of Total Marks)

Article to read: **Pasquier, H. (2008).** "Acteurs, stratégies et lieux de 'recherche et développement' dans l'industrie horlogère suisse, 1900–1970." [*Actors, strategies and 'places of research' in the Swiss watchmaking industry*], *Entreprises et histoire*, 52, pp. 76–84.

Supplementary notes:

Explanation for p. 79, first paragraph: François Caron (1997) makes a distinction between test and research laboratories. According to François Caron the former rely on an exchange of knowledge and savoir faire with the production workshops. They favor an approach based on experience and the application of theoretical science. The latter carry out institutional, identified, programmed and planned research. Research laboratories have access to significant funding, and employ scientists and specialists.

The rest of the story:

The first electronic watches were launched in 1970. Omega showed several cutting-edge products at the Basel Fair, and over the following years developed many quartz movements. However, during the 1970s and 1980s the Swiss watch industry was plunged into an unprecedented crisis. Most of the manufacturers either disappeared or were subsumed into SMH, the *Société suisse de Microélectronique et d'Horlogerie*, created by Nicolas Hayek. Omega was absorbed into SMH when it was created in 1983. SMH was renamed the Swatch group in 1998 as a tribute to the little watch that had made the company so successful at the end of the 1980s.

Seiko 35 SQ Astron, First quartz watch on the market (Tokyo, Christmas 1969, $1250)

Swatch watch at the end of the 1980s. The watch is non repairable. In a departure from the usual architecture, the case also supports the movement, allowing more compact watches.

Using the accompanying text, develop the following points:

1. According to the author, the Swiss watch industry embodies a regional R&D model. Try to characterize it by answering the following questions:

   - Is there a representation of the object shared at the industrial sector level? If so, what is it? What is the associated logic of innovation?
   - What is the performance model for the development of new items?
   - What is the organizational model for the firms?

- What is the organizational model for the sector?
- In conclusion: characterize the design regime for the Swiss watch industry at the end of the 1950s.

2. The author describes the transformations the sector made to confront the issues of the "non-mechanical watch". Try to characterize this new regime by answering the following questions:

- Does the object have a crisis of identity? If so, in what dimension(s)?
- What are the organizations adopted by the four companies?
- Compare the strategies employed by Tissot, Longines, Le Coultre and Omega. What strategy seems to you the most robust from the innovative design point of view?
- Based on these elements, how would you explain the difficulties of the Swiss watch industry in the face of the rise in Japanese competitors marketing quartz watches in the 1970s? After examining the photograph of one of the first Swatch watches, what hypotheses can you offer regarding the innovative design strategy developed by Swatch in the 1980s?

Hints for III (2009)

1. According to the author, the Swiss watch industry embodies a regional R&D
   model. Try to characterize it by answering the following questions:

   - Is there a representation of the object shared at the industrial sector level? If
     so, what is it? What is the associated logic of innovation?
     *See the definitions of "manufacture", "prototype house" and
     "établisseurs": the movement and its components on the one hand, and the
     casing on the other. (p. 77–78). See also the product performance:
     "Exceptional aesthetic and functional performance" p. 80. Finally, see the
     skills developed at LSRH: alloys, magnetism, rust, etc.)*
   - What is the performance model for the development of new items?
     *See bottom of p. 80–81 :the text shows that the performance logic is that of
     writing off R&D costs by playing on the potential returns due to economies
     of scale—hence the logic is that each development has to pay for itself, i.e.
     maximizing static returns.*
   - What is the organizational model for the firms?
     *An engineering department with progressive sophistication of languages
     (Mech. dept., Process & planning dept., Inspection dept.)*
   - What is the organizational model for the sector?
     *The sector comprises companies with "engineering departments" which
     split production within the branch; due to agreements with the cartel, no
     information is exchanged with the outside world. Training is accomplished
     by technical schools and a common research laboratory.*
   - In conclusion: characterize the design regime for the Swiss watch industry at
     the end of the 1950s.
     Rule-based regime!

2. The author describes the transformations the sector made to confront the issues
   of the "non-mechanical watch". Try to characterize this new regime by
   answering the following questions:

   - Does the object have a crisis of identity? If so, in what dimension(s)?
     Disruption qualified as "technical"
   - What are the organizations adopted by the four companies?
     *Rather, "ambidextrous" with separation of rule-based and innovative
     regimes*
   - Compare the strategies employed by Tissot, Longines, Le Coultre and
     Omega. What strategy seems to you the most robust from the innovative
     design point of view?
     *The first three companies are faced with something of a technical challenge,
     and seek knowledge they do not possess from outside sources; Omega
     explores the three paths with a logic that internalizes the knowledge they
     lack.*

– Based on these elements, how would you explain the difficulties of the Swiss watch industry in the face of the rise in Japanese competitors marketing quartz watches in the 1970s? After examining the photograph of one of the first Swatch watches, what hypotheses can you offer regarding the innovative design strategy developed by Swatch in the 1980s?

For the first 3: no learning, risky strategy; for Omega, they would have been able to win out but other factors (cost?) had to appear;

For the first 3: no learning, risky strategy; for Omega, they would have been able to win out but other factors (cost?) had to appear;

For Swatch, not just a technical disruption! Extend the crisis of the object = mode, non-repairable, ...

Appendix G
Knowledge Control 2010—Product Design and Innovation

G.1. Course Question (1/3 of Total Marks)

1. Mme Laroche (Verallia Saint-Gobain) presented several means by which the cost of buying can be reduced. State two of these methods requiring an effort of design.
2. Give a major difference between prototype designed under a rule-based system and a prototype designed under an innovative design system.
3. Explain why systematic design (Pahl and Beitz, 1977) is more effective than parametric design.
4. In C-K theory, what is a restrictive partition and an expansive partition?
5. What criteria can you use to establish whether the objects in an industrial sector are in a crisis situation? Give two examples.

G.2. Summary of Workshops (1/3 of Total Marks)

G.2.1. Analysis of the Design Regime in an Industrial Sector

Nils Saclier (Renault) presented the case of vehicle navigation for all. Laurence Laroche (Saint-Gobain Verralia) presented the case of new glass ovens in the bottling domain. Alain Dieulin (Vallourec) presented innovation in the oil-well drill-tube sector. Bernard Vaudeville (T/E/S/S) presented several cases of innovative buildings. Pascal Daloz (Dassault Systèmes) presented the evolution of the sector producing computer aided design tools.

1. Choose two of these cases on which you must answer the following questions:

    a. Give some rules that are characteristic of the dominant design in these sectors.
    b. Do the objects have a crisis of identity? Justify your answer using examples given by the speakers.

2. Using examples from the five presentations and from the cases covered in the course (Baldwin, Tefal, Gehry, etc.) characterize customer relations in a rule-based situation and innovative situation respectively.

© Springer International Publishing AG 2017
P. Le Masson et al., *Design Theory*,
DOI 10.1007/978-3-319-50277-9

G.2.2. Rule-Based and Innovative Design Reasoning

Let the concept C0 be the "smart vacuum-cleaner".

1. Give two functions of the domestic vacuum-cleaner and justify your answer.
2. Give a conceptual model of the vacuum-cleaner.
3. Using this knowledge base, suggest two expansive partitions on the "Smart vacuum-cleaner" concept (provide justification for the expansive nature of these partitions). Indicate the knowledge bases you might turn to in order to pursue the design reasoning on these expansive partitions.

G.3. Case Study (1/3 of Total Marks)

Article to read: **Lenfle, S. (2010).** "Managing Parallel Strategy in Projects with Unforeseeable Uncertainty: The Manhattan Case in Retrospect." European Academy of Management, Rome, Italy.

Reading guide: the detailed case study is on pp. 5–13; you will find certain items analyzing parallel strategies on pp. 13–15; organizational aspects are analyzed on pp. 15–18.

Using the accompanying text, develop the following points:

1. *Characterization of the design situation (rule-based vs. innovative)*

What is the classical doctrine for project management (see course and hints in the document, p. 2)? This type of project relies on the existence of design rules. What conditions on these design rules make traditional project organization possible?

Characterize the *initial concept* of the Manhattan project and the state of knowledge of the designers involved in the project. How did the project distance itself from the conditions allowing a rule-based project?

2. *Analysis of the design reasoning*

Covering the production of fissile material, the text reconstructs a C-K reasoning, finishing up with the conjunction indicated on p. 10, "S50 to K25 to Y12". Indicate the concepts dealt with over the course of the reasoning and the associated learning processes.

The economic logic of managing the portfolio of technologies generally leads to choosing the best technology as soon as possible. Under what conditions may this logic be implemented? Is this logic adopted here? Why? What logic is followed?

3. *Organization of design*

The organization adopted contrasts with traditional project management. Give two examples.

Nevertheless, the Manhattan project was built on a small number of well defined management rules. State which these were. Compare with other innovative project organizations encountered in the course.

4. *Assessment of the project*

More generally, did the Manhattan project consist of managing a set of *uncertain* alternatives in the hope of finding one that worked? Justify your answer using the arguments put forward in the text.

How would you analyze the Manhattan project as an innovative design project (expansion in concepts, expansion in knowledge)? What additional items of analysis would you require to answer with greater accuracy?

Hints for correction III (2010)

1. What is the classical doctrine for project management (see course and hints in the document, p. 2)? This type of project relies on the existence of design rules. What conditions on these design rules make traditional project organization possible?

Characterize the *initial concept* of the Manhattan project and the state of knowledge of the designers involved in the project. How did the project distance itself from the conditions allowing a rule-based project?

Clear CQT target; coordination of tasks
existence of systematic design languages: F/MC/E/D with coherence between these languages → ensures conjunction and expansion along the cone of performance (dominant design)
"Decisive weapon based on exothermic fission reaction, before the Germans". Gaps in K on the production of equipment and design of the bomb. Crudely: almost no rules!
Manhattan: hardly any F; varied and poorly understood MC; E unknown; D unknown.

2. Covering the production of fissile material, the text reconstructs a C-K reasoning, finishing up with the conjunction indicated on p. 10, "S50 to K25 to Y12". Indicate the concepts dealt with over the course of the reasoning and the associated learning processes.

The economic logic of managing the portfolio of technologies generally leads to choosing the best technology as soon as possible. Under what conditions may this logic be implemented? Is this logic adopted here? Why? What logic is followed?

C0: fissile material for the bomb
C_U11 and C_U12 = U235: gaseous diffusion and EM separation; also $C_{Pu}1x$ paths for Pu.
dK = poor effectiveness (that said, there is a lack of information on the learning that actually took place and on the generation of concepts throughout the text! AND there is a need for uranium for the bomb (result of explosions on the bomb: gun design does not work for Pu and Pu bomb (implosion) risks not working.
C_U13: thermal diffusion → dK → C_U131= only/C_U132= with C_U12 and C_U11 Economic logic = choose one of the 2 (or 3) initial paths provided NPV (cost/return) is known; here = maximization of learning → choose the paths that are richest for learning: when the learning drops off, take another path and DON'T develop it too soon since there is still much to learn; fine-tuning might not be the best route to knowledge.

3. The organization adopted contrasts with traditional project management. Give two examples.

Nevertheless, the Manhattan project was built on a small number of well defined management rules. State what these were. Compare with other innovative project organizations encountered in the course.

Division of labor redefined over the course of the route; reallocation of resources; organization of tasks not defined at the outset; parallelism,... no stable structure.

Weekly symposia; global view of the part played by project leaders (Groves & Oppenheimer : strong centralization → VM highly informed; who defines the DS (division of labor sufficiently clear?)

4. More generally, did the Manhattan project consist of managing a set of *uncertain* alternatives in the hope of finding one that worked? Justify your answer using the arguments put forward in the text.

How would you analyze the Manhattan project as an innovative design project (expansion in concepts, expansion in knowledge)? What additional items of analysis would you require to answer with greater accuracy?

Portfolio: probabilistic → each path is tried in the hope that one will work; not the case since there is a recombination (for fissile material) and a redefinition of certain paths (implosion → hence path not given at the outset.

CI: one would wish to know what are the new concepts and what are the learning processes that lead to these concepts; indeed, one has the impression that the concepts are "already there", but the work on conceptual development cannot be seen; does the uranium path resemble an algebraic combination rather than an innovative concept? Question: is this a dC project or rather, a DC project? To answer this question, would it be necessary to have a better understanding of the rules that have been broken? (in particular rules on industrialization? research? etc.)

Appendix H
Knowledge Control 2011—Product Design and Innovation

H.1. Course Question (1/3 of Total Marks)

1. State the first axiom of the theory of axiomatic design. In what way is an upper triangular matrix interesting?
2. A common definition of innovation is: "An idea that has found a market". Give a limitation of this definition in the organization of the process of innovative design.
3. Among those given in the courses on rule-based design, give an example of a generative model.
4. Why is statistical decision theory (Savage 1954) not a theory of design?
5. In C-K theory, what is a concept? What is an item of knowledge?

H.2. Summary of Workshops (1/3 of Total Marks)

H.2.1. Analyze Disruptions to the Identity of Objects

Each of the five speakers in the week presented projects involving innovation: Nils Saclier (Renault) presented the case of the Twizy and vehicle navigation for all; Bernard Vaudeville (T/E/S/S) presented two cases of innovative construction; Yves Parlier (Beyond the Sea) presented a maritime transport project using a kite; Pascal Daloz (Dassault Systèmes) showed how the sector for computer aided design tools has evolved; Michel Lescanne (Nutriset) presented innovations in the fight against severe malnutrition.

For each of these speakers, identify a disruption in the identity of the objects, pointing out the reference identity and explaining the disruption.

H.2.2. Rule-Based and Innovative Design Reasoning

Let the concept C0 be: "all-weather umbrella".

1. Give two functions of the umbrella, and justify your answer
2. Give a conceptual model of the umbrella.
3. Using this knowledge base and C-K theory, suggest two expansive partitions for the concept "all-weather umbrella" (justify the expansive nature of these partitions). Indicate the knowledge bases you might turn to in order to pursue the design reasoning on these expansive partitions.

© Springer International Publishing AG 2017
P. Le Masson et al., *Design Theory*,
DOI 10.1007/978-3-319-50277-9

H.3. Case Study (1/3 of Total Marks)

Article to read:**Hughes, Thomas P. (1983).** "L'électrification de l'Amérique." [*The electrification of America*] *Culture technique*, No. 10, June 1983, "USA", pp. 21–41.

Reading guide: Only the chapter on Edison, p. 21–29, is relevant to the questions.

Historical notes:

1. The Yablochkoff candle is an arc lamp. In 1876, one year before Edison's incandescent bulb, Yablochkoff candles were used to light the streets in front of the great stores in Paris and London. These systems gave rise to numerous innovations. They tended to use high currents at low voltages.
2. During the years 1887-1888, Edison built his own "invention factory", West Orange, where a great many electrical innovations would be developed (electric motors, tramways, electric vehicles, batteries and cells for railway use, a universal electric motor, etc.), sound (phonograph, talking doll, recording and studio systems, dictaphone, etc.), cinema (movie camera, camera manufacture, "coin slot" machines for viewing short films, etc.), etc.

Using the accompanying text, develop the following points:

1. *Characterization of the design reasoning*

   a. What was Edison's initial concept? Was this original in the context of the age?
   b. What knowledge did Edison combine for the first reasoning processes?
   c. What reasoning step does the author consider to be critical? Why is this an expansive partition?
   d. For this phase of reasoning, characterize the types of knowledge of which Edison made use.
   e. What followed on from this reasoning?
   f. In this reasoning, identify those elements that appear to involve innovative design, and those that appear to involve rule-based design.

2. *Characterization of the organization adopted at Menlo Park*

   a. Why can the organization at Menlo Park be likened almost to a rule-based design project?
   b. Under what conditions can a "project" type organization can be successfully adopted?
   c. How did Edison resolve the contradiction between a question requiring an innovative design reasoning and a rule-based organization? What skills were presupposed on his part?
   d. More generally, what is the logic behind Edison's performance? How is this distinguishable from the logic of the project undertaken at Menlo Park?

Hints for correction III (2011)

1. Characterization of the design reasoning

   a. What was Edison's initial concept? Was this original in the context of the age?
   Electrical lighting system—not original: cf. Yablochkoff in Europe

   b. What knowledge did Edison assemble for the first reasoning processes?
   Economic models for gas and arc lighting; modeling electrical systems (main components and laws governing the system) – Ohm's law at resistance of a copper cable.

   c. What reasoning step does the author consider to be critical? Why is this an expansive partition?
   High voltage system, while existing systems preferred high current, low voltage.

   d. For this phase of reasoning, characterize the types of knowledge of which Edison made use.
   Conceptual model

   e. What followed on from this reasoning?
   Based on high voltage and economic equations → rapid definition of the main elements of the systems and division of labor (generators, cables, etc.); also combine the necessary conditions: concession.

   f. In this reasoning, identify those elements that seem to involve innovative design, and those that seem to involve rule-based design.
   Upstream part: clearly an expansive partition, "definition of the problem"; Downstream part: quasi SD: many components can be re-used (and improved); problem solved (filament)

2. *Characterization of the organization adopted at Menlo Park*

   a. Why can the organization at Menlo Park be likened almost to a rule-based design project?
   Skills are combined, work is split, target is clear; test equipment and learning processes are already identified.

   b. Under what conditions can a "project" type organization can be successfully adopted?
   Combine resources (skills, experts), ability to assess the risks and benefits before the event. Avoid expansive partitions during the process!

   c. How did Edison resolve the contradiction between a question requiring an innovative design reasoning and a rule-based organization? What skills were presupposed on his part?
   No CQT! On the contrary: maximize exploratory capabilities! and as a result, this maximization occurs via the ability to define locally "rule-based" projects.

d. More generally, what is the logic behind Edison's performance? How is this distinguishable from the logic of the project undertaken at Menlo Park?
No CQT! On the contrary: maximize exploratory capabilities! and as a result, this maximization occurs via the ability to define locally "rule-based" projects.

Index of Cited Authors

© Springer International Publishing AG 2017
P. Le Masson et al., *Design Theory*,
DOI 10.1007/978-3-319-50277-9

Index of Companies, Organisations and Products

© Springer International Publishing AG 2017
P. Le Masson et al., *Design Theory*,
DOI 10.1007/978-3-319-50277-9

Index of Notions

© Springer International Publishing AG 2017
P. Le Masson et al., *Design Theory*,
DOI 10.1007/978-3-319-50277-9

Printed in the United States
By Bookmasters